Transmission of Electrical Energy

Transmission of Electrical Energy

Overhead Lines

Ailson Pereira de Moura
Adriano Aron Freitas de Moura
Ednardo Pereira da Rocha

CRC Press is an imprint of the
Taylor & Francis Group, an **informa** business

CRC Press
Taylor & Francis Group
6000 Broken Sound Parkway NW, Suite 300
Boca Raton, FL 33487-2742

First issued in paperback 2022

ISBN-13: 978-0-367-47709-7 (hbk)
ISBN-13: 978-1-03-233623-7 (pbk)
DOI: 10.1201/9781003038597

Publisher's Note

The publisher has gone to great lengths to ensure the quality of this reprint but points out that some imperfections in the original copies may be apparent.

Library of Congress Cataloging-in-Publication Data

Names: Moura, Ailson P. de (Ailson Pereira), author. | Moura, Adriano Aron F. de (Adriano Aron Freitas), author. | Rocha, Ednardo P. da (Ednardo Pereira), author.
Title: Transmission of electrical energy : overhead lines / Ailson P. de Moura, Adriano Aron F. de Moura, Ednardo P. da Rocha.
Description: First edition. | Boca Raton, FL : CRC Press, 2020. | Includes bibliographical references and index.
Identifiers: LCCN 2019060113 (print) | LCCN 2019060114 (ebook) | ISBN 9780367477097 (hardback) | ISBN 9781003038597 (ebook)
Subjects: LCSH: Overhead electric lines. | Electric power transmission--Technique. | Electric power distribution--Alternating current--Computer simulation.
Classification: LCC TK3226 .M68 2020 (print) | LCC TK3226 (ebook) | DDC 621.319/22--dc23
LC record available at https://lccn.loc.gov/2019060113
LC ebook record available at https://lccn.loc.gov/2019060114

Visit the Taylor & Francis Web site at
http://www.taylorandfrancis.com

and the CRC Press Web site at
http://www.crcpress.com

Contents

Preface

This book aims to provide an introduction to alternating current power transmission, deep enough to provide undergraduate and graduate students in electrical engineering at various higher education institutions with a solid theoretical background on the subject, specifically for students wishing to pursue their studies in electrical power systems. The book presents the transmission of electricity, particularly with a focus on computer use through the ATPdraw program and algorithms developed in MATLAB.

Chapter 1 introduces the ATPdraw program. This program, used worldwide, serves as a reference for all computer simulations performed in the text.

Chapter 2 presents the physical constitution of the transmission lines and the corona effect.

Chapter 3 presents the generalized calculation of the transmission line parameters.

In Chapter 4, the modeling of the steady-state transmission lines is detailed and exemplified. An introduction to the electrical design of the transmission lines is introduced.

Chapter 5 presents the modeling of transient transmission lines.

Finally, in Chapter 6, an introduction to the modal propagation theory is presented.

Chapters 1, 5, and 6 may be taught in postgraduate courses in electrical engineering.

Ailson Pereira de Moura
Adriano Aron Freitas de Moura
Ednardo Pereira da Rocha

Acknowledgments

The authors would like to thank God and his beloved son Jesus for allowing the creation of this work. We thank our family members for allowing us to be absent during the endless hours for the completion of this work. To all the students, who also provided us with treats and learning on the subject, here is our thanks, because without a doubt it is by teaching, learning, and using the imagination that we can realize and transform our dreams into reality.

For Aluísio de Oliveira Moura
(Ailson's father, Adriano's grandfather), (In memoriam)

About the Authors

Ailson Pereira de Moura was born in Brazil in August 1955. He got a degree in Electrical Engineering at the Federal University of Ceara—UFC in July 1979. From 1979 to 1980 he worked as an engineer in the textile industry. In October 1981, he joined the staff of the Department of Electrical Engineering at the Federal University of Ceara as teaching and research auxiliary. He taught electrical installations and electrical measurement disciplines. In 1982 and 1983, he completed specializations in distribution systems and power systems from the universities of Fortaleza and Federal of Itajuba, respectively. In 1986, he completed his MSc at the Federal University of Campina Grande. In 1996, he completed his PhD at the Federal University of Campina Grande, having reached the highest grade. In December 2003, he got a degree in accounting at the Ceara State University. From September 2013 until May 2014, he worked as a researcher at the Institute for Systems and Computer Engineering Technology and Science—Porto and University of Porto, with Prof. João Abel Peças Lopes, where he was awarded his post-doctorate. He teaches graduation and postgraduation courses, and is responsible for the following disciplines:

- Electrical circuits;
- Electric power transmission;
- Power systems analysis;
- Distribution systems analysis (MSc);
- Distributed generation network analysis (MSc).

Dr. Ailson Moura was responsible for several research projects, including:

- Impact of wind power plants on a sub-transmission-distribution system in steady-state;
- Methods for voltage regulation in wind-generated distribution networks;
- Models of wind farm generators for studies of steady-state power systems;
- Impact of wind farms on the steady-state operation of three-phase modeling distribution systems;
- New method and models for power and short-circuit studies;
- Short-circuit calculations using phase components in distribution systems;
- Newton-Raphson power flow with constant matrices.

He is a reviewer of the following journals:

- 2007—*Renewable Power Generation;*
- 2013—*International Journal of Electrical Power & Energy Systems;*
- 2019—*IEEE Transactions on Power Systems.*

He is the author of six books:

- *Power systems engineering—load flow exercises.* 1st ed. Sao Paulo: Artliber, 2018. 176p.
- *Power systems engineering—load flow analysis in power systems.* 1st ed. Sao Paulo: Publisher Artliber, 2018. 314p.
- *Power systems engineering—analysis of alternating-current circuits for power systems.* 1st ed. Sao Paulo: Publisher Artliber, 2018. 316p.
- *Power systems engineering—transmission of electric power in alternating current.* 1st ed. Fortaleza: Publisher of the Federal University of Ceará—UFC, 2019. v. 1. 353p.
- *Power systems engineering—hydroelectric and wind power generation.* 1st ed. Fortaleza: Publisher of the Federal University of Ceará—UFC, 2019. v. 1. 291p.
- *Power systems engineering—computer models for three-phase load flow, continued power flow and external equivalents.* Fortaleza: Publisher of the Federal University of Ceará—UFC, 2020. v. 1. 300p.

Dr. Ailson Moura has presented over 60 papers at various national and international workshops and conferences and in international technical journals. He has been responsible for the supervision of two PhD students and eight MSc students.

He has also been responsible for the development of the following computer programs:

- Moura, Adriano Aron Freitas de; de Moura, A. P. Three phase load flow program with a wind turbine model, 2008.
- Santos, Lucélia Alves dos; de Moura, A. P. Short circuit program using phase components, 2007.
- de Moura, A. P.; Lima, Siomara Peixoto. Rapidly decoupled load flow program with wind power plant models, 2006.
- de Moura, A. P. Port of winds program, 2003.
- de Moura, A. P. Program for calculating transmission line parameters, 1998.
- de Moura, A. P. Program to calculate a three-phase load flow based on the equivalent power method, 1996.

- de Moura, A. P. Program for load flow calculation based on the equivalent power method, 1995.
- de Moura, A. P. Program to calculate a charge flow based on the moment method, 1994.
- de Moura, A. P. Three-phase load flow calculation program based on the moment method, 1994.
- de Moura, A. P. Program with constant tangents, 1991.
- de Moura, A. P. Direct current link load flow program, 1986.

Dr. Ailson Moura's complete curriculum can be accessed on the CNPQ platform: http://lattes.cnpq.br/7100954284070403.

Adriano Aron Freitas de Moura was born in Brazil in December 1983. He graduated in electrical engineering from the Federal University of Ceará (2006). He has written a monograph in planning the growth of a distribution system with the integration of wind generators. He has a master's degree in distribution systems with a three-phase load flow and integration of wind turbines in the power grid (2009). He completed his PhD at the Federal University of Ceará (2013) with the thesis entitled new method and models for studies of power flow and short circuit. In 2010, he was contracted as assistant professor at the Federal Rural University of Semiarid. He taught solar energy engineering, electrical and magnetics materials, power system analysis, and wind energy conversion systems. He is experienced in electrical engineering, focusing on electrical systems, generation, transmission, electricity distribution, and renewable energy sources. Interested in computer simulation of power systems in MATLAB language, he is the creator and leader of the research group at UFERSA GESPER—Renewable Energy Power Systems Study Group. He has research interests in electrical power systems, involving power flow, short circuit, electromechanical stability, electromagnetic transients, power electronics for power systems, and control of microgrids with renewable energy sources. He teaches in graduate and postgraduate courses, and is responsible for the following disciplines:

- Power system analysis;
- Wind energy;
- Transients and harmonics (MSc);
- Modern power system analysis (M Sc).

Dr. Adriano Moura was responsible for several research projects such as:

- Models of wind farm generators for studies of steady-state power systems;
- Impact of wind farms on the steady-state operation of three-phase modeling distribution systems;

- New method and models for power and short-circuit studies;
- Short-circuit calculations using phase components in distribution systems;
- Newton-Raphson power flow with constant matrices;
- Didactic simulator of wind energy systems in power systems;
- Stability analysis models of wind energy systems connected to power systems.

He is the author of six books:

- *Power systems engineering—load flow exercises.* 1st ed. São Paulo: Artliber, 2018. 176p.
- *Power systems engineering—load flow analysis in power systems.* 1st ed. Sao Paulo: Publisher Artliber, 2018. 314p.
- *Power systems engineering—analysis of alternating-current circuits for power systems.* 1st ed. Sao Paulo: Publisher Artliber, 2018. 316p.
- *Power systems engineering—transmission of electric power in alternating current.* 1st ed. Fortaleza: Publisher of the Federal University of Ceará - UFC, 2019. v. 1. 353p.
- *Power systems engineering—hydroelectric and wind power generation.* 1st ed. Fortaleza: Publisher of the Federal University of Ceará—UFC, 2019. v. 1. 291p.
- *Power systems engineering—computer models for three-phase load flow, continued power flow and external equivalents.* Publisher of the Federal University of Ceará—UFC, 2020. v. 1. 300p.

Dr. Adriano Moura has presented 42 papers at various national and international workshops and conferences and in international technical journals. He has been responsible for the supervision of seven MSc students, 16 graduate student monographs, and 11 starting science students.

He has also been responsible for the development of the following computer programs:

- Moura, Adriano Aron Freitas de; de Moura, A. P. Three phase load flow program with a wind turbine model. 2008—ANAREDGEE. Runs IEEE 13, 34, 37, and 123 test systems.
- Moura, Adriano Aron Freitas de; de Moura, A. P. Three phase load flow program with a wind turbine model. 2008—ANAREDGEE object oriented programmed.
- Moura, Adriano Aron Freitas de; de Moura, A. Generalized power flow software. 2013—FLUPOTGEN.

Dr. Adriano Moura's complete curriculum can be accessed on the CNPQ platform: http://buscatextual.cnpq.br/buscatextual/visualizacv.do?id=K4211383T5.

Ednardo Pereira da Rocha is an electrical engineer and has a bachelor's degree in science and technology from the Federal Rural University of Semi-Arid and a masters in communication and automation systems—PPGSCA/UFERSA. He is an assistant professor at the Federal Rural University of Semi-Arid. He has interests in the following areas of study: electrical power systems; alternative energy sources; and industrial electrical installations.

Prof. Ednardo Pereira was responsible for several research projects such as:

- Study, analysis, and control of low power DC wind generators;
- New techniques and algorithms for short-circuit processing in-phase components and continuous power flow with a three-phase modeling;
- Development of a unified computational platform for simulation and analysis of distributed generation and three-phase modeling of permanent power transmission and distribution systems;
- Development of a computational platform for simulation and analysis of distributed generation permanent electricity distribution systems.

He is the author of six books:

- *Power systems engineering—load flow exercises.* 1st ed. Sao Paulo: Artliber, 2018. 176p.
- *Power systems engineering—load flow analysis in power systems.* 1st ed. Sao Paulo: Publisher Artliber, 2018. 314p.
- *Power systems engineering—analysis of alternating-current circuits for power systems.* 1st ed. Sao Paulo: Publisher Artliber, 2018. 316p.
- *Power systems engineering—transmission of electric power in alternating current.* 1st ed. Fortaleza: Publisher of the Federal University of Ceará—UFC, 2019. v. 1. 353p.
- *Power systems engineering—hydroelectric and wind power generation.* 1st ed. Fortaleza: Publisher of the Federal University of Ceará—UFC, 2019. v. 1. 291p.
- *Power systems engineering—computer models for three-phase load flow, continued power flow and external equivalents.* Publisher of the Federal University of Ceará—UFC, 2020. v. 1.300p.

Prof. Ednardo Pereira has presented 11 papers at various national and international workshops and conferences and in international technical journals. He has been responsible for the supervision of 23 graduate student monographs.

Prof. Ednardo Pereira's complete curriculum can be accessed on the CNPQ platform: http://buscatextual.cnpq.br/buscatextual/visualizacv.do?id=K4465423J0.

1

ATPdraw Program

1.1 Introduction

ATPdraw is a preprocessor for the alternative transients program (ATP), resulting from cooperation between the Bonneville power administration (BPA) of Canada and the Norwegian Electric Power Research Institute of Norway, which developed ATP for the Windows operating system environment. ATPDraw output is a file that is used as input to the ATP program.

1.2 Resolution Algorithms

The first studies related to the digital simulation of traveling wave problems in power systems were conducted in the early 1960s using two techniques: the Bewley diagram and the Bergeron method. These developed techniques were applied in the analysis of small networks with lumped, linear or non-linear parameters and distributed parameters. The extension to multi-node networks was done by Hermann W. Dommel, which resulted in BPA's electromagnetic transients program (EMTP) and later alternative transients program (ATP). The scheme developed by Dommel combines the Bergeron method and the trapezoidal rule in a transient calculation algorithm in multi-phase networks with concentrated and distributed parameters.

The calculation of electromagnetic transients using the trapezoidal rule is done in the time domain. Other techniques have been developed to perform this calculation in the frequency domain or using the Z transform. Programs based on the trapezoidal rule are mostly used in the simulation of transients in power systems. This is because of the simplicity of the trapezoidal rule. However, this method has important limitations: it uses a fixed integration step and produces numerical oscillations. The integration step can be set in an aba called ATP settings and is represented by Δt. The selection of the integration interval is greatly influenced by the phenomenon under investigation. Simulations involving high frequencies require tiny integration steps, while low frequency phenomena can be calculated with larger integration steps—for example, switching transients (25 – 100 ms) $\rightarrow \Delta t \leq \tau / 10$, where τ is the transit time of the shortest analyzed transmission line (TL) length,

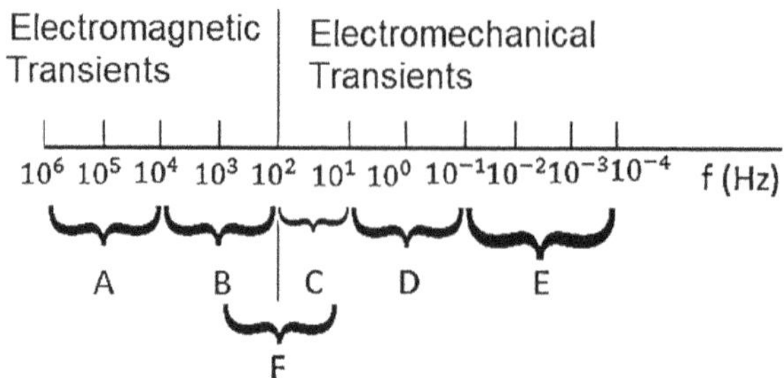

FIGURE 1.1
Transient frequency spectrum in power systems.

A = lightning overvoltage
B = switching overvoltage
C = short-circuit transients and sub-synchronous resonance
D = stability
E = thermal machine control transients
F = harmonics because of transformer saturation

and lightning phenomena $(0,01 - 0,1\ ms) \rightarrow \Delta t \leq 1/2f$, where f is the frequency of interest of the phenomenon under analysis. Figure 1.1 shows the frequency spectrum of transients in power systems.

In the aba ATP settings, it is still possible to adjust the maximum simulation time (t_{max}), which depends on the phenomenon studied and the results obtained during the analysis, for example, TL energization (50 ms) and lightning phenomena (20 ms). It is still possible to change Xopt and Copt (when Xopt = 0, L must be given in mH, when Copt = 0, C must be given in μF).

1.3 Trapezoidal Integration Rule

Starting from the initial conditions at $t = 0$, the bus voltages are determined at $t = \Delta t, 2\Delta t, 3\Delta t, \ldots,$ up to the maximum time (t_{max}). In calculating the stresses of the bars at time t, it is necessary to know the voltages of these nodes up to a certain previous time, $t = t - \Delta t,\ t - 2\Delta t,\ t - 3\Delta t, \ldots, t - \tau$. It is also necessary to know the previous values of the system voltages for the calculation at time t.

Numerical processes are used to calculate the integral value within a defined range. Thus, to calculate the area under the curve that defines the function $f(x)$ in the range x_n to x_{n+1}, is shown in Figure 1.2.

From Figure 1.2, we have the trapezoidal area. Thus:

$$A = \int_{x_n}^{x_{n+1}} f(x)dx = \frac{\Delta x}{2}\left[f(x_{n+1}) + f(x_n)\right] \tag{1.1}$$

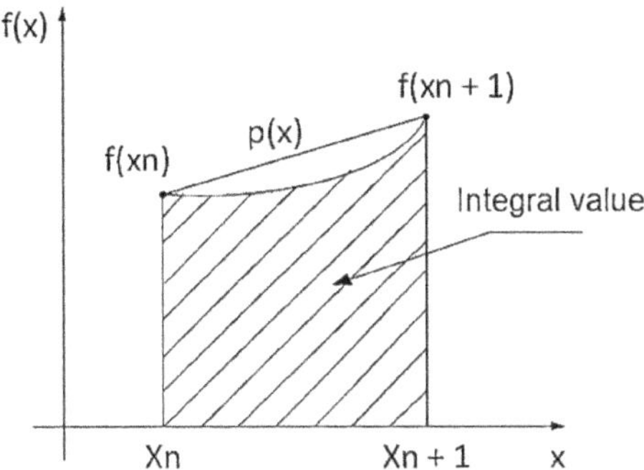

FIGURE 1.2
Trapezoidal integration method.

1.3.1 Lumped Resistance Model

The equivalent circuit of a resistor is shown in Figure 1.3.

We can write the equation that defines the branch as:

$$i_{jk}(t)=\frac{1}{R}\left[v_j(t)-v_k(t)\right]=\frac{1}{R}v_{jk}(t) \tag{1.2}$$

1.3.2 Lumped Inductance Model

Let L be an inductance connected between the *j* and *k* nodes of a circuit, as shown in Figure 1.4.

The inductance equation that relates voltage and current at its terminals is expressed as:

$$v_{jk}(t)=L\frac{d}{dt}[i_{jk}(t)] \tag{1.3}$$

$$\int_{t-\Delta t}^{t} di_{jk}(t)=\frac{1}{L}\int_{t-\Delta t}^{t} v_{jk}(t)dt \tag{1.4}$$

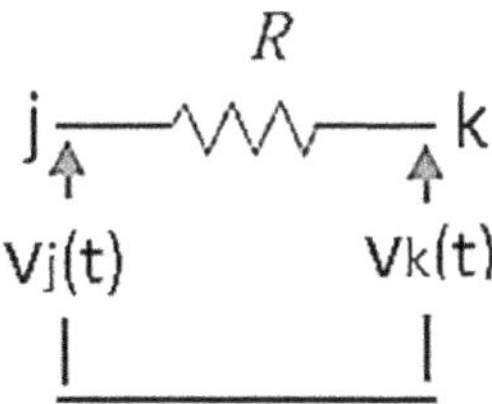

FIGURE 1.3
Equivalent resistor circuit.

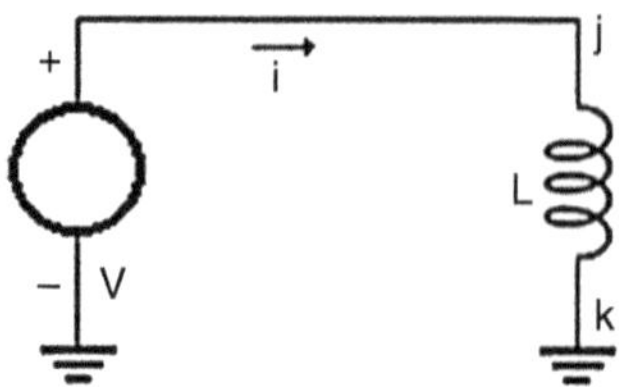

FIGURE 1.4
Inductance circuit.

Using the trapezoidal integration law, we have:

$$i_{jk}(t) - i_{jk}(t-\Delta t) = \frac{\Delta t}{2L}\left[v_{jk}(t-\Delta t) + v_{jk}(t)\right] \tag{1.5}$$

$$i_{jk}(t) = \frac{\Delta t}{2L}v_{jk}(t) + \frac{\Delta t}{2L}v_{jk}(t-\Delta t) + i_{jk}(t-\Delta t) = \frac{\Delta t}{2L}v_{jk}(t) + I_L(t-\Delta t) \tag{1.6}$$

where:

$$I_L(t-\Delta t) = \frac{\Delta t}{2L}v_{jk}(t-\Delta t) + i_{jk}(t-\Delta t) \tag{1.7}$$

Current $I_L(t-\Delta t)$ is interpreted as a current source dependent on past values of voltage and current over the inductor.

Therefore, the equation that defines the branch is expressed as:

$$i_{jk}(t) = \frac{1}{R_L}\left[v_{jk}(t)\right] + I_L(t-\Delta t) \tag{1.8}$$

where:

$$R_L = \frac{2L}{\Delta t} \tag{1.9}$$

From equation (1.6) to $t = t - \Delta t$, we arrive at equation (1.10).

$$i_{jk}(t-\Delta t) = \frac{\Delta t}{2L}v_{jk}(t-\Delta t) + \frac{\Delta t}{2L}v_{jk}(t-2\Delta t) + i_{jk}(t-2\Delta t) \tag{1.10}$$

Replacing (1.10) in (1.7), we have:

$$I_L(t-\Delta t) = \frac{\Delta t}{2L}v_{jk}(t-\Delta t) + \frac{\Delta t}{2L}v_{jk}(t-\Delta t) + \frac{\Delta t}{2L}v_{jk}(t-2\Delta t) + i_{jk}(t-2\Delta t) \tag{1.11}$$

$$I_L(t-\Delta t) = \frac{2v_{jk}(t-\Delta t)}{R_L} + I_L(t-2\Delta t) \tag{1.12}$$

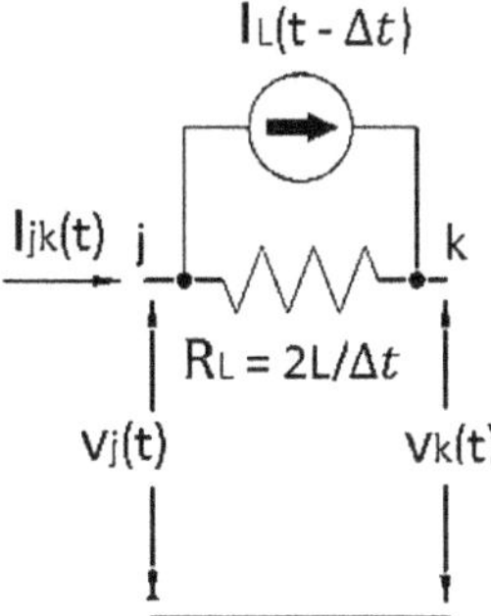

FIGURE 1.5
Inductance equivalent circuit.

where:

$$I_L\left(t-2\Delta t\right)=\frac{\Delta t}{2L}v_{jk}\left(t-2\Delta t\right)+i_{jk}\left(t-2\Delta t\right) \tag{1.13}$$

The equivalent circuit is shown in Figure 1.5.

Example 1.1

Calculate $V_L(t)$ for $t=2\Delta t$ in the circuit of Figure 1.6. The frequency is 60 Hz and $\Delta t=1\ \mu seg$. The initial conditions are given by:

$$V_F\left(0^-\right)=0;\ I_F\left(0^-\right)=0;\ V_L\left(0^-\right)=0;\ I_L\left(0^-\right)=0.$$

SOLUTION:

Circuit model:

$$R_L=\frac{2L}{\Delta t}=\frac{2x10x10^{-3}}{10^{-6}}=20000\ \Omega$$

$$I_L\left(t-\Delta t\right)=\frac{2V_L\left(t-\Delta t\right)}{R_L}+I_L(t-2\Delta t)$$

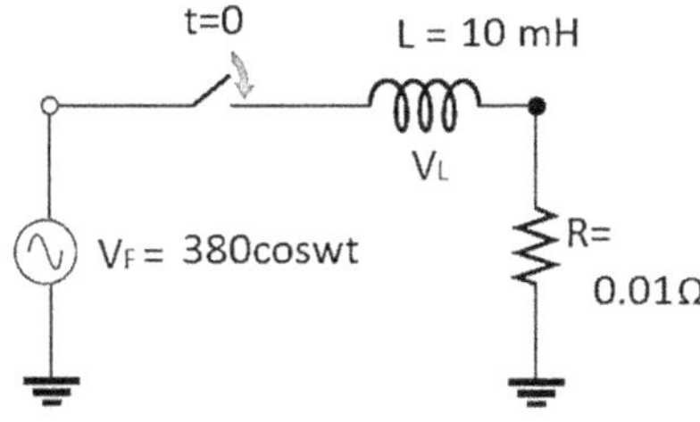

FIGURE 1.6
Circuit RL.

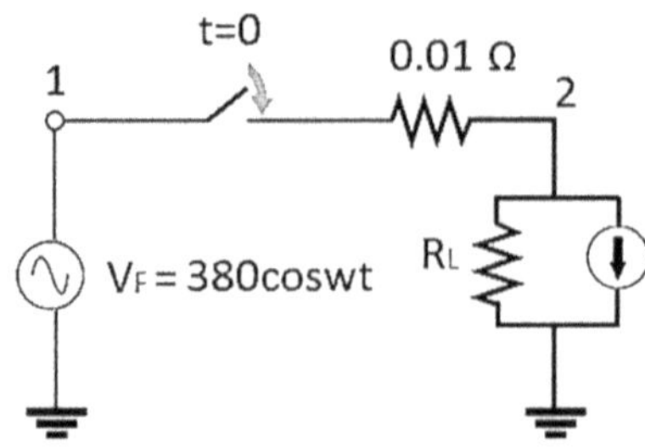

FIGURE 1.7
Equivalent circuit RL.

First time interval: $t = \Delta t$

From the circuit of Figure 1.7, we can write:

$$V_L(t) = V_F(t) - RI_F(t)$$

But

$$I_F(t) = I_L(t - \Delta t) + \frac{V_L(t)}{R_L}$$

Overriding the value of $I_F(t)$, we have:

$$V_L(t) = V_F(t) - R\left[I_L(t - \Delta t) + \frac{V_L(t)}{R_L}\right]$$

Grouping $V_L(t)$, we have:

$$V_L(t)\left[1 + \frac{R}{R_L}\right] = V_F(t) - RI_L(t - \Delta t)$$

The equation for voltage calculation is:

$$V_L(t)\left[1 + \frac{0.01}{20000}\right] = V_F(t) - RI_L(t - \Delta t)$$

$$1.0000005V_L(t) = V_F(t) - RI_L(t - \Delta t)$$

Calculating $V_L(t)$ for $t = \Delta t$:

$$V_F(\Delta t) = 380cos\left(2\pi 60x10^{-6}\right) = 379.9999729967627$$

It comes: $I_L(\Delta t - \Delta t) = I_L(0) = \frac{2V_L(0)}{R_L} + I_L(-\Delta t) = 0$

$$1.0000005V_L(\Delta t) = 379.9999729967627 - 0.01x0$$

$$1.0000005V_L(\Delta t) = 379.9999729967627$$

$$V_L(\Delta t) = 379.99978299687112$$

Second time interval: $t = 2\Delta t$

$$V_F(2\Delta t) = 380cos\left(2x2\pi 60x10^{-6}\right) = 379.9998919870545$$

$$I_L(2\Delta t - \Delta t) = \frac{2V_L(2\Delta t - \Delta t)}{R_L} + I_L(2\Delta t - 2\Delta t)$$

$$I_L(\Delta t) = \frac{2V_L(\Delta t)}{R_L} + I_L(0) = \frac{2x379.99978299687112}{20000} = 0.037999978299687$$

Therefore:

$$1.0000005V_L(2\Delta t) = 379.9998919870545 - 0.01x0.037999978299687$$

$$V_L(2\Delta t) = \frac{379.9995119872716}{1.0000005} = 379.9993219876106$$

MATLAB language code is shown next, and we show the graphs of the inductor voltage and current in Figure 1.8.

```
clc
clear all
format long
%Initial conditions
IL0=0;% Inductor Start Current
VL=[];%VL(0)=0, Inductor start voltage
IF = []; %Source initial current
```

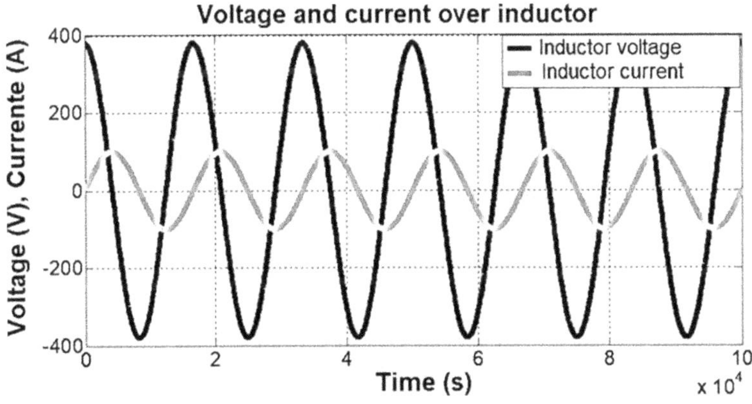

FIGURE 1.8
Inductor voltage and current.

```
at=input ('Enter the interaction step deltat
(seconds)\n');% used in the exercise was 1e-6s
R=input('enter the resistance value (Ohms) \n');% used in
the exercise was 0.01ohm;
L=input('enter the inductance value (Henry)\n');% used in
the exercise was 10e-3H;
timetotal=input('' enter total simulation time value
(seconds) \n');% used in the exercise was 0.1s
RL1=2*L/(at);% resistance for inductor model
intime=round(temptotal/at);% number of interactions
possible at the indicated time (the nearest integer of
interactions)
timetotal=intime*at;% time at which the inductor voltage
will be calculated
for t=1:intime
VF(t) =380*cos(2*pi*60*at*t);
VL(t)=(VF(t)-R*IL0)/(1+(R/RL1));
IL0=(2*VL(t)/RL1)+IL0;
IF(t)=(VF(t)-VL(t))/R;
end
fprintf('% D interactions were made and the values of
inductor voltage and source current \ nfor instant of% ds
(% d * interaction step) \ nbeat was% 2.19f V and% 2.19f
A respectively \ n', intime , timetotal, intime, VL
(intime), IF (intime))
%%%%%%%%%%%%%%%%%%%%%%% Graphics Plot
%%%%%%%%%%%%%%%%%%%%%%%%
t=linspace(1,intime,intime);
plot(t,VL,'k',t,IF,'b')
legend ('Inductor voltage', 'Inductor current')
title('Voltage and current over inductor')
xlabel('Time(s)')
ylabel('Voltage (V), Current (A)')
```

1.3.3 Nodal Equations

In the ATP program, all elements are replaced by their equivalent circuits, and then the nodal conductance matrix of the network under analysis is assembled. The result is a system of linear equations that in computational terms is solved using backward substitution triangularization.

$$[G][v(t)]=[i(t)]-[I] \tag{1.14}$$

where:

$[G]$ = the nodal conductance matrix, which is symmetrical and real and will be constant if the time interval Δt is constant

$[v(t)]$ = the vector of nodal tensions at each instant of time. These are the unknowns of equation (1.12)

$[i(t)]$ = the vector of injected currents
$[I]$ = the vector of known currents, which contains the values of the currents of past time

1.3.3.1 Conductance Matrix Assembly by Inspection

Let us take electrical circuit of Figure 1.9 next.

As already described in the previous item, one can write in matrix form, according to Kirchhoff's first law, the system of equations (1.15):

$$\begin{aligned} i_1 &= g_{12}(v_1 - v_2) + g_{14}(v_1 - v_4) \\ i_2 &= g_{12}(v_2 - v_1) + g_{23}(v_2 - v_3) \\ i_3 &= g_{23}(v_3 - v_2) + g_{34}(v_3 - v_4) \\ i_4 &= g_{14}(v_4 - v_1) + g_{34}(v_4 - v_3) \end{aligned} \tag{1.15}$$

In matrix form:

$$\begin{bmatrix} i_1 \\ i_2 \\ i_3 \\ i_4 \end{bmatrix} = \begin{bmatrix} g_{12}+g_{14} & -g_{12} & 0 & -g_{14} \\ -g_{12} & g_{21}+g_{23} & -g_{23} & 0 \\ 0 & -g_{32} & g_{32}+g_{34} & -g_{34} \\ -g_{41} & 0 & -g_{43} & g_{41}+g_{43} \end{bmatrix} \begin{bmatrix} v_1 \\ v_2 \\ v_3 \\ v_4 \end{bmatrix} \tag{1.16}$$

Assuming that:

$$\begin{aligned} G_{11} &= g_{12} + g_{14} \\ G_{22} &= g_{21} + g_{23} \\ G_{33} &= g_{32} + g_{34} \\ G_{44} &= g_{41} + g_{43} \end{aligned} \tag{1.17}$$

$$\begin{gathered} G_{12} = -g_{12} \; G_{21} = -g_{21} \quad G_{13} = 0 \quad G_{14} = -g_{14} \\ G_{31} = 0 \quad G_{23} = -g_{23} \; G_{32} = -g_{32} \quad G_{24} = 0 \\ G_{41} = -g_{41} \quad G_{42} = 0 \quad G_{34} = -g_{34} \; G_{43} = -g_{43} \end{gathered} \tag{1.18}$$

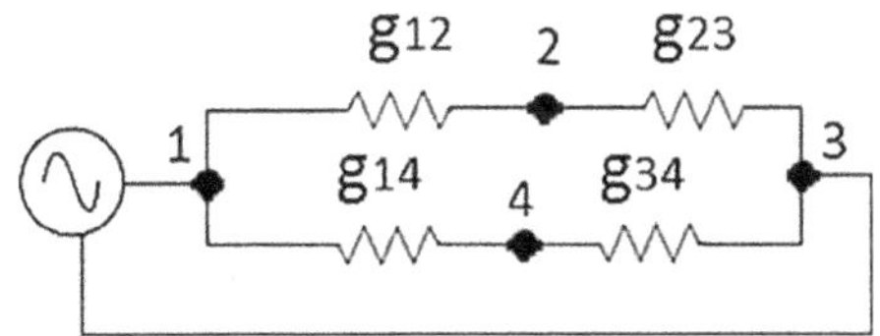

FIGURE 1.9
Four-node circuit.

In equation (1.16), we note that the main diagonal elements are given by the sum of the conductances directly connected to each node.

In equation (1.18), we note that the elements outside the main diagonal are given by the conductance connected between the corresponding nodes, with the opposite sign.

Thus, the nodal conductance matrix can be easily constructed by inspection of:

a. Elements of the main diagonal: the conductances of indices *jj* are given by the sum of all conductances directly connected to node j.
b. Elements outside the main diagonal: the conductances of indices *jk* are given by the equivalent physical conductance (if there are over one branches between two nodes) connected between nodes *j* and *k*, with the opposite sign.

Example 1.2

Solve example 1.1 by mounting the nodal conductance matrix.

MATRIX SOLUTION:

According to Figure 1.7, we have:

Conductance matrix assembly:

$$G_{11} = g_{12} = \frac{1}{0.01} = 100\ S$$

$$G_{12} = G_{21} = -g_{12} = -\frac{1}{0.01} = -100\ S$$

$$G_{22} = g_{20} + g_{12} = \frac{1}{20000} + 100 = 100.00005\ S$$

Nodal equation:

$$\begin{bmatrix} G_{11} & G_{12} \\ G_{21} & G_{22} \end{bmatrix} \begin{bmatrix} V_F(t) \\ V_L(t) \end{bmatrix} = \begin{bmatrix} I_F(t) \\ -I_L(t-\Delta t) \end{bmatrix}$$

Overriding the values:

$$\begin{bmatrix} 100 & -100 \\ -100 & 100.00005 \end{bmatrix} \begin{bmatrix} 380\cos(2\pi 60\Delta t) \\ V_L(t) \end{bmatrix} = \begin{bmatrix} I_F(t) \\ -I_L(t-\Delta t) \end{bmatrix}$$

$$\begin{bmatrix} 380\cos(2\pi 60x10^{-6}) \\ V_L(t) \end{bmatrix} = \begin{bmatrix} 20000.00999933607 & 19999.99999933607 \\ 19999.99999933607 & 19999.99999933607 \end{bmatrix} \begin{bmatrix} I_F(t) \\ -I_L(t-\Delta t) \end{bmatrix}$$

From the first equation, we have:

$$380\cos(2\pi f\Delta t) = 20000.00999933607 I_F(t) - 19999.99999933607 I_L(t-\Delta t)$$

$$I_F(t) = \frac{380\cos(2\pi 60 x \Delta t) + 19999.99999933607 I_L(t-\Delta t)}{20000.00999933607}$$

From the second equation:

$$V_L(t) = 19999.99999933607 I_F(t) - 19999.99999933607 I_L(t-\Delta t)$$

Overriding the value of $I_F(t)$, it comes:

$$V_L(t) = 19999.99999933607\left(\frac{380\cos(2\pi 60 x \Delta t) + 19999.99999933607 I_L(t-\Delta t)}{20000.00999933607}\right) - 19999.99999933607 I_L(t-\Delta t)$$

Gathering $I_L(t-\Delta t)$, we have:

$$V_L(t) = 379.9998100000950\cos(2\pi 60 x \Delta t) - 0.009999995003454 I_L(t-\Delta t)$$

For $t = \Delta t = 10^{-6}$, we have:

$$I_L(\Delta t - \Delta t) = I_L(0) = 0$$

$$V_L(\Delta t) = 379.9997829968712\ (V)$$

For $t = 2\Delta t$

$$I_L(2\Delta t - \Delta t) = \frac{2V_L(2\Delta t - \Delta t)}{R_L} + I_L(2\Delta t - 2\Delta t)$$

$$I_L(\Delta t) = \frac{2V_L(\Delta t)}{R_L} + I_L(0) = \frac{2x379.9997829968712}{20000} = 0.037999978299687$$

$$V_L(t) = 379.9998100000950\cos(2\pi 60 x 2\Delta t) - 0.009999995003454 I_L(\Delta t)$$

$$V_L(t) = 3.799993219876104\ (V)$$

MATLAB code is shown next:

```
clc
clear all;
format long;
%Initial conditions
IL0=0;% Inductor start current
VL=[];%VL(0)=0, Inductor start voltage
IF = []; %Source initial current
```

```
at=input('Enter the interaction step deltat
(seconds)\n');% used in the exercise was 1e-6s
R=input('enter the resistance value (Ohms) \n');% used in
the exercise was 0.01ohm;
L=input('enter the inductance value (Henry)\n');% used in
the exercise was 10e-3H;
timetotal=input('' enter total simulation time value
(seconds) \n');% used in the exercise was 0.1s
RL1=2*L/(at);% resistance for inductor model
intime=round(temptotal/at);% number of interactions possible
at the indicated time (the nearest integer of interactions)
timetotal=intime*at;% time at which the inductor voltage
will be calculated
G=[1/R -1/R;-1/R (1/R + 1/RL1)];
Y=inv(G);
for t=1:intime
VF(t) = 380*cos(2*pi*60*at*t);
IF1(t)=(VF(t)+IL0*Y(1,2))/Y(1,1);
I=[IF1(t);-IL0];
res=Y*I;
VL(t)=res(2,1);
IL0=(2*VL(t)/RL1)+IL0;
end
fprintf('% D interactions were made and the values of
inductor voltage and source current \ nfor instant of% ds
(% d * interaction step) \ nbeat was% 2.19f V and% 2.19f
A respectively \ n', intime , timetotal, intime, VL
(intime), IF (intime))
%%%%%%%%%%%%%%%%%%%%%% Graphics Plot %%%%%%%%%%%%%%%%%%%%%%%%
t=linspace(1,intime,intime);
plot(t,VL,'k',t,IF,'b')
legend ('Inductor voltage', 'Inductor current')
title('Voltage and current over inductor')
xlabel('Time(s)')
ylabel('Voltage (V), Current (A)')
```

1.3.4 Lumped Capacitance Model

Let C be a capacitance connected between the j and k nodes of a circuit, as shown in Figure 1.10.

The capacitance equation that relates voltage and current at its terminals is expressed as:

$$i_{jk}(t) = C\frac{d}{dt}\left[v_{jk}(t)\right] \tag{1.19}$$

$$\int_{t-\Delta t}^{t} dv_{jk}(t) = \frac{1}{C}\int_{t-\Delta t}^{t} i_{jk}(t)dt \tag{1.20}$$

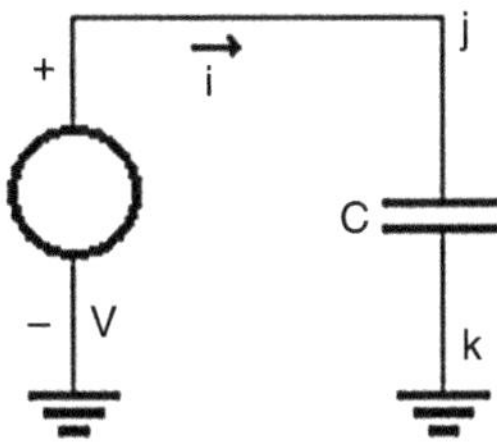

FIGURE 1.10
Capacitance circuit.

Using the trapezoidal integration law, we have:

$$v_{jk}(t) - v_{jk}(t-\Delta t) = \frac{\Delta t}{2C}\left[i_{jk}(t) + i_{jk}(t-\Delta t)\right] \tag{1.21}$$

The equation that defines the branch is expressed as:

$$i_{jk}(t) = \frac{1}{R_C}\left[v_{jk}(t)\right] - \frac{1}{R_C}\left[v_{jk}(t-\Delta t)\right] - i_{jk}(t-\Delta t) = \frac{1}{R_C}\left[v_{jk}(t)\right] - I_C(t-\Delta t) \tag{1.22}$$

where:

$$R_C = \frac{\Delta t}{2C} \tag{1.23}$$

$$I_C(t-\Delta t) = \frac{1}{R_C}\left[v_{jk}(t-\Delta t)\right] + i_{jk}(t-\Delta t) \tag{1.24}$$

Using equation (1.22) with $t = t - \Delta t$, we have:

$$i_{jk}(t-\Delta t) = \frac{1}{R_C}\left[v_{jk}(t-\Delta t)\right] - \frac{1}{R_C}\left[v_{jk}(t-2\Delta t)\right] - i_{jk}(t-2\Delta t) \tag{1.25}$$

Substituting equation (1.25) in (1.24), we have:

$$\begin{aligned} I_C(t-\Delta t) = &\frac{1}{R_C}\left[v_{jk}(t-\Delta t)\right] + \frac{1}{R_C}\left[v_{jk}(t-\Delta t)\right] - \frac{1}{R_C}\left[v_{jk}(t-2\Delta t)\right] \\ &- I_{jk}(t-2\Delta t) = \frac{2}{R_C}\left[v_{jk}(t-\Delta t)\right] - \frac{1}{R_C}\left[v_{jk}(t-2\Delta t)\right] \\ &- i_{jk}(t-2\Delta t) = \frac{2}{R_C}\left[v_{jk}(t-\Delta t)\right] - I_C(t-2\Delta t) \end{aligned} \tag{1.26}$$

Therefore:

$$I_C(t-\Delta t) = \frac{2}{R_C}\left[v_{jk}(t-\Delta t)\right] - I_C(t-2\Delta t) \tag{1.27}$$

The capacitance equivalent circuit is shown in Figure 1.11.

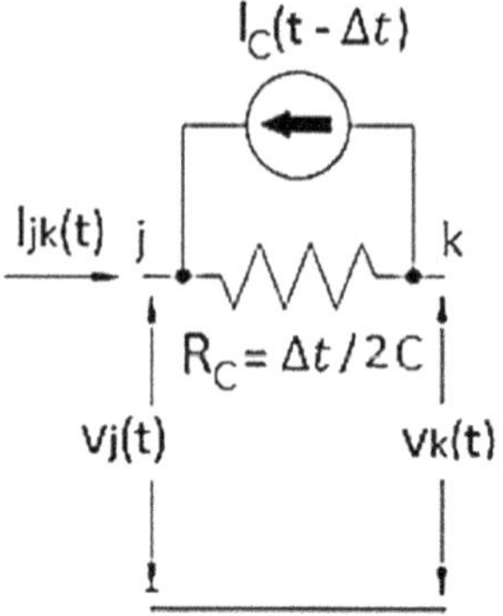

FIGURE 1.11
Capacitance equivalent circuit.

Example 1.3

Calculate $V_C(t)$ for $t = 2\Delta t$ in the circuit of Figure 1.12. The frequency is 60 Hz and $\Delta t = 1$ μseg. The initial conditions are given by:

$$V_F\left(0^-\right)=0;\ I_F\left(0^-\right)=0;\ V_C\left(0^-\right)=0;\ I_C\left(0^-\right)=0.$$

SOLUTION:

Circuit model:

$$R_C=\frac{\Delta t}{2C}=\frac{10^{-6}}{2x10^{-6}}=0.5\ \Omega$$

$$I_C\left(t-\Delta t\right)=\frac{2V_C\left(t-\Delta t\right)}{R_C}-I_C(t-2\Delta t)$$

First time interval: $t = \Delta t$

From the circuit of Figure 1.13, we can write:

$$V_C\left(t\right)=V_F\left(t\right)-RI_F\left(t\right)$$

$$I_F\left(t\right)=-I_C\left(t-\Delta t\right)+\frac{V_C\left(t\right)}{R_C}$$

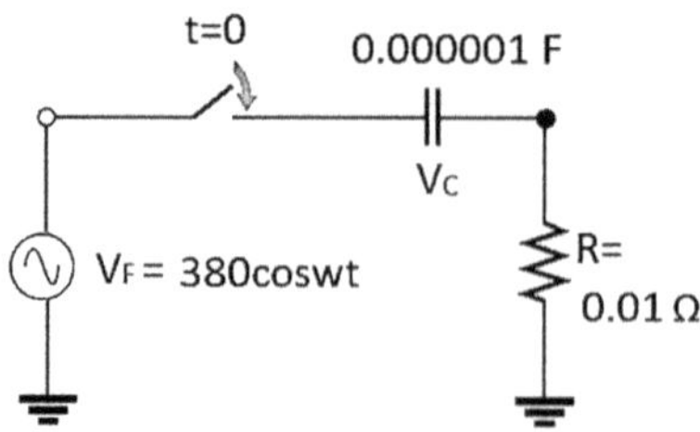

FIGURE 1.12
RC circuit.

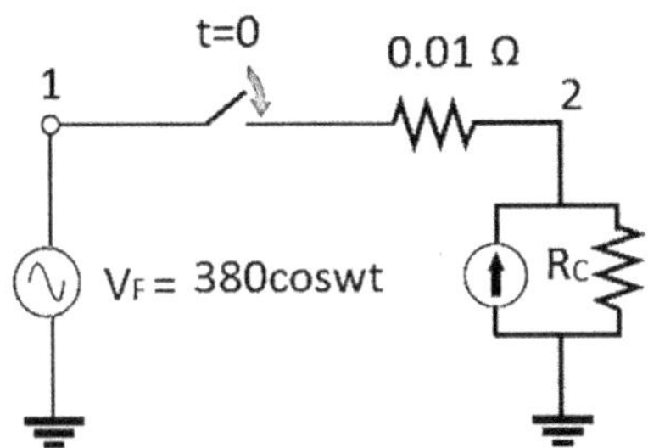

FIGURE 1.13
RC equivalent circuit.

Overriding the value of $I_F(t)$, we have:

$$V_C(t) = V_F(t) - R\left[-I_C(t-\Delta t) + \frac{V_C(t)}{R_C}\right]$$

Grouping $V_C(t)$, we have:

$$V_C(t)\left[1 + \frac{R}{R_C}\right] = V_F(t) + RI_C(t-\Delta t)$$

The equation for voltage calculation is:

$$V_C(t)\left[1 + \frac{0.01}{0.5}\right] = V_F(t) + RI_C(t-\Delta t)$$

$$1.02V_C(t) = V_F(t) + RI_C(t-\Delta t)$$

$$V_F(\Delta t) = 380cos\left(2\pi 60x10^{-6}\right) = 379.9999729967627$$

And: $I_C(\Delta t - \Delta t) = I_C(0) = \frac{2V_C(0)}{R_C} + I_C(-\Delta t) = 0$

$$1.02V_C(t) = 379.9999729967627$$

$$V_C(t) = 372.5489931340811$$

Second time interval: $t = 2\Delta t$

$$V_F(2\Delta t) = 380cos\left(2x2\pi 60x10^{-6}\right) = 379.9998919870545$$

$$I_C(2\Delta t - \Delta t) = \frac{2V_C(2\Delta t - \Delta t)}{R_C} - I_C(2\Delta t - 2\Delta t)$$

$$I_C(\Delta t) = \frac{2V_C(\Delta t)}{R_C} - I_C(0) = \frac{2x372.5489931340811}{0.5} = 1490.195972536325$$

$$1.02V_C(2\Delta t) = 379.9998919870545 + 0.01x1490.195972536325$$

$$V_C(2\Delta t) = 387.1586781494292\ (V)$$

MATLAB code is shown next:

```
clc
clear all
format long
%Initial conditions
IC0=0;% Capacitor start current
VC=[];%VC(0)=0, Capacitor start voltage
IF = []; %Source start current
at=input('Enter the interaction step deltat
(seconds)\n');% used in the exercise was 1e-6s
R=input('enter the resistance value (Ohms) \n');% used in
the exercise was 0.01ohm;
C=input('enter the capacitance value (Faraday)\n');% used
in the exercise was 10e-6H;
temptotal=input('' enter total simulation time value
(seconds) \n');% used in the exercise was 0.1s
RC1=at/(2*C);% resistance for capacitor model
intime=round(temptotal/at);% number of interactions
possible at the indicated time (the nearest integer of
interactions)
timetotal=intime*at;% time at which the capacitor voltage
will be calculated
for t=1:intime
VF(t) =380*cos(2*pi*60*at*t);
VC(t)=(VF(t)+R*IC0)/(1+(R/RC1));
IC0=(2*VC(t)/RC1)-IC0;
IF(t)=(VF(t)-VC(t))/R;
end
fprintf('%d interactions were made and the values of
capacitor voltage and source current \ nfor instant of% ds
(% d * interaction step) \ nbeat was% 2.19f V and% 2.19f A
respectively \ n', intime , timetotal, intime, VL (intime),
IF (intime))
```

Example 1.4

Solve example 1.3 by mounting the nodal conductance matrix.

SOLUTION:

Conductance matrix assembly:

$$G_{11} = g_{12} = \frac{1}{0.01} = 100\ S$$

$$G_{12} = G_{21} = -g_{12} = -100\ S$$

$$G_{22} = g_{20} + g_{12} = \frac{1}{0.5} + 100 = 102\ S$$

Nodal equation:

$$\begin{bmatrix} G_{11} & G_{12} \\ G_{21} & G_{22} \end{bmatrix} \begin{bmatrix} V_F(t) \\ V_C(t) \end{bmatrix} = \begin{bmatrix} I_F(t) \\ I_C(t-\Delta t) \end{bmatrix}$$

Overriding the values:

$$\begin{bmatrix} 100 & -100 \\ -100 & 102 \end{bmatrix} \begin{bmatrix} 380\cos(2\pi 60\Delta t) \\ V_C(t) \end{bmatrix} = \begin{bmatrix} I_F(t) \\ I_C(t-\Delta t) \end{bmatrix}$$

$$\begin{bmatrix} 380\cos(2\pi 60x10^{-6}) \\ V_C(t) \end{bmatrix} = \begin{bmatrix} 0.51 & 0.50 \\ 0.50 & 0.50 \end{bmatrix} \begin{bmatrix} I_F(t) \\ I_C(t-\Delta t) \end{bmatrix}$$

From the first equation, we have:

$$380\cos(2\pi f\Delta t) = 0.51 I_F(t) + 0.5 I_C(t-\Delta t)$$

$$I_F(t) = \frac{380\cos(2\pi 60x\Delta t) - 0.5 I_C(t-\Delta t)}{0.51}$$

From the second equation, we have:

$$V_C(t) = 0.5 I_F(t) + 0.5 I_C(t-\Delta t)$$

Overriding the value of $I_F(t)$, comes:

$$V_C(t) = 0.5\left(\frac{380\cos(2\pi 60x\Delta t) - 0.5 I_C(t-\Delta t)}{0.51}\right) + 0.5 I_C(t-\Delta t)$$

Gathering $I_C(t-\Delta t)$, we have:

$$V_C(t) = 372.5490196078432\cos(2\pi 60x\Delta t) + 0.009803921568627 I_C(t-\Delta t)$$

For $t = \Delta t = 10^{-6}$, we have:

$$I_C(\Delta t - \Delta t) = I_C(0) = 0$$

$$V_C(\Delta t) = 372.5490196078432\cos(2\pi 60x10^{-6}) = 372.5489931340811\ (V)$$

For $t = 2\Delta t$

$$I_C(2\Delta t - \Delta t) = \frac{2V_C(2\Delta t - \Delta t)}{R_C} - I_C(2\Delta t - 2\Delta t)$$

$$I_C(\Delta t) = \frac{2V_C(\Delta t)}{R_C} + I_C(0) = \frac{2x372.5489931340811}{0.5} = 1490.195972536325$$

$$V_C(2\Delta t) = 372.5490196078432\cos(2\pi 60x2\Delta t) + 0.009803921568627 I_C(\Delta t)$$

$$V_C(2\Delta t) = 387.1586781494286\ (V)$$

1.4 Transmission Line Models

We can divide existing models into three types:

- Models with lumped parameters;
- Models with distributed parameters;
- Models with frequency-dependent parameters.

1.4.1 Models with Lumped Parameters

For single-phase modeling, the model used is the π circuit, as shown in Figure 1.14.

For the three-phase modeling, the model used is the coupled π circuit, as shown in Figure 1.15.

The impedance matrix is assembled using Carson's equations and a Kron reduction when the lines have one or two ground wire cables.

For both single-phase and three-phase lines, modeling can be done using multiple cascaded π circuits. For example, each portion of a transposed TL may be represented by a circuit π, as shown in Figure 1.16.

1.4.2 Models with Distributed Parameters

Bergeron's model uses the energy propagation phenomenon of a TL and the disturbance propagates subject to attenuation taking a finite time to be reflected by one end of the line, with a delay of voltages and currents at opposite ends. Although the Bergeron model takes into account the distributed characteristic of the parameters, they are considered frequency invariant. With the application of the Bergeron characteristics method and the trapezoidal method for single-phase modeling, one can have two models: lossless and lossy.

For a lossless line, the transit time τ for a wave to travel the distance l is shown in Figure 1.17.

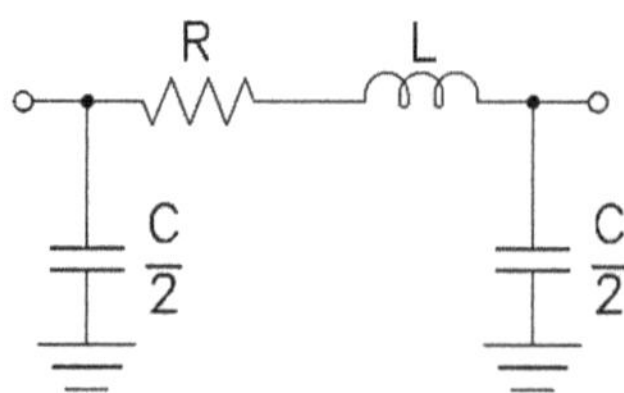

FIGURE 1.14
Circuit for single-phase TL modeling.

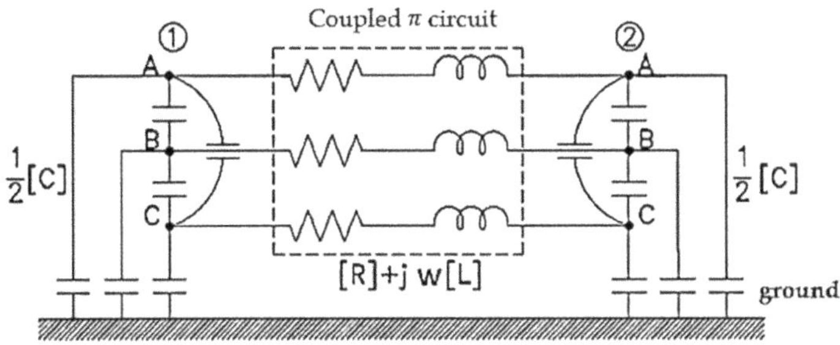

FIGURE 1.15
Three-phase TL modeling circuit.

$$\tau = \frac{l}{v} = \frac{l}{\frac{1}{\sqrt{LC}}} = l\sqrt{LC}\ (s) \tag{1.28}$$

where:
v = the wave propagation speed in (m/s)
L = the inductance of TL in (H)
C = the capacitance of TL
l = the length of TL

The equation that relates voltage and current between the j and k ends of a line length with transit time τ is given by:

$$v_j(t-\tau) + Zi_{jk}(t-\tau) = v_k(t) - Zi_{kj}(t) \tag{1.29}$$

where the positive current is from node j to node k.
Similarly, for a wave moving from k to j.

$$v_k(t-\tau) + Zi_{kj}(t-\tau) = v_j(t) - Zi_{jk}(t) \tag{1.30}$$

From (1.30):

$$i_{jk}(t) = \frac{1}{Z} v_j(t) - i_j(t-\tau) \tag{1.31}$$

$$i_j(t-\tau) = i_{kj}(t-\tau) + \frac{1}{Z} v_k(t-\tau) \tag{1.32}$$

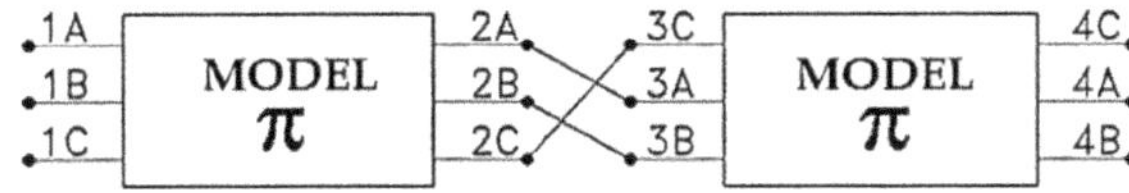

FIGURE 1.16
Cascade π circuits.

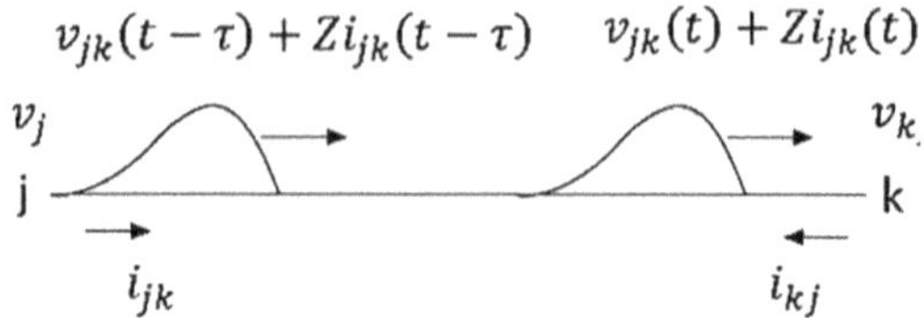

FIGURE 1.17
Wave shift between two ends of TL.

From (1.29):

$$i_{kj}(t)=\frac{1}{Z}v_k(t)-i_k(t-\tau) \tag{1.33}$$

$$i_k(t-\tau)=i_{jk}(t-\tau)+\frac{1}{Z}v_j(t-\tau) \tag{1.34}$$

Replacing (1.31) in (1.34) for $t=t-\tau$, we get:

$$i_k(t-\tau)=\frac{1}{Z}v_j(t-\tau)-i_j(t-2\tau)+\frac{1}{Z}v_j(t-\tau)=\frac{2}{Z}v_j(t-\tau)-i_j(t-2\tau) \tag{1.35}$$

Replacing (1.33) in (1.32), for $t=t-\tau$, we get:

$$i_j(t-\tau)=\frac{1}{Z}v_k(t-\tau)-i_k(t-2\tau)+\frac{1}{Z}v_k(t-\tau)=\frac{2}{Z}v_k(t-\tau)-i_k(t-2\tau) \tag{1.36}$$

The equivalent circuit is shown in Figure 1.18.

Example 1.5

A voltage source $v(t)=230cos377t$ energizes an inductance of 0.1 H through a 246 Ω surge impedance TL. A 1000 Ω preinsertion resistor is inserted during switching.

Calculate the voltage transients at nodes j and k to $t=\Delta t$. The integration step is $\Delta t=0.1$ ms. The line transit time is 1.5 ms.

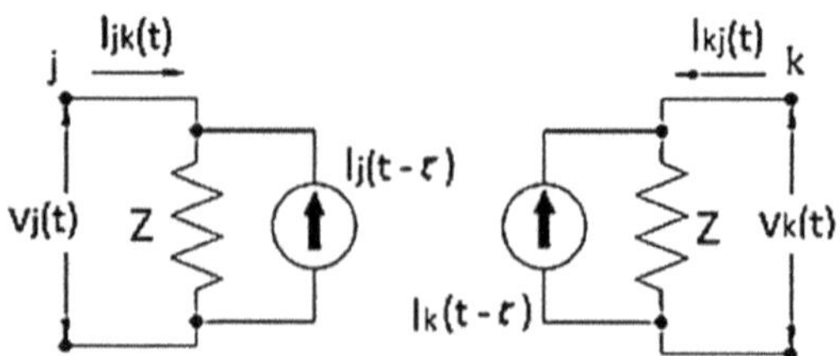

FIGURE 1.18
Lossless transmission line equivalent circuit.

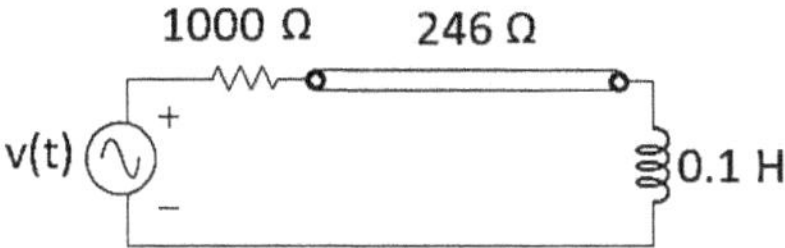

FIGURE 1.19
Circuit for example 1.5.

Currents and voltages are zero for times less than zero. The problem circuit is shown in Figure 1.19 with the resistor already connected.

SOLUTION:

The equivalent circuit for the example is shown in Figure 1.20, where the voltage source has been transformed into a current source.

Inductor resistance calculation:

$$R_L = \frac{2L}{\Delta t} = \frac{2x0.1}{0.1x10^{-3}} = 2000\ \Omega$$

Conductance matrix assembly:

$$G_{11} = g'_{10} + g''_{10} = \frac{1}{1000} + \frac{1}{246} = 0.00506504065$$

$$G_{22} = g'_{20} + g''_{20} = \frac{1}{2000} + \frac{1}{246} = 0.00456504065$$

$$G_{12} = G_{21} = -g_{12} = 0$$

Therefore:

$$\begin{bmatrix} 0.00506504065 & 0 \\ 0 & 0.00456504065 \end{bmatrix} \begin{bmatrix} v_j(t) \\ v_k(t) \end{bmatrix} = \begin{bmatrix} 0.23cos377t + i_j\left(t - 1.5x10^{-3}\right) \\ i_k\left(t - 1.5x10^{-3}\right) - I_L(t - \Delta t) \end{bmatrix}$$

$$\begin{bmatrix} v_j(t) \\ v_k(t) \end{bmatrix} = \begin{bmatrix} 197.4318 & 0 \\ 0 & 219.0561 \end{bmatrix} \begin{bmatrix} 0.23cos377t + i_j\left(t - 1.5x10^{-3}\right) \\ i_k\left(t - 1.5x10^{-3}\right) - I_L(t - \Delta t) \end{bmatrix}$$

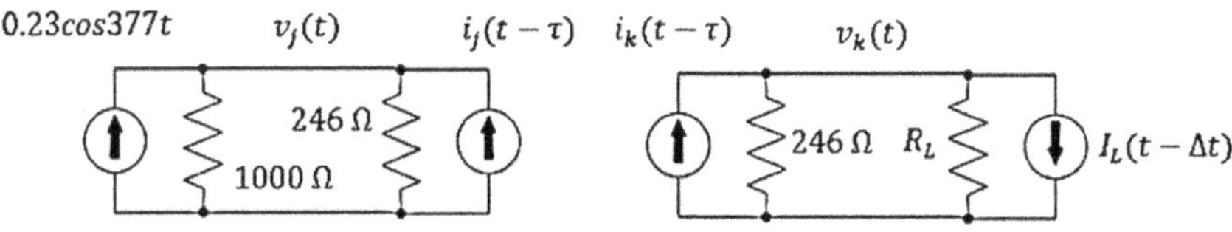

FIGURE 1.20
Equivalent circuit.

The iteration equations are given by:

$$v_j(t) = 197.4318\left[0.23cos377t + i_j\left(t - 1.5x10^{-3}\right)\right]$$

$$v_k(t) = 219.0561\left[i_k\left(t - 1.5x10^{-3}\right) - I_L(t - \Delta t)\right]$$

Using equations (1.28) and (1.29), we have:

$$i_k(t - \tau) = \frac{2}{246}v_j(t - \tau) - i_j\left(t - 2x1.5x10^{-3}\right)$$

$$i_j(t - \tau) = \frac{2}{246}v_k(t - \tau) - i_k\left(t - 2x1.5x10^{-3}\right)$$

For the inductor current source, we will use equation (1.12).

$$I_L(t - \tau) = \frac{2v_k(t - \tau)}{2000} + I_L(t - 2x0.1x10^{-3})$$

We can get the current equations at time t:

$$i_k(t) = \frac{2}{246}v_j(t) - i_j\left(t - 1.5x10^{-3}\right)$$

$$i_j(t) = \frac{2}{246}v_k(t) - i_k\left(t - 1.5x10^{-3}\right)$$

$$I_L(t) = \frac{2v_k(t)}{2000} + I_L\left(t - 0.1x10^{-3}\right)$$

For $t = 0$

$$i_j\left(-1.5x10^{-3}\right) = 0$$

$$v_j(0) = 197.4318x0.23cos377x0 + 0 = 45.409314\ (V)$$

$$i_k(0) = \frac{2}{246}v_j(0) - i_j\left(0 - 1.5x10^{-3}\right) = \frac{2x45.409314}{246} + 0 = 0.369181414$$

$$I_L\left(-0.1x10^{-3}\right) = 0$$

$$v_k(0) = 219.0561\left[i_k\left(0 - 1.5x10^{-3}\right) - I_L\left(0 - 0.1x10^{-3}\right)\right] = 0$$

Para $t = \Delta t$

$$v_k\left(0.1x10^{-3}\right) = 219.0561\left[i_k\left(0.1x10^{-3} - 1.5x10^{-3}\right) - I_L(0)\right] = 0$$

$$i_j\left(0.1x10^{-3}\right) = \frac{2}{246}v_k\left(0.1x10^{-3}\right) - i_k\left(0.1x10^{-3} - 1.5x10^{-3}\right) = 0$$

$$v_j\left(0.1x10^{-3}\right) = 197.4318x0.23cos377x0.1x10^{-3} + i_j\left(0.1x10^{-3} - 1.5x10^{-3}\right)$$
$$= 45.40930417 + 0$$

$$i_k\left(0.1x10^{-3}\right)=\frac{2}{246}v_j\left(0.1x10^{-3}\right)-i_j\left(0.1x10^{-3}-1.5x10^{-3}\right)=0$$

$$I_L\left(0.1x10^{-3}\right)=\frac{2v_k\left(0.1x10^{-3}\right)}{2000}+I_L(0)=0$$

MATLAB code is shown next:

```
clc
clear all;
format long;
%Initial conditions
Ij=[];
Ik=[];
Vk=[];
Vj=[];
IL=0;
R = 1000;
tal=1.5e-3;
at=0.1e-3;
Rz=246;
L=0.1;
timetotal=10e-3;
Tal = round(tal/at);
Ij(1:Tal)=0;
Ik(1:Tal)=0;
RL1=2*L/(at);                  % resistance for inductor
                                 model
intime=round(timetotal/at);    % number of interactions
                                 possible at the indicated
                                 time (nearest number of
                                 interactions)
timetotal=intime*at;           % time at which the
                                 inductor voltage will be
                                 calculated
G=[(1/R + 1/Rz),0 ;0 , (1/RL1 + 1/Rz)];
Y=inv(G);
for t=1:intime+1
Vj(t) = Y(1,1)*(0.23*cos(377*at*(t-1)) + Ij(t));
Vk(t) = Y(2,2)*(Ik(t) - IL);
if t>=Tal
Ij(t+1) = (2/Rz)*Vk(t-Tal+1) - Ik(t-Tal+1);
Ik(t+1) = (2/Rz)*Vj(t-Tal+1) - Ij(t-Tal+1);
IL = (2/RL1)*Vk(t-Tal+1) + IL;
end
end
fprintf(% d interactions were made and the voltage value
Vj and Vk \ nfor instant of% ds (% d * interaction step) \
nbound was% 2.19f V and% 2.19f V respectively \n',intime,
timetotal,intime,Vj(intime+1),Vk(intime+1))
```

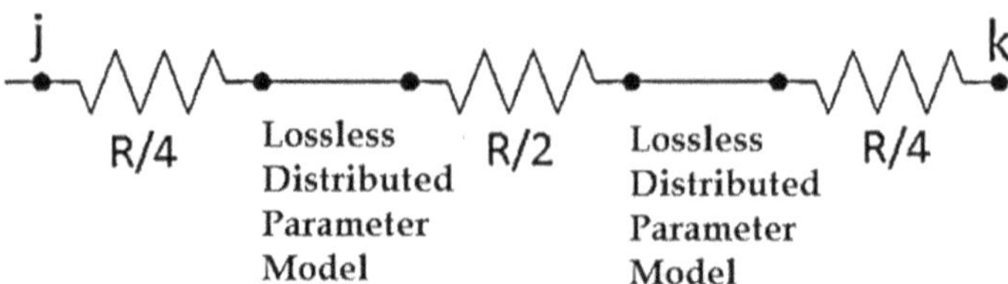

FIGURE 1.21
Circuit with lossy distributed parameters.

Losses are inserted into the previous model by placing concentrated resistors as shown in Figure 1.21.

Three-phase modeling uses modal quantities, which is equivalent to breaking down the three-phase system into three single-phase systems. At the end of the study, the inverse process is performed and the quantities of the three-phase system are obtained.

When the three-phase TL is not transposed, it is necessary to calculate eigenvalues and eigenvectors to get the transformation matrix from phase components to modal components.

1.4.3 Models with Frequency-Dependent Parameters

ATPdraw uses the following frequency-dependent models:

a. Semlyen;
b. JMarti;
c. Noda (Taku Noda).

These models use the convolution process.

1.5 Models of Lightning Discharge Current Sources

There are four different type-15 fonts in ATPdraw software to represent an atmospheric surge. The sources are Double Exponential, Heidler, Standler and Cigré, and all can be chosen as voltage or current sources.

1.5.1 Double Exponential Source

Figure 1.22 represents this source.

To implement a Double Exponential font in ATPdraw, some values must be entered to define the discharge, indicated by α, β, and A.

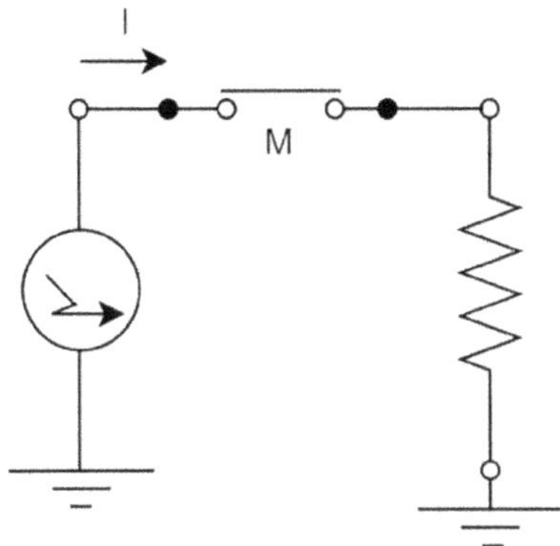

FIGURE 1.22
Double Exponential source representation.

As initial conditions, it can be assumed that β is much larger than α, so X will be much smaller than 1. Therefore, a low value greater than zero must be set for X. Then an iterative process is started until it is reached the desired response for a given error value. At the end of this process, the values of the variables should be entered in the source model. This model depends on the peak discharge values and the crest and tail times.

The function representing this source is shown in the following equation:

$$f(t) = Amp\left(e^{At} - e^{Bt}\right) \tag{1.37}$$

where:

Amp = constant in [A] or [V]. It does not correspond exactly to the maximum value of the wave increase
A = negative number specifying the downward slope
B = negative number specifying the upward slope

Figure 1.23 shows the Double Exponential type font curve.

When the ratio of tail time to crest time is greater than 3, the parameters found for the curve are accurate and represent well the source. Otherwise, such values are not accurate and there is an error in at least one of these values. To represent a wave of the standard current form, 8/20 μs, using this source is not a good option.

The original font of two exponential types-15 has many disadvantages:

1. It is just a rough approximation of the measured lightning currents because the highest time of the function occurs at time zero. This does not correspond to the pulse shape proposed by CIGRÉ Study Committee No 33.

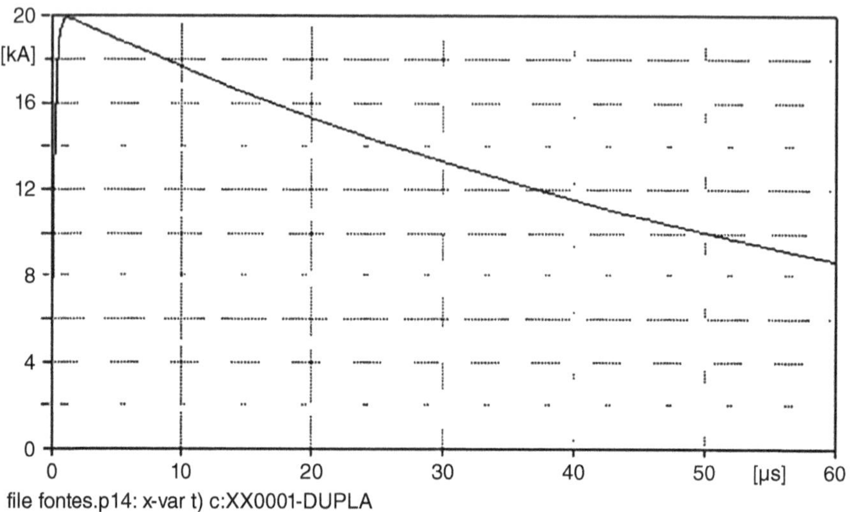

FIGURE 1.23
Current surge waveform for Double Exponential source (1.2/50 μs).

2. It is complicated for the user that the Amplitude field does not correspond to the maximum value.
3. Finally, because of the numerical instability of the formula (subtraction of "two exponentials"), numerical oscillations may occur.

1.5.2 Heidler Source

Bernd Stein of FGH Mannheim West Germany introduced this model, and it is shown in Figure 1.24.

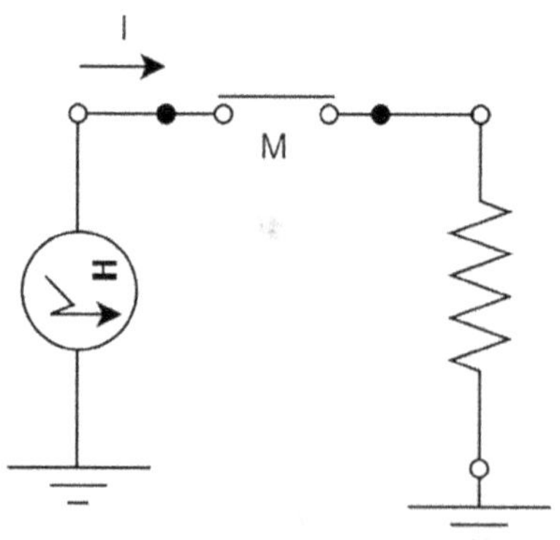

FIGURE 1.24
Heidler source representation.

The Heidler source has three parameters for its definition: A, which is the peak value of the curve; T_f, which is the wavefront time given in seconds; and τ, which is the time in seconds when the amplitude of the surge drops to 37% of the peak value. Equation 1.38 defines its curve.

$$f(t) = A\left(\left(\frac{t}{T_f}\right)^{\left(\frac{n}{\left(1+\left(\frac{t}{T_f}\right)^n\right)e^{-\left(\frac{t}{\tau}\right)}}\right)}\right) \tag{1.38}$$

where n is a factor of influence of the growth rate of the function, which increases proportionally with the slope of the surge wave.

Figure 1.25 shows the Heidler source curve.

Since the known values of a discharge are the wave front, tail, and peak current or voltage times, two parameters for this surge source are already defined. The value of τ does not correspond to the tail time, since this is the time for the amplitude of the curve to be half of its maximum value, but the two values are close. The most practical way to determine τ is to perform successive simulations, with the tail time as the initial value, until the shape got adequately represents the desired discharge.

The model is reasonable and accurate for "standard overvoltage functions" (e.g., 1.2/50 μs).

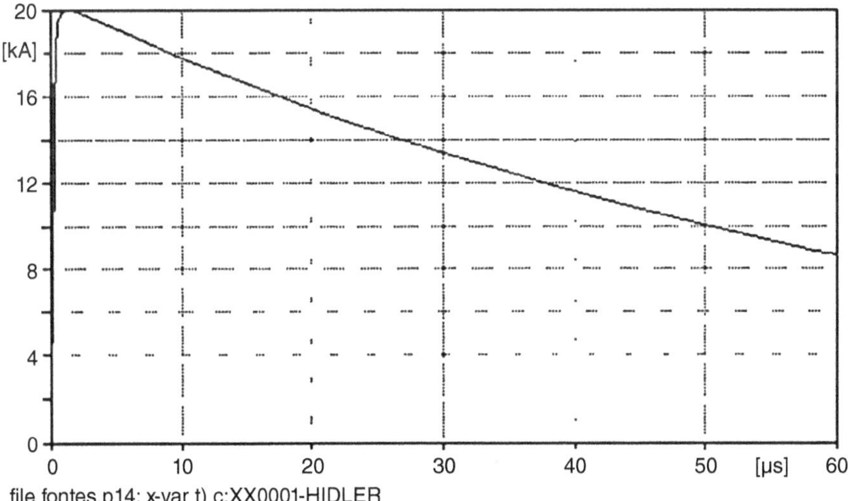

FIGURE 1.25
Current surge waveform for Heidler-type source (1.2/50 μs).

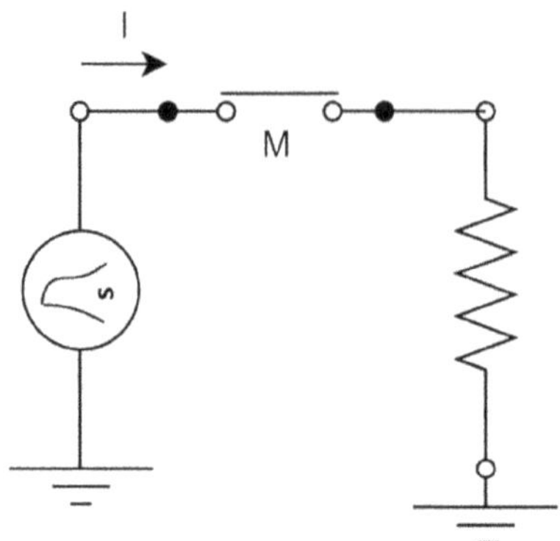

FIGURE 1.26
Standler source representation.

1.5.3 Standler Source

Standler font is shown in Figure 1.26.

The Standler source is got from a function close to the previous source, but the variables represent different parameters: τ is the tail time; *Amp* is a constant; and *n* is an exponent. Standler's source equation is given by:

$$f(t) = Amp\left(\frac{t}{\tau}\right)^n e^{-\left(\frac{t}{\tau}\right)} \tag{1.39}$$

This wave type presents a more complex parameterization form than the Heidler type, since only the tail time data is entered directly into this source model. The waveform representation is shown in Figure 1.27.

1.5.4 CIGRÉ Source

CIGRÉ source is shown in Figure 1.28.

CIGRÉ source (Figure 1.29) presents as variables: *A*, which is its amplitude; T_f, which is the wavefront time; T_h, which is the tail time; S_{max}, which is the maximum rate of increase, amps per second for a current source and volts per second for a voltage source.
The parameters are:

U/I = 0: Voltage source.

–1: Actual source.

Amp = Amplitude of the function in [A] or [V].

Tf = The front time constant in seconds.

Th = Tail time constant in seconds.

Smax = Maximum slope (amps or volts per second).

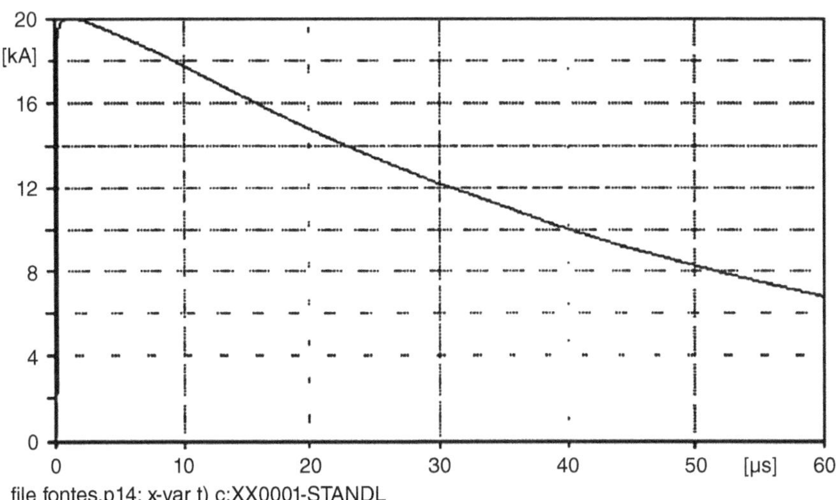

FIGURE 1.27
Current surge waveform for Standler type source (1.2/50μs).

Tsta = Start time in seconds. Zero font value for T < Tsta.

Tsto = End time in seconds. Zero font value for T > Tsto.

Node: CIGRÉ = Positive node of overvoltage function.

The negative node is grounded.

Since the variables that define an outbreak are the peak values and the crest, or wavefront, and tail times, it is not common to have the rate of change of the curve as the initial data of the simulation. This makes getting the parameters for this font type more complex.

The choice of one of the four sources should be based on their comparison.

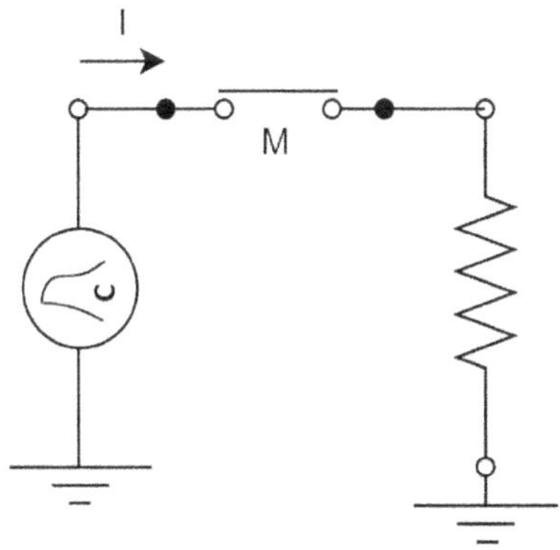

FIGURE 1.28
Cigré source representation.

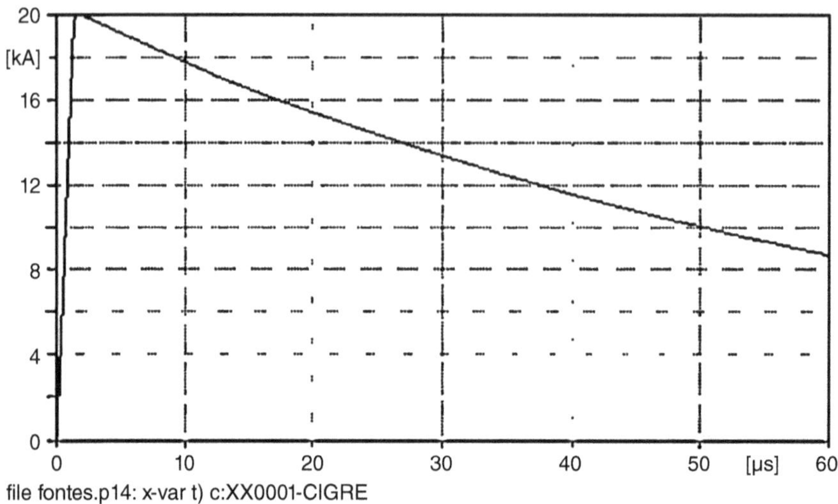

FIGURE 1.29
Current surge waveform for CIGRÉ type source (1.2/50 μs).

1.6 Numerical Oscillations

Several methodologies that take advantage of the trapezoidal rule have been proposed. However, this method has numerical oscillations. The integration step chosen determines the maximum frequency that can be simulated. This means that the user must know in advance the frequency range of the transient process being simulated, as there may be transients at different frequencies in the various power system bars under study.

Most times, the trapezoidal rule causes deviations in the solution of the transient process and introduces sustained oscillations of no physical significance. Several techniques have been proposed to control or reduce these oscillations: additional damping, time modification of the solution method, critical damping change procedure, and interpolation.

1.7 Using ATPdraw

ATPdraw is a great tool for users who are just starting to use electrical power system programs. Both single-phase and complex three-phase circuits and particularly TLs with concentrated, distributed and frequency-dependent parameters can be constructed and simulated using ATPdraw.

The capacity of the standard distributed program is as follows: 6,000 nodes, 10,000 branches, 900 sources, 2,250 nonlinear elements, 1,200 keys, and 90 synchronous machines.

All ATPdraw components can be seen in Figure 1.30.

ATPdraw can be got from different sources around the world as shown in Figure 1.31.

ATPdraw group contacts can be found on the Internet in the following link: https://www.eeug.org/index.php/about-eeug/who-we-are/user-groups. Examples: Canadian/American EMTP User Group. Drs. W. Scott Meyer and Tsu-huei Liu, Co-chairmen. 3179 Oak Tree Court West Linn, Oregon 97068. USA. E-mail: canam@emtp.org. European EMTP-ATP Users Group Assoc. (EEUG). Prof. Dr.-Ing. Harald Wehrend Chairman (EEUG). University of Applied

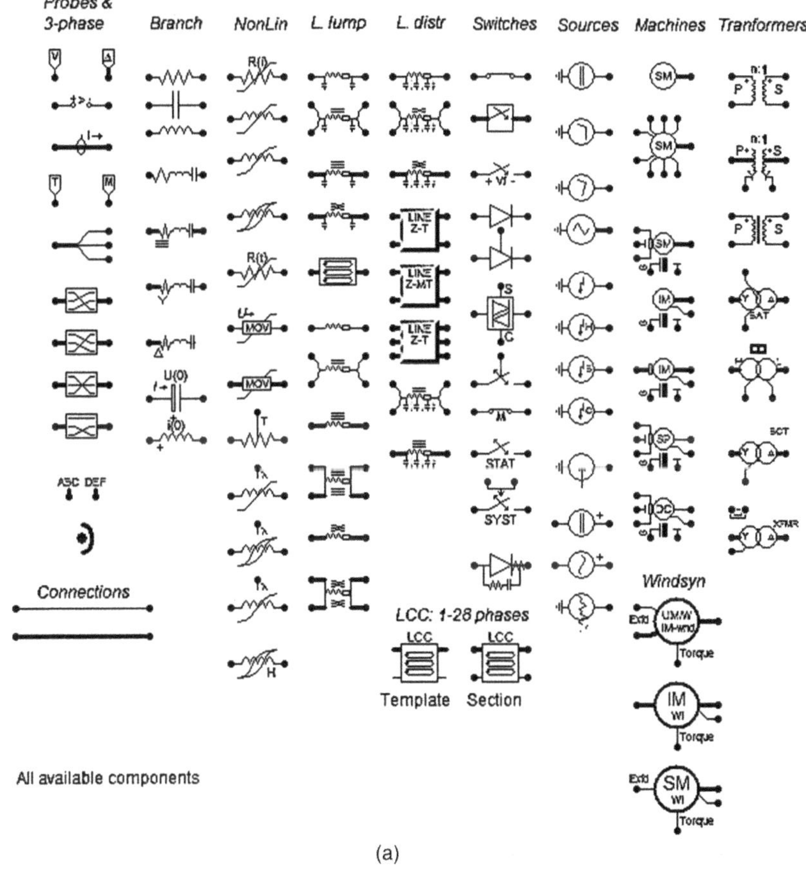

FIGURE 1.30
(a) ATPdraw components. (b) Continuation for ATPdraw components.

(Continued)

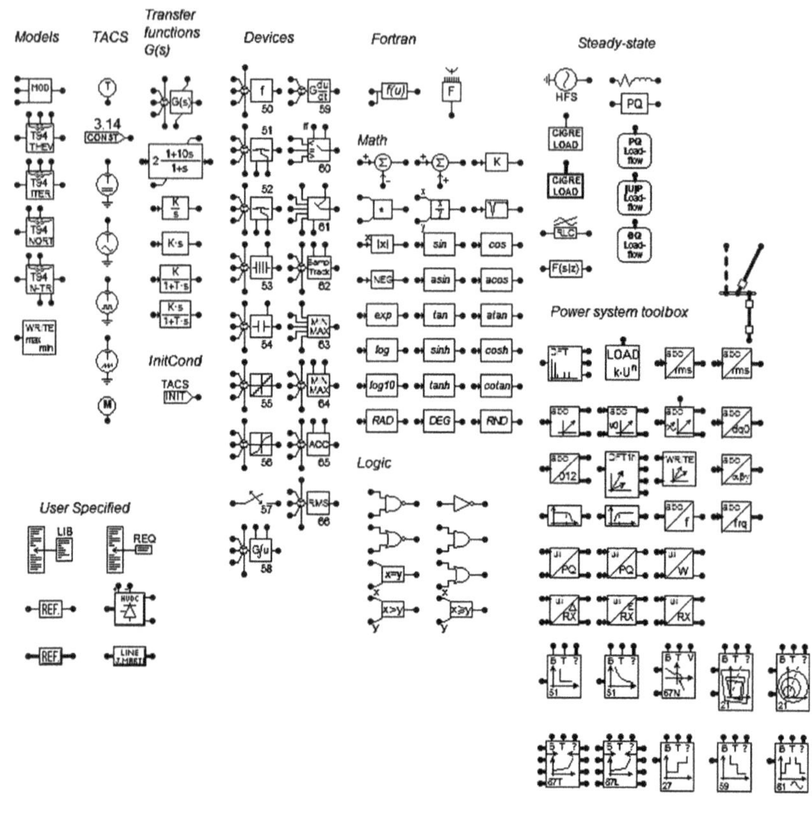

FIGURE 1.30
(Continued)

Sciences Kiel Faculty of Computer Science and Electrical Eng. Grenzstraße 5 D-24149 Kiel Germany. WWW: http://www.eeug.org. E-mail: eeug@emtp.org

After the program installation, ATPdraw configuration should be performed, making the procedure trigger Tools - Options - Preferences and configure the program according to Figure 1.32.

The basic elements for electrical circuit simulation in ATPdraw are shown in Figures 1.33 to 1.36 and described below. The respective figures are obtained by pressing the right mouse key on the ATPdraw screen.

Standard measurement components (Probes & 3-phase):

Probe Volt: voltmeter (voltage to ground);

Probe Branch Voltage: voltmeter (voltage between two circuit points);

Probe Current: ammeter (current at one branch of the circuit);

Splitter: transforms a single-phase node into three phases or vice versa.

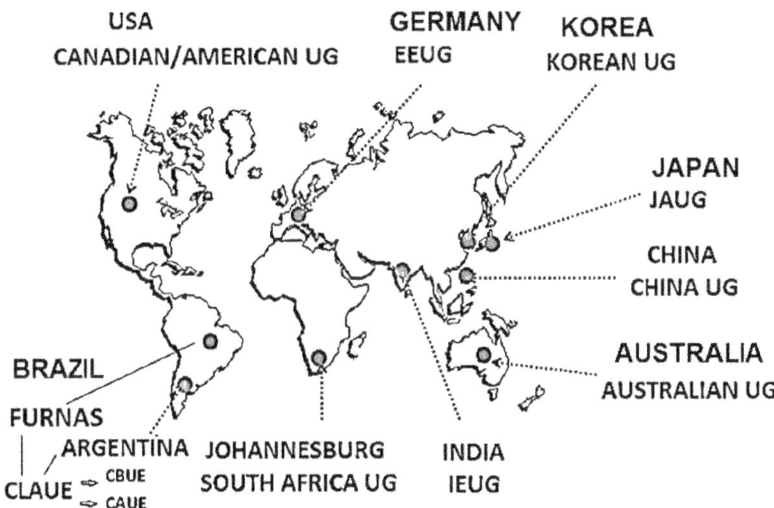

FIGURE 1.31
Location of ATPdraw groups.

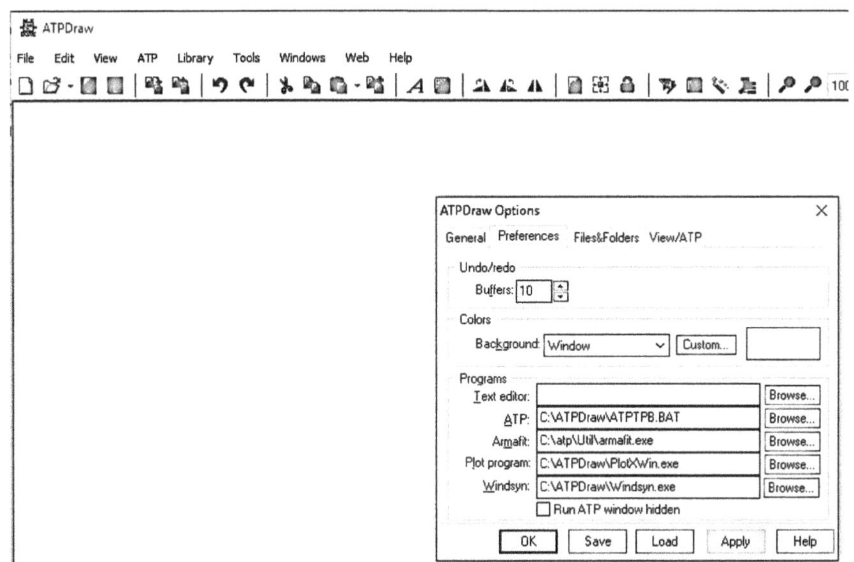

FIGURE 1.32
ATPdraw configuration.

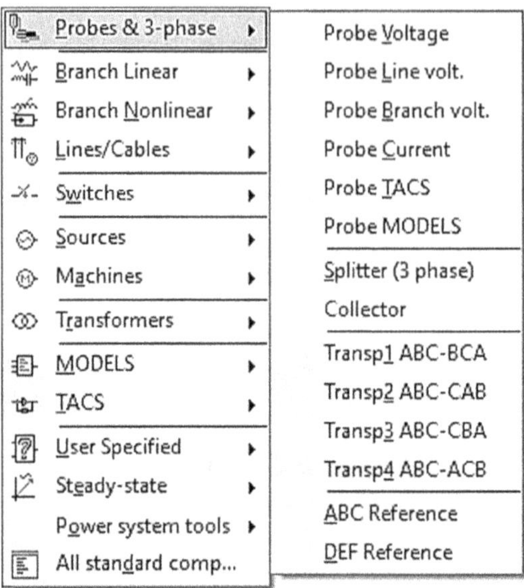

FIGURE 1.33
Standard measurement components.

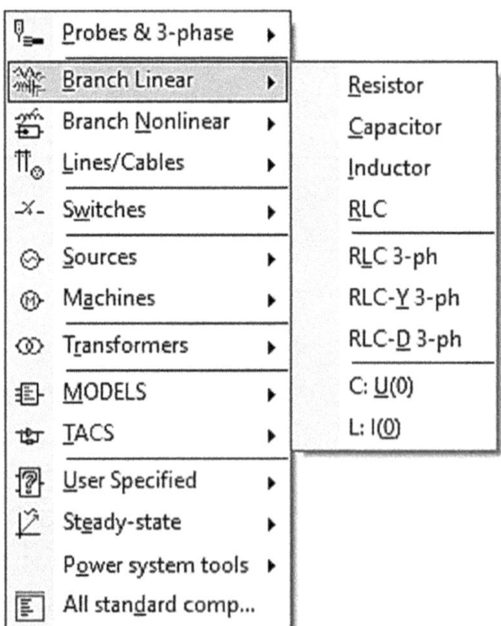

FIGURE 1.34
Linear branch components.

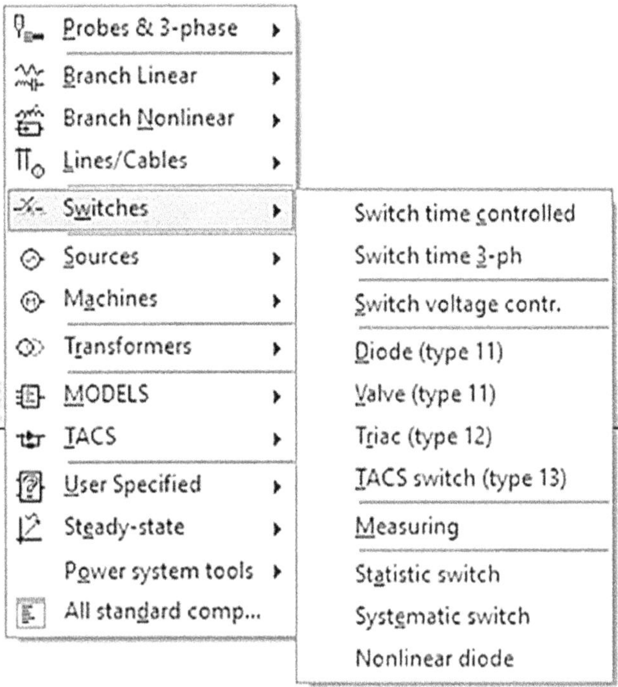

FIGURE 1.35
Switches.

Linear Branch:

Resistor: ideal resistor;
Capacitor: capacitance;
Inductor: inductance;
RLC: single phase extension / load with R-L-C in series;
3-ph RLC: three-phase extension with R-L-C in series;
RLC-Y 3-ph: three-phase star connected load;
RLC-D 3-ph: delta connected three-phase load;
C: U (0): capacitor with initial load;
L: I (0): self-induction with initial load.

Switches:

Switch time controlled: time controlled single-phase switch;
Switch time 3-ph: time controlled three phase switches (independent phases).

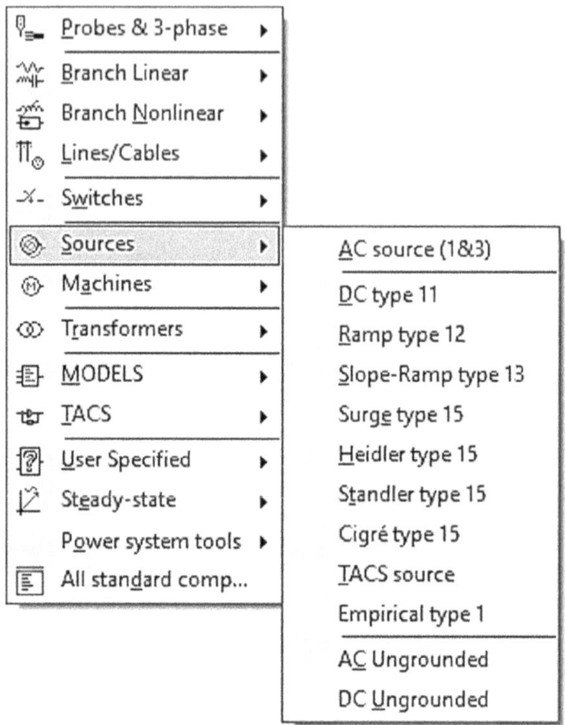

FIGURE 1.36
Sources.

Sources:

AC source: current / voltage source, 1/3 phase, grounded/ungrounded;
AC Ungrounded: ungrounded AC voltage source.

Graphical Result Viewers (PlotXY) (Figure 1.37):
In this viewer, you can represent up to eight curves in the same graph and represent in the same sheet curves from three different files. Curves can be well differentiated when different colors are used.

1.7.1 First ATPdraw Circuit

Using ATPdraw, solve an RL circuit with R = 0.01 Ω and L = 10mH, subjected to a voltage of $V = 380\angle 0°\ V$, 60 Hz. The maximum simulation time is 0.1 s and the integration step is from 1.e-8. Draw the energy curve dissipated by the resistor.

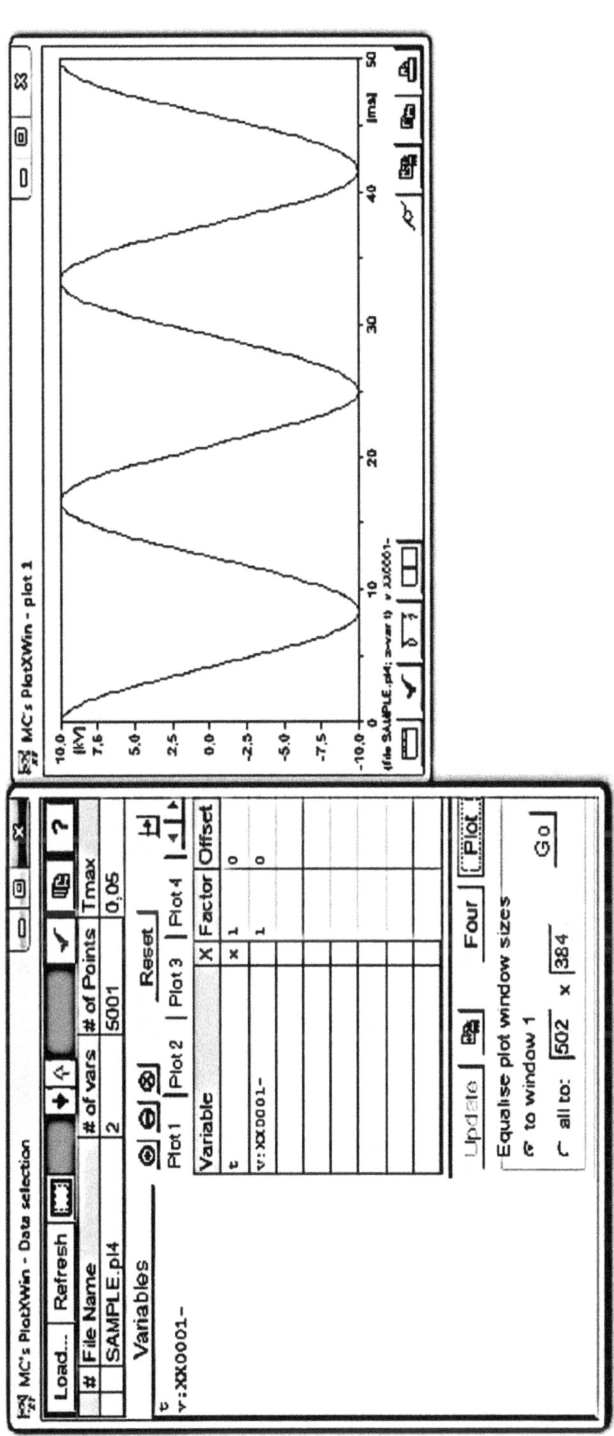

FIGURE 1.37
PlotXY viewer.

SOLUTION:

Click on the ATPdraw icon. Then, with the mouse, press file - new and the following screens will appear (Figures 1.38 and 1.39).

Right-click of the mouse and move the cursor to source (Figure 1.40).

Double-click the right side of the mouse: the window for entering the source values will appear and the value given in the statement should be placed, replacing the default values (Figure 1.41).

Repeat the previous procedure to place the resistor in the circuit (Figure 1.42).

Drag the resistor with the mouse and connect it with the source (Figure 1.43).

Repeat the procedure to place the inductor in the circuit. Rotate the inductor by right-clicking and then clicking rotate. Place the mouse over the inductor knot and tighten its right side. A window will appear to ground the node, checking the ground box (Figure 1.44).

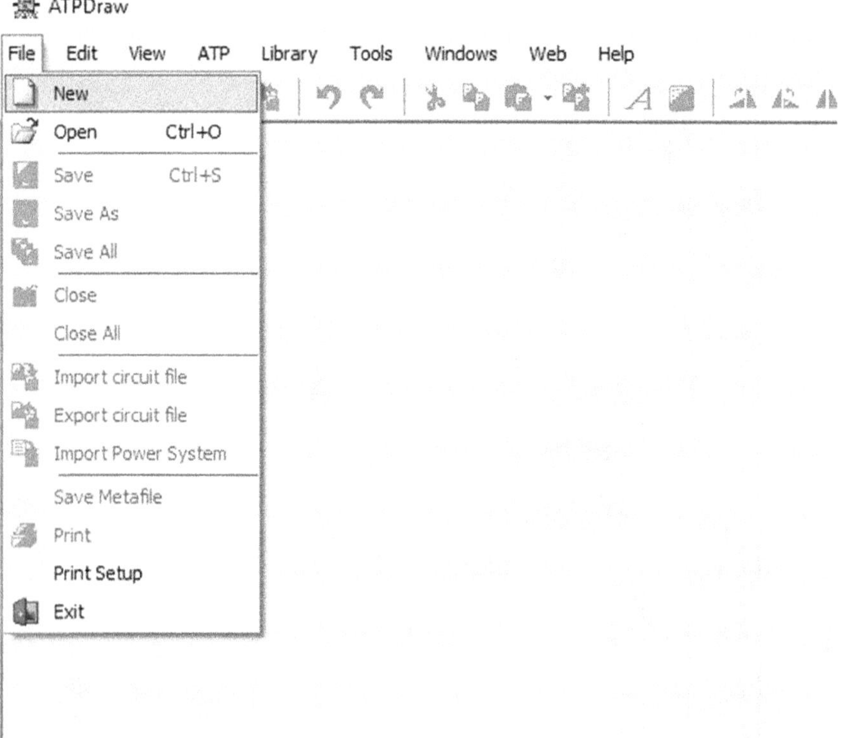

FIGURE 1.38
ATPdraw screen.

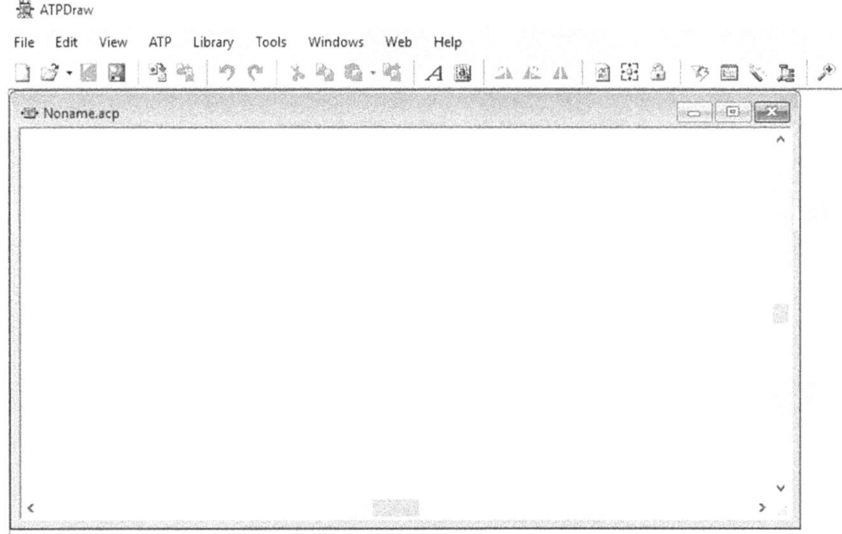

FIGURE 1.39
ATPdraw blank screen.

Separate the source, place the current probe and two voltage probes, and the circuit will be ready for simulation. Double-click on each probe to configure them. Save the file with a name. Then move the mouse to ATP Settings and set the simulation by changing the default values (Figure 1.45).

Then move the mouse to ATP → run ATP and then to run Plot, mark the curves you want to plot and the screen with the circuit simulation results will show the next screen (Figure 1.46).

Note that the current is behind the voltage by 89.8^0.

Then double-click on the resistor and check the Power & Energy option. Rerun the circuit and plot the energy graph using the voltage and current on the resistor (Figure 1.47).

Resistor power curve (Figure 1.48).

1.7.2 Using the Line Constant Routine

The Line Constant routine (Line Cable Constant or LCC) is a reference when calculating TL parameters.

To use the Line Constant routine, press File - New and then right-click the LCC template block as shown in Figure 1.49.

After inserting the LCC block, the screen will be as shown in Figure 1.50.

By pressing the block twice, the routine screen is displayed on the "Model" tab, as shown in Figure 1.51.

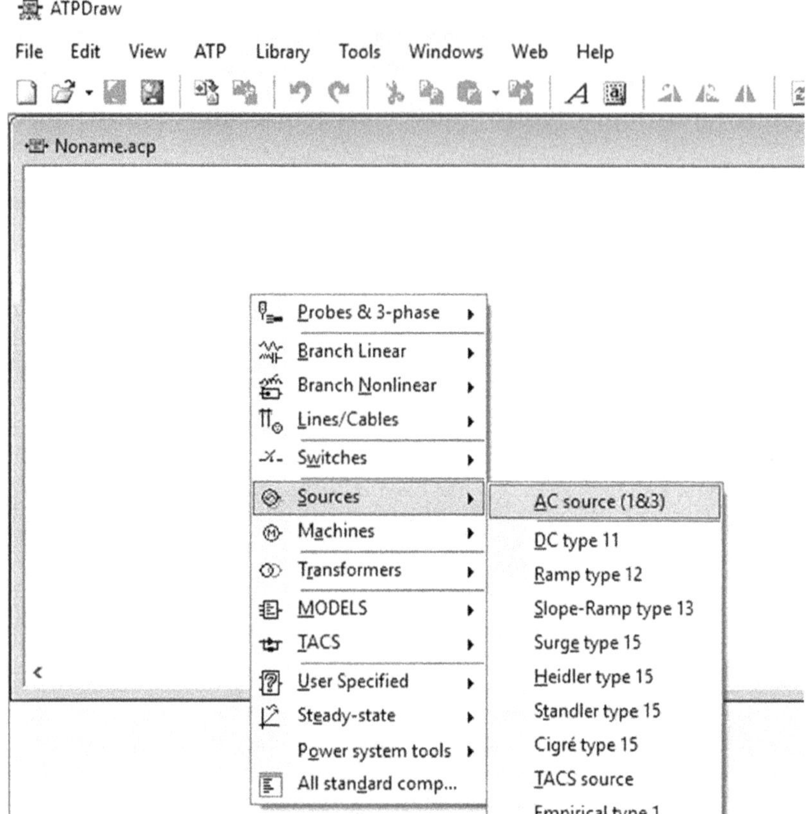

FIGURE 1.40
ATPdraw screen for choosing sources.

In the screen presented in Figure 1.51, it is possible to carry out several studies and choose several network characteristics for both overhead lines and underground cables:

a. System type: place of choice of TL type, whether overhead or underground. For airlines, the option chosen should be Overhead Line. In this section it is also possible to determine the number of phases (equal to 3 for single three-phase lines or 6 for three-phase double-circuit lines), incorporate Skin Effect, Autobunding, Segmented ground, informing that the ground wires are grounded in each structure, choose the unit of measurement, among others;
b. Model: place of choice of the output or model used in the calculations. The π line model is widely used (Figures 1.14 and 1.15);
c. Standard data: nominal line characteristics such as ground resistivity, grid frequency and line length.

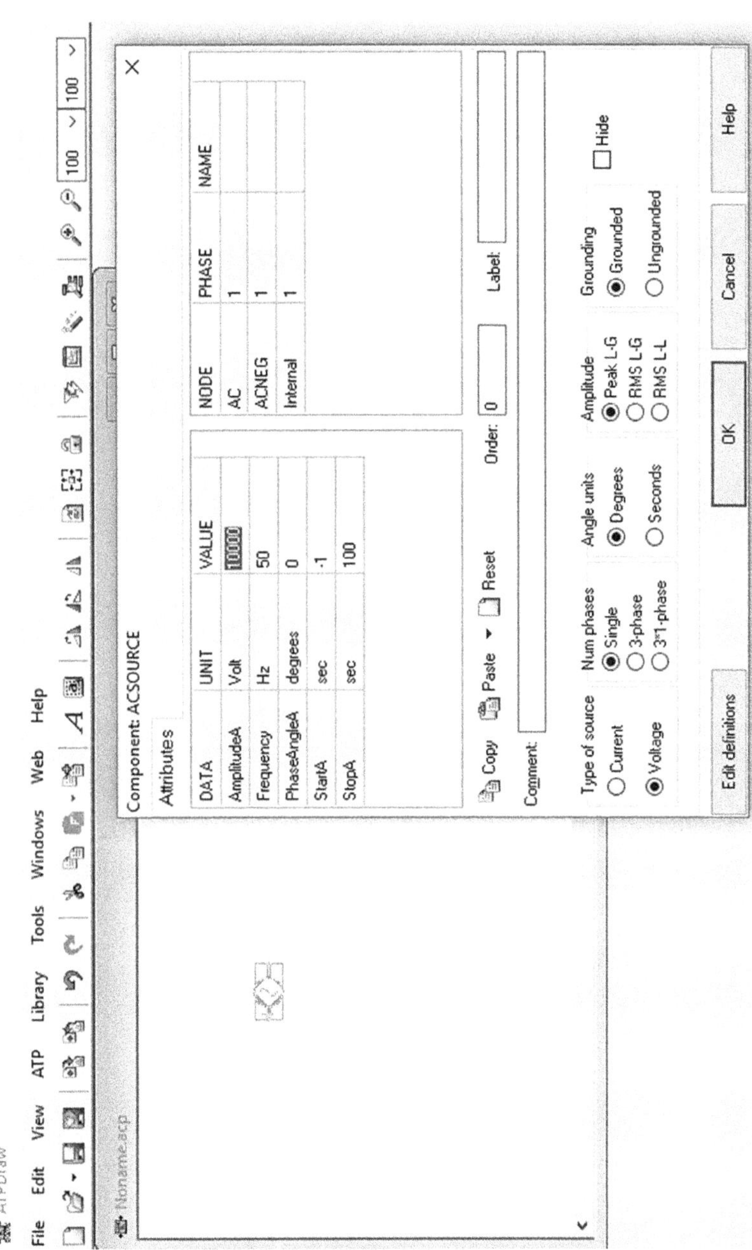

FIGURE 1.41
ATPdraw screen for entering source values.

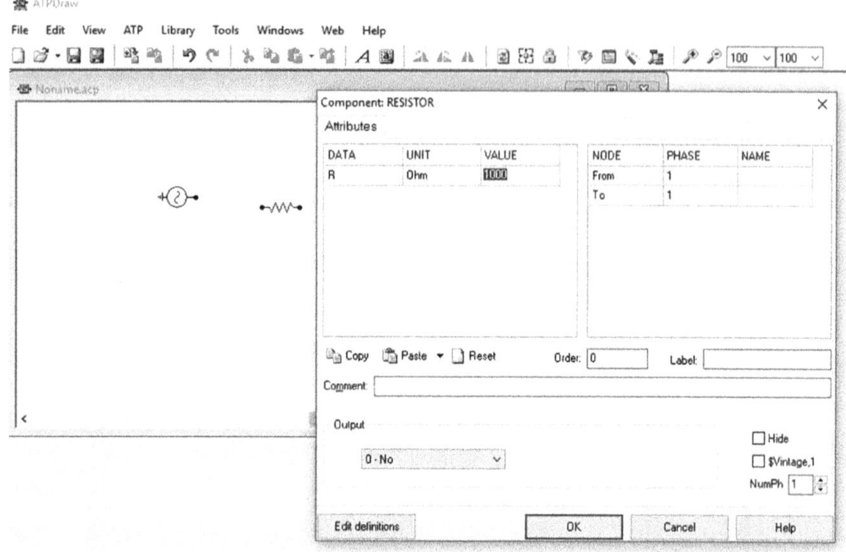

FIGURE 1.42
ATPdraw screen for resistor value placement.

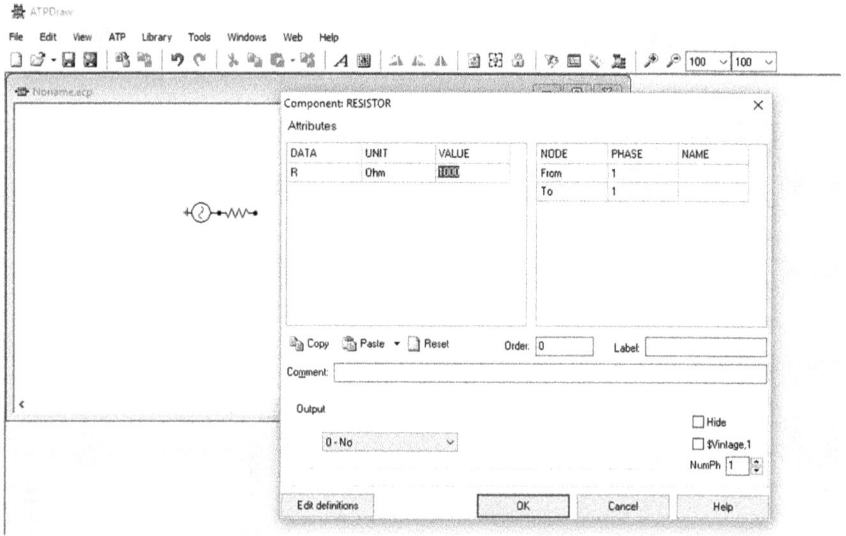

FIGURE 1.43
ATPdraw screen showing resistor connection.

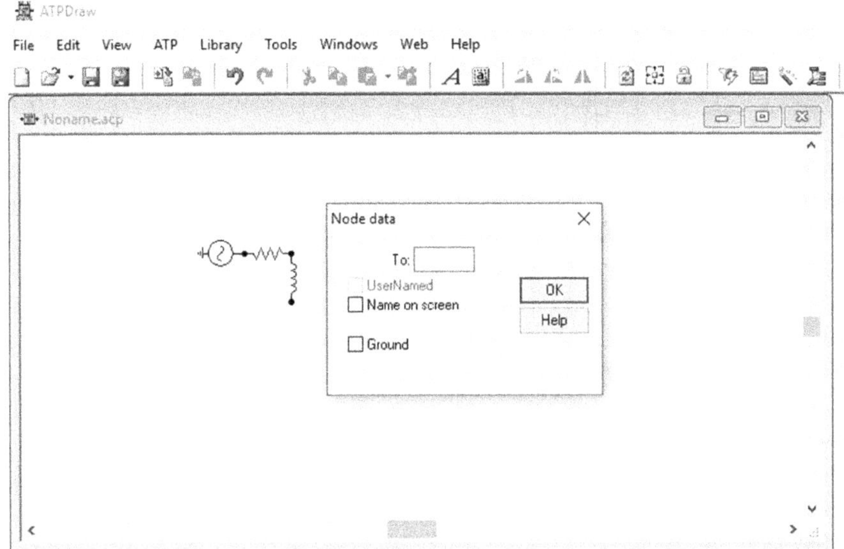

FIGURE 1.44
ATPdraw screen showing inductor rotation and connection.

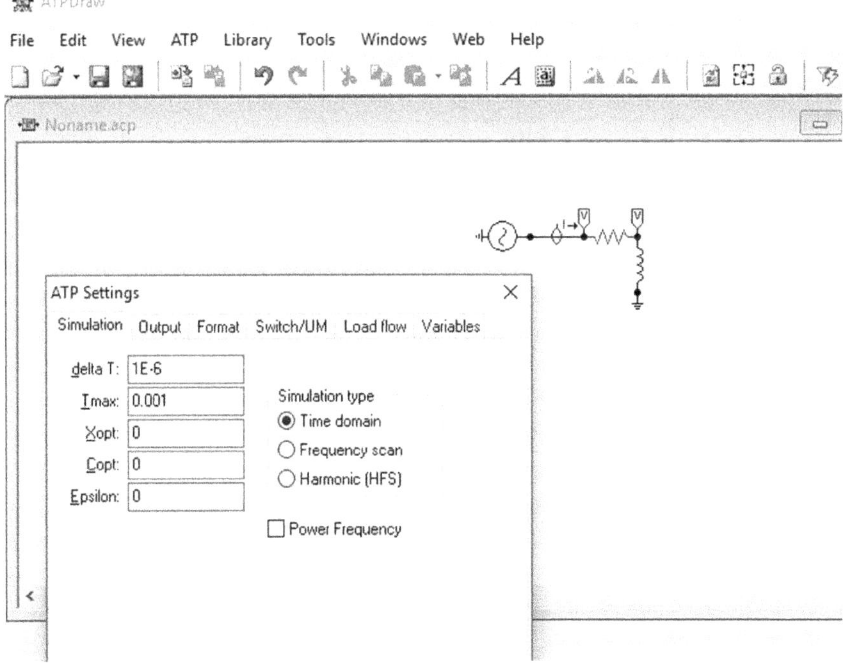

FIGURE 1.45
ATPdraw screen showing settings screen for value placement.

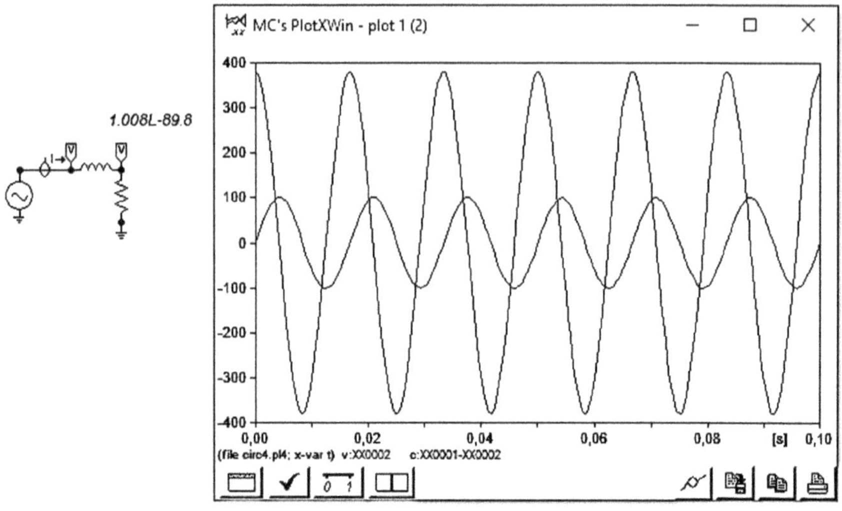

FIGURE 1.46
ATPdraw PlotXY screen showing graph with voltage and current in circuit.

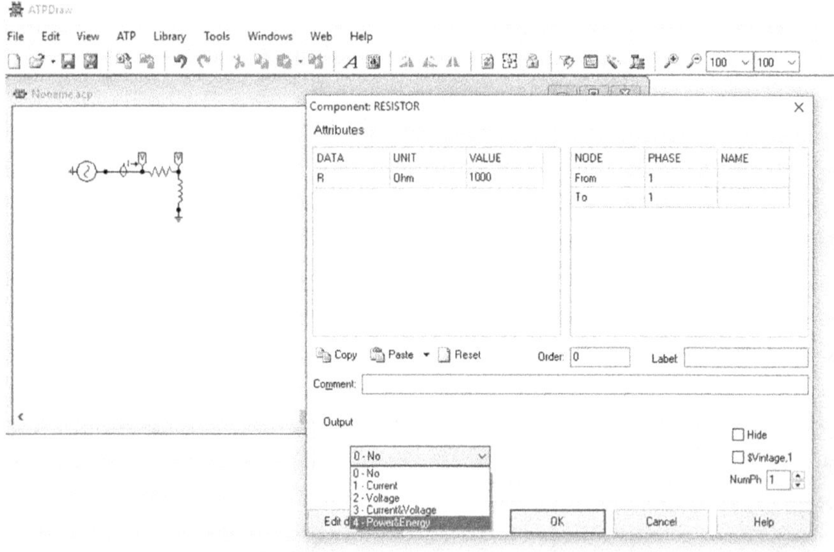

FIGURE 1.47
ATPdraw screen for resistor energy measurement.

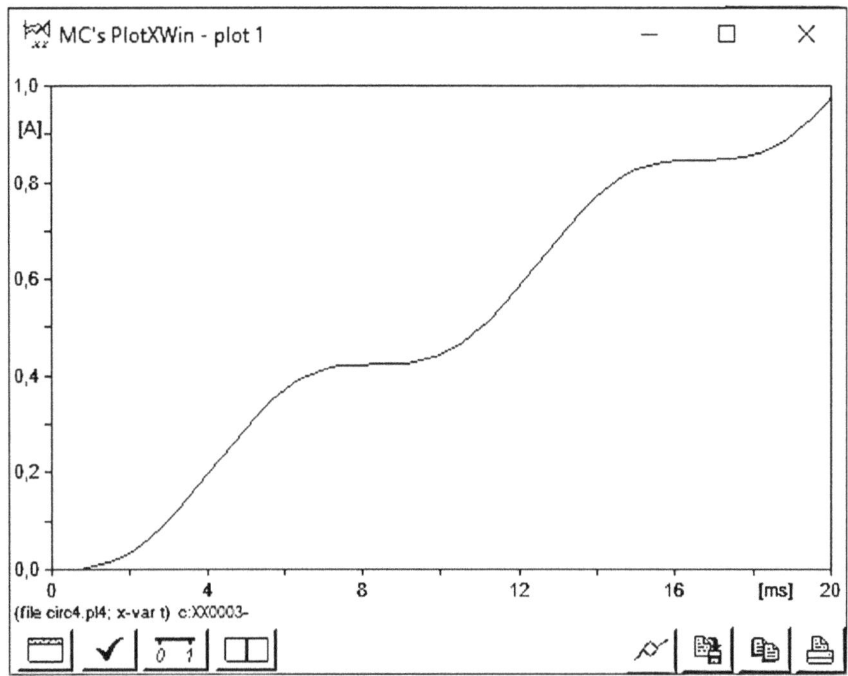

FIGURE 1.48
PlotXY screen with resistor energy.

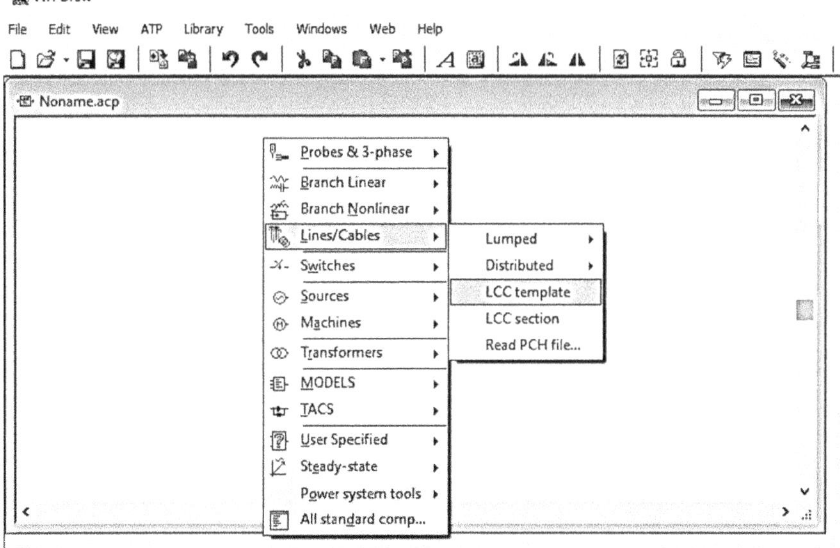

FIGURE 1.49
Using the LCC routine.

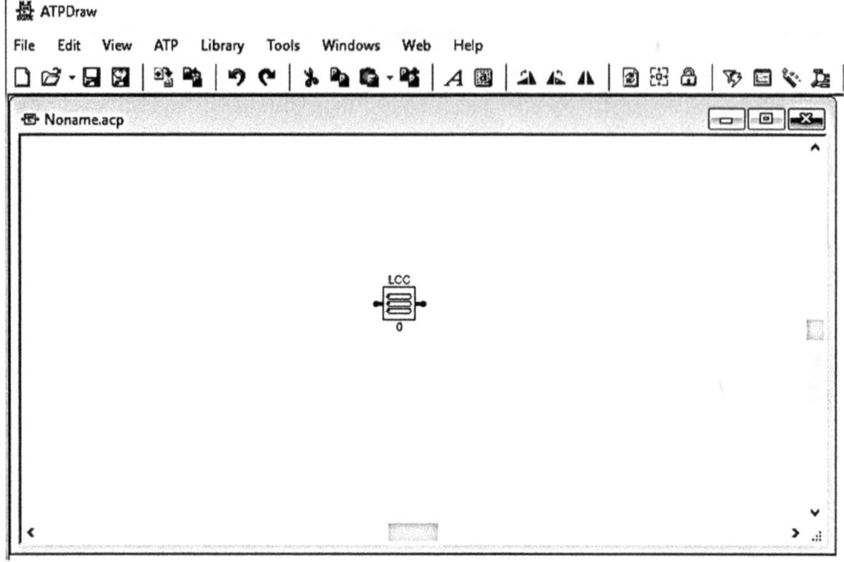

FIGURE 1.50
ATPdraw LCC block for transmission line parameter study.

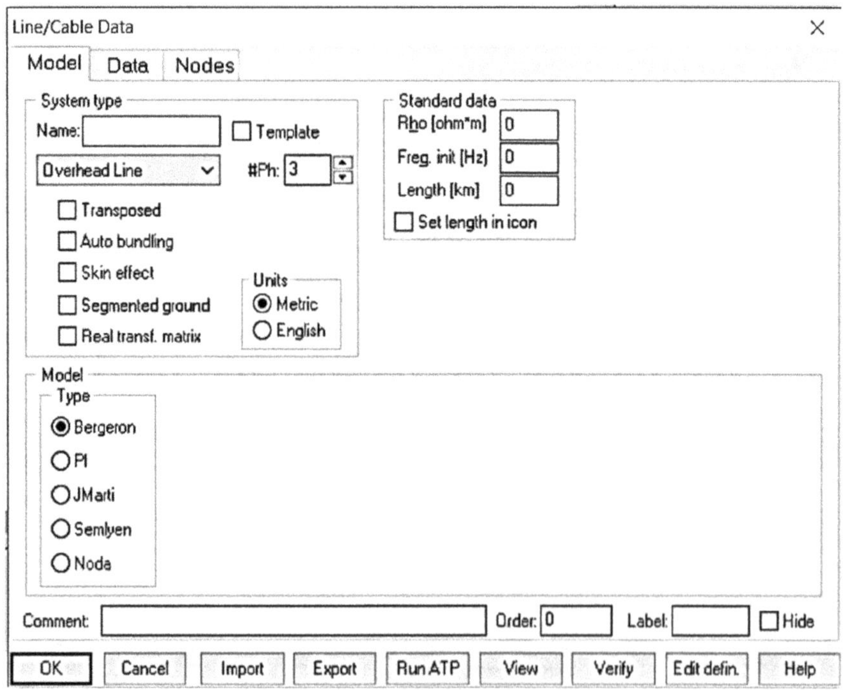

FIGURE 1.51
Aba model on the LCC block line parameters study design screen.

The buttons at the bottom of the window shown in Figure 1.51 highlight:

a. Import: used to import data from a study already carried out previously in the extension ".alc";
b. Export: save the present study with all its data in a file with extension ".alc";
c. Run ATP: executes the study generating a text output card with ".lis" format;
d. View: preview the structure during data entry.

In the Data tab, shown in Figure 1.52, you can enter the data of the conductor cable or lightning conductor:

Ph.no - phase conductor number. When the number is zero, we have a ground wire;

Rout - external radius of conductor;

Horiz - horizontal distance from a conductor or lightning conductor to center of a tower. If the conductor or lightning conductor is on the left side of the tower, the distance is negative and on the right side positive;

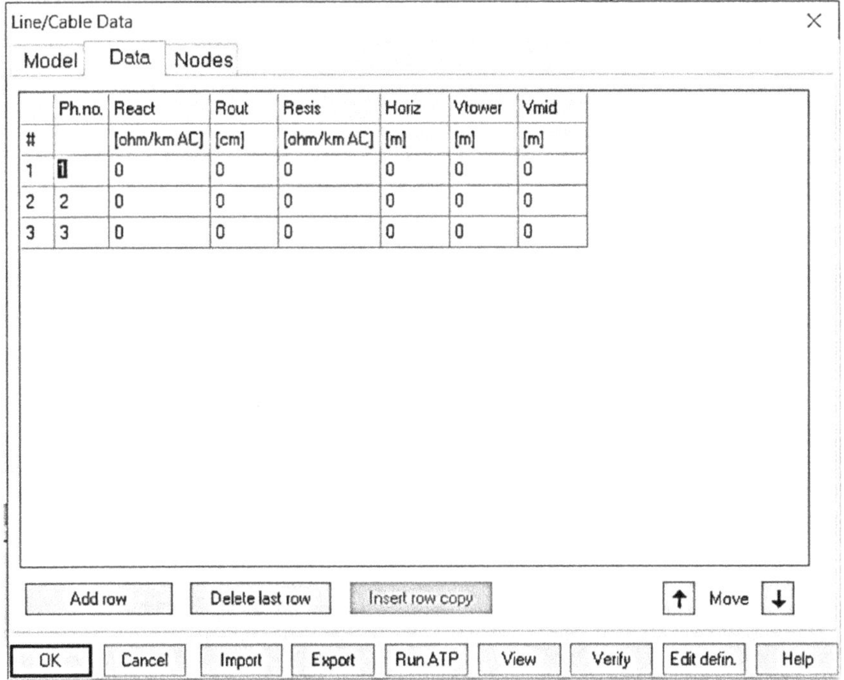

FIGURE 1.52
Aba Data in the LCC block line parameters study preparation screen.

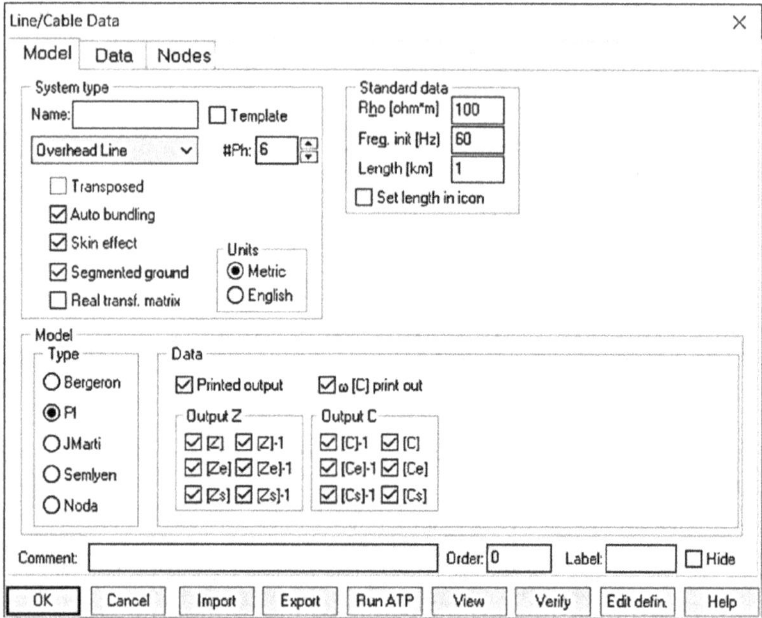

FIGURE 1.53
Study of 500 kV TL parameters.

Vtower - height of a conductor or lightning conductor to ground;

Vmid - conductor height or lightning conductor cable in the middle of the span. Calculated from conductor arrow or ground wire conductor.

Figures 1.53 and 1.54 show the previous screens filled with data from a 500-kV double circuit TL.

After triggering run ATP, a .lis file is generated, where the TL results under analysis are printed. To access this file, press Tools - Text Editor, as in Figure 1.55.

When the text editor screen opens, press file - open and choose the .lis files as shown in Figure 1.56.

An example of .lis file is shown in Figure 1.57 (a) and (b).

Figure 1.57 (b) shows an impedance matrix. We can observe, for example, where it is written:

1. 3.001544E-01
   9.517302E-0.1

We must understand this as: impedance matrix element at position (1,1) and value of$Z_{11} = 0.3001544 + j0.9517302\ \Omega$. The other elements must be understood in a similar way.

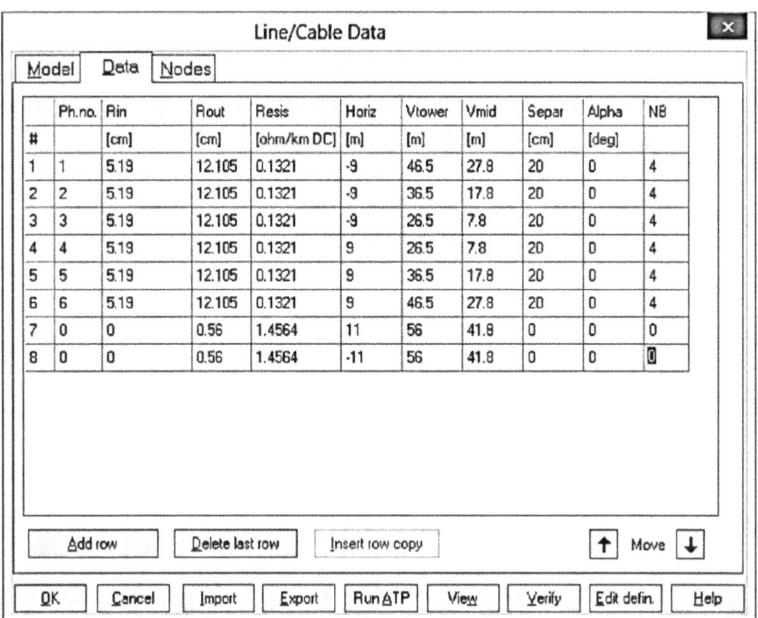

#	Ph.no.	Rin [cm]	Rout [cm]	Resis [ohm/km DC]	Horiz [m]	Vtower [m]	Vmid [m]	Separ [cm]	Alpha [deg]	NB
1	1	5.19	12.105	0.1321	-9	46.5	27.8	20	0	4
2	2	5.19	12.105	0.1321	-9	36.5	17.8	20	0	4
3	3	5.19	12.105	0.1321	-9	26.5	7.8	20	0	4
4	4	5.19	12.105	0.1321	9	26.5	7.8	20	0	4
5	5	5.19	12.105	0.1321	9	36.5	17.8	20	0	4
6	6	5.19	12.105	0.1321	9	46.5	27.8	20	0	4
7	0	0	0.56	1.4564	11	56	41.8	0	0	0
8	0	0	0.56	1.4564	-11	56	41.8	0	0	0

FIGURE 1.54
500-kV double circuit TL parameters.

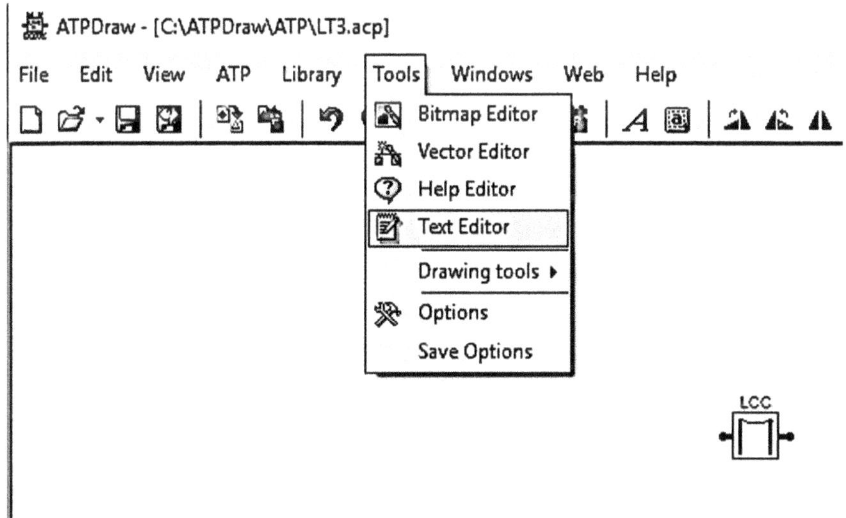

FIGURE 1.55
Print file access.

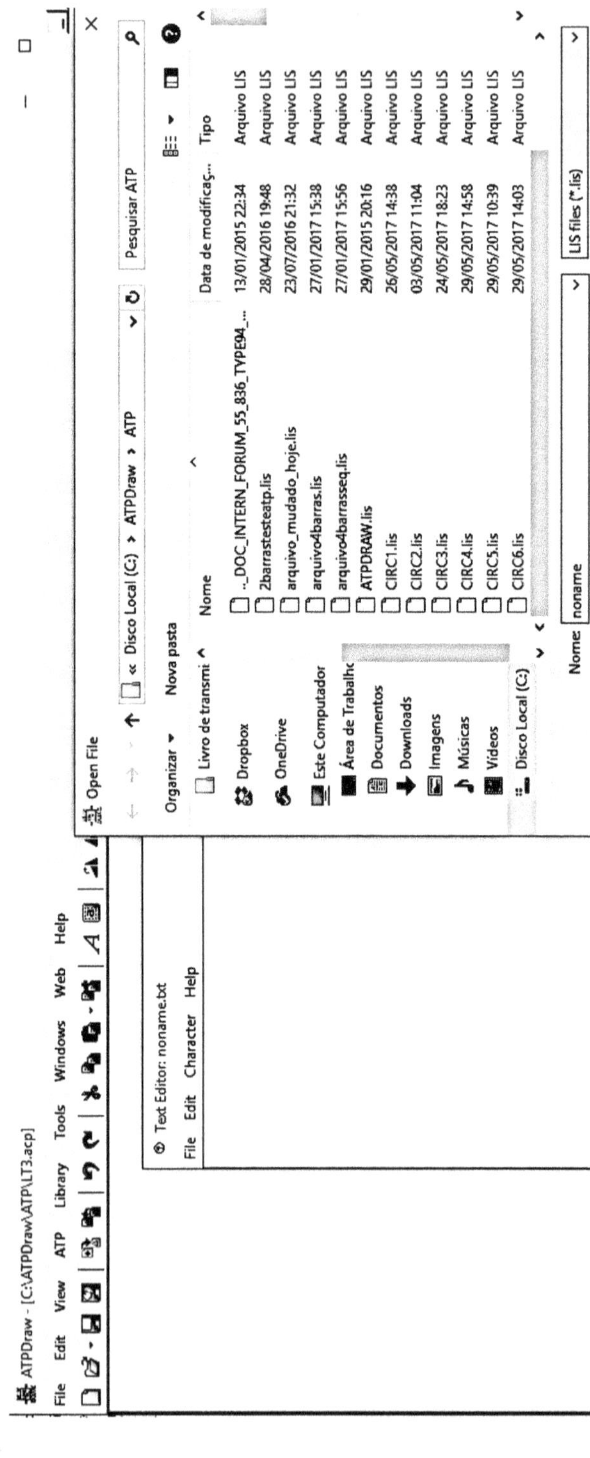

FIGURE 1.56
The archives .lis no ATPdraw.

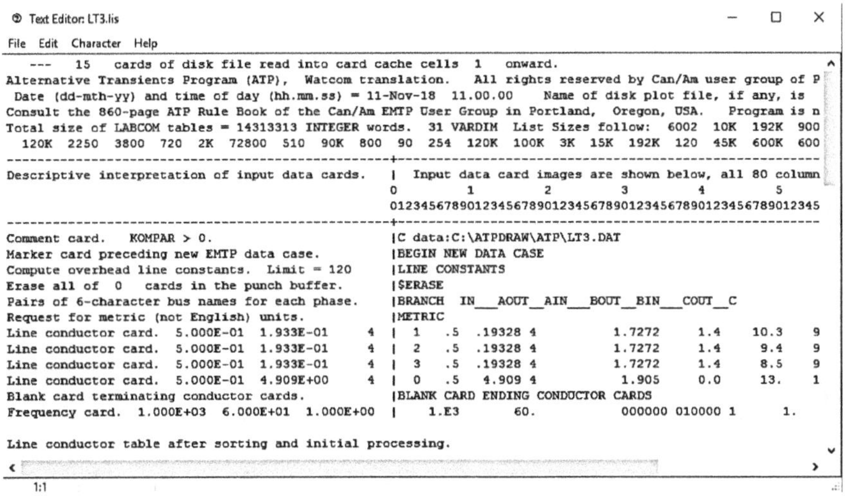

Text Editor: LT3.lis

File Edit Character Help

```
   ---   15   cards of disk file read into card cache cells  1   onward.
Alternative Transients Program (ATP),  Watcom translation.   All rights reserved by Can/Am user group of P
 Date (dd-mth-yy) and time of day (hh.mm.ss) = 11-Nov-18  11.00.00    Name of disk plot file, if any, is
Consult the 860-page ATP Rule Book of the Can/Am EMTP User Group in Portland,  Oregon, USA.   Program is n
Total size of LABCOM tables = 14313313 INTEGER words.  31 VARDIM  List Sizes follow:  6002  10K  192K  900
  120K  2250  3800  720  2K  72800  510  90K  800  90  254  120K  100K  3K  15K  192K  120  45K  600K  600
--------------------------------------------------+-------------------------------------------------------
Descriptive interpretation of input data cards.   |  Input data card images are shown below, all 80 column
                                                  0         1         2         3         4         5
                                                  012345678901234567890123456789012345678901234567890123456789012345
--------------------------------------------------+-------------------------------------------------------
Comment card.   KOMPAR > 0.                       |C data:C:\ATPDRAW\ATP\LT3.DAT
Marker card preceding new EMTP data case.         |BEGIN NEW DATA CASE
Compute overhead line constants.  Limit = 120     |LINE CONSTANTS
Erase all of  0   cards in the punch buffer.      |$ERASE
Pairs of 6-character bus names for each phase.    |BRANCH  IN___AOUT__AIN___BOUT__BIN___COUT__C
Request for metric (not English) units.           |METRIC
Line conductor card.  5.000E-01  1.933E-01      4 |  1    .5  .19328 4           1.7272      1.4    10.3     9
Line conductor card.  5.000E-01  1.933E-01      4 |  2    .5  .19328 4           1.7272      1.4     9.4     9
Line conductor card.  5.000E-01  1.933E-01      4 |  3    .5  .19328 4           1.7272      1.4     8.5     9
Line conductor card.  5.000E-01  4.909E+00      4 |  0    .5   4.909 4            1.905      0.0     13.     1
Blank card terminating conductor cards.           |BLANK CARD ENDING CONDUCTOR CARDS
Frequency card.  1.000E+03  6.000E+01  1.000E+00  |    1.E3       60.            000000 010000 1       1.

Line conductor table after sorting and initial processing.
```

1:1

(a)

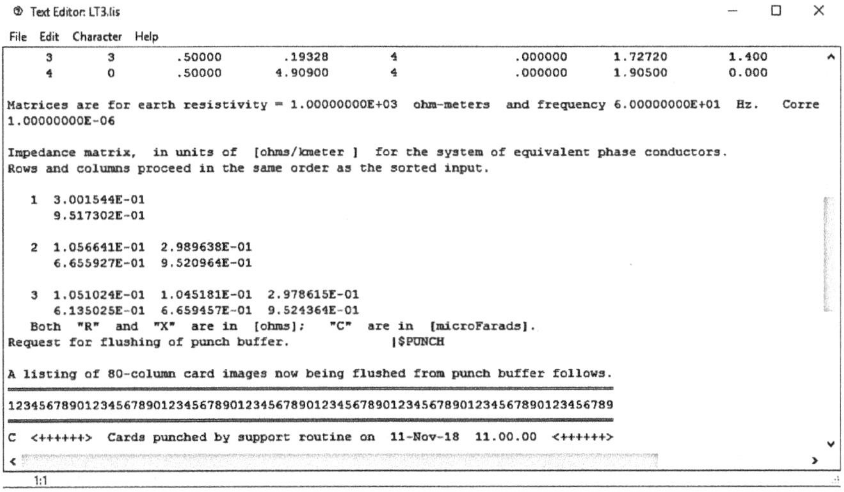

Text Editor: LT3.lis

File Edit Character Help

```
     3        3          .50000          .19328          4                .000000      1.72720        1.400
     4        0          .50000         4.90900          4                .000000      1.90500        0.000

Matrices are for earth resistivity = 1.00000000E+03  ohm-meters  and frequency 6.00000000E+01  Hz.   Corre
1.00000000E-06

Impedance matrix,  in units of  [ohms/kmeter ]  for the system of equivalent phase conductors.
Rows and columns proceed in the same order as the sorted input.

   1  3.001544E-01
      9.517302E-01

   2  1.056641E-01  2.989638E-01
      6.655927E-01  9.520964E-01

   3  1.051024E-01  1.045181E-01  2.978615E-01
      6.135025E-01  6.659457E-01  9.524364E-01
   Both  "R"  and  "X"  are in  [ohms];   "C"  are in  [microFarads].
Request for flushing of punch buffer.              |$PUNCH

A listing of 80-column card images now being flushed from punch buffer follows.
=======================================================================================
12345678901234567890123456789012345678901234567890123456789012345678901234567890123456789
=======================================================================================
C  <++++++>  Cards punched by support routine on  11-Nov-18  11.00.00  <++++++>
```

1:1

(b)

FIGURE 1.57
(a), (b) .lis file screens.

1.8 Problems

1.8.1 Talk about the resolution algorithms for calculating electromagnetic transients.

Answer: See item 1.2 Resolution Algorithms.

1.8.2 What is the frequency range for switching transients?

Answer: See item 1.2 Resolution Algorithms.

1.8.3 Calculate $V_L(t)$ for $t = \Delta t$ in the circuit of Figure 1.58. The frequency is 60 Hz and e $\Delta t = 1\ \mu seg$. The initial conditions are given by:

$$V_F\left(0^-\right) = 0;\ I_F\left(0^-\right) = 0;\ V_L\left(0^-\right) = 0;\ I_L\left(0^-\right) = 0.$$

Answer: $V_L\left(t\right) = 379.9983724955027400000\ V$.

1.8.4 Explain current source models to represent a lightning discharge in ATPdraw.

Answer: See item 1.5.

1.8.5 Explain how to calculate the parameters of a TL using the ATP program.

Answer: See item 1.7.2.

1.8.6 The following instruction lines are an excerpt from an alternative transient program (ATP) software input file. They refer to the model data with constant distributed parameters with the frequency of a 100-km-long 230-kV three-phase transmission line between two power transmission substations called *SE*1 and *SE*2.

–1_*SE*1_*A*	*SE*1_*A*	0.38	1.09	2.23	100.0
–1_*SE*1_*B*	*SE*2_*B*	0.09	0.52	3.12	100.0
–1_*SE*1_*C*	*SE*2_*C*				

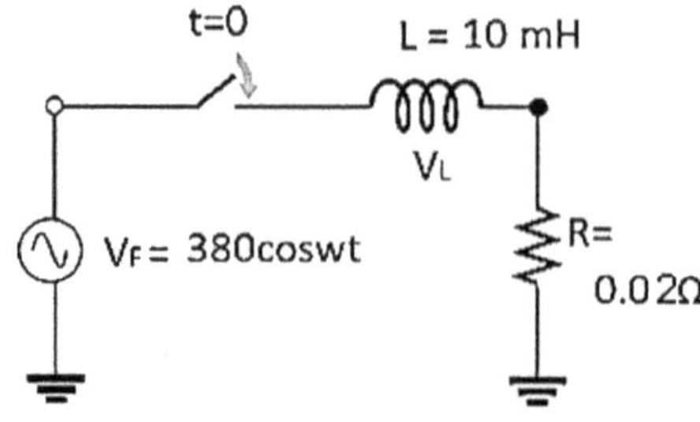

FIGURE 1.58
RL circuit for Exercise 1.8.3.

In these instruction lines, the resistance, inductive reactance and capacitive susceptibility values, per unit length of the transmission line, relative to the frequency 60 Hz are reported. Judge the following items:

a. The zero-sequence impedance of the transmission line is equal to $9 + j52\ \Omega$.

b. The positive sequence impedance of the transmission line is $38 + j109\ \Omega$.

c. The negative sequence impedance of the transmission line is $38 + j109\ \Omega$.

Answers: a) Wrong, b) Wrong, c) Wrong.

2

Transmission Lines: Physical Aspects

2.1 Introduction

Transmission lines (TLs) are essential for transporting large blocks of energy over long distances in a technically and economically viable manner.

According to historical narrative, long and high voltage TLs were used to carry the energy generated at hydroelectric plants far from the consumption centers.

Nowadays, with the increasing use of the interconnected system, it is increasingly important to have a transmission network that guarantees the quality of transport and energy supply.

2.2 Transmission System

The first lines built in the world are described in Table 2.1 (see page 56).

Currently, the interconnected transmission system can cross a country like Brazil or a continent like Europe (Figure 2.1) (see page 56).

Standardized transmission voltages in North America and Europe are shown in Table 2.2 (see page 56).

The Brazilian system, according to the national electric system operator (ONS), is formed by over 125,000 km of TLs, with voltages between 69 kV and 750 kV. Table 2.3 and Figure 2.2 (see page 57) show the lines of the basic grid, which work at voltages greater than or equal to 230 kV.

The TLs in Brazil carry the energy generated in the various existing plants in the country, as shown in Figure 2.3 (see page 57).

2.3 Components of a Transmission Line

The main components of a TL are support structures, conductors, ground wires, insulators, and fittings.

TABLE 2.1
First Transmission Lines

	AC/DC	Length (km)	Voltage (kV)	Year	Country
First line	CC	50	2.4	1882	Germany
Single-phase first line	CA	21	4	1889	US
First three-phase line	CA	179	12	1891	Germany

TABLE 2.2
Standard Voltages

North America Transmission (kV)	Europe Transmission (kV)
69	60
115	110
138	132
161	220
230	275
345	400
500	–
735 – 765	765

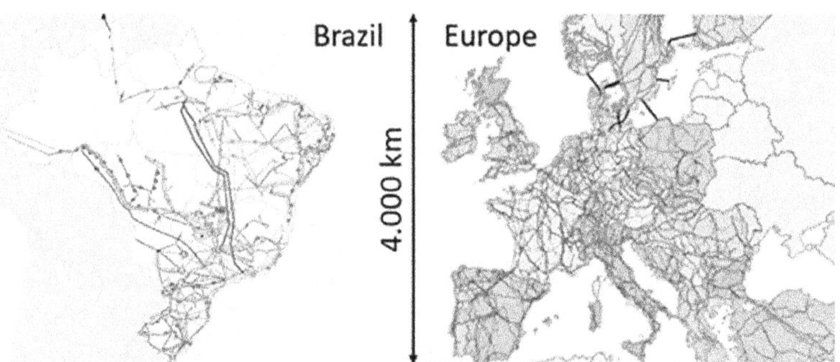

FIGURE 2.1
Maps of Brazil and Europe with electrical systems.

2.3.1 Supporting Structures

TL structures are one of the most visible elements of the electric transmission system. They support the conductors used to transport electricity from load-generating sources to consumers and mechanically withstand the stresses transmitted by the insulators. Concrete, metal structures with galvanized steel profiles or steel poles are used. The choice of the structures to be used in a

TABLE 2.3
Transmission Lines in Brazil

	Length of SIN TLs - km					
Voltage kV	**2010**	**2011**	**2012**	**2013**	**2014**	**Var % 14/13**
230	43.184,5	45.708,7	47.893,5	49.969,0	52.449,8	4.96
345	10.060,5	10.061,9	10.223,9	10.272,3	10.303,2	0.30
440	6.670,5	6.680,7	6.728,2	6.728,2	6.728,2	-
500	34.356,2	35.003,4	35.726,2	39.123,1	40.659,4	3.93
600 DC	3.224,0	3.224,0	3.224,0	7.992,0	12.816,0	60.36
750	2.683,0	2.683,0	2.683,0	2.683,0	2.683,0	-
SIN	100.178,7	103.361,7	106.478,8	116.767,7	125.639,6	7.60

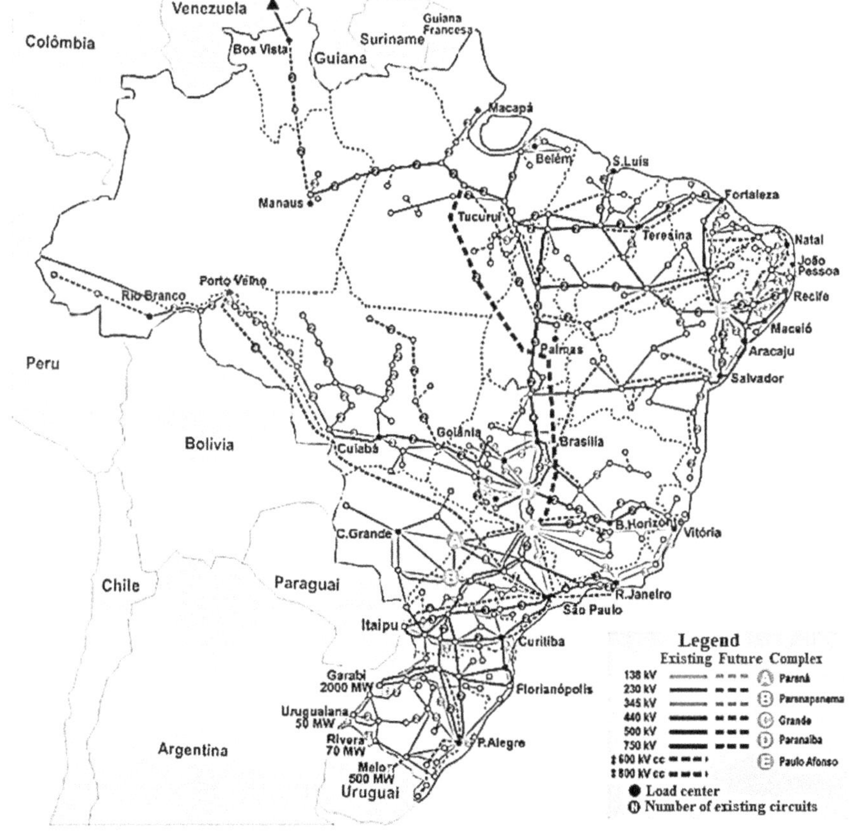

FIGURE 2.2
Map of Brazil with TLs.

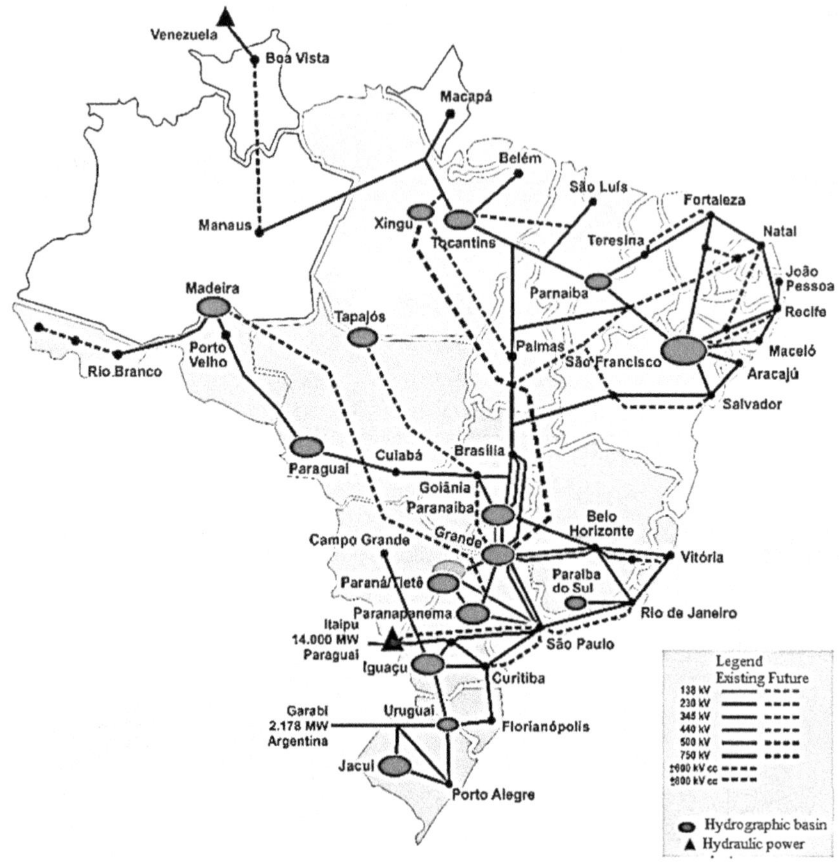

FIGURE 2.3
Map with river basins.

TL depends on several factors, such as maximum and minimum phase spacing; insulators configuration; protection angle of the lightning conductor cable (usually the protection zone is assumed to be contained at an angle of 30° on either side of the conductor); minimum electrical distances between energized pairs and towers; conductors arrow; number of circuits and safety height.

They are classified as self-supporting (Figure 2.4 (a), (b)), which are supported by the structure itself, and cable-stayed (Figure 2.5 (a), (b)), which are supported by tensioned cables in the ground.

We can classify structures, according to their function in the line, into:

2.3.1.1 Suspension Structure

It is the most common, even because it is the simplest and most economical. Its function is to support the conductor and lightning conductor cables,

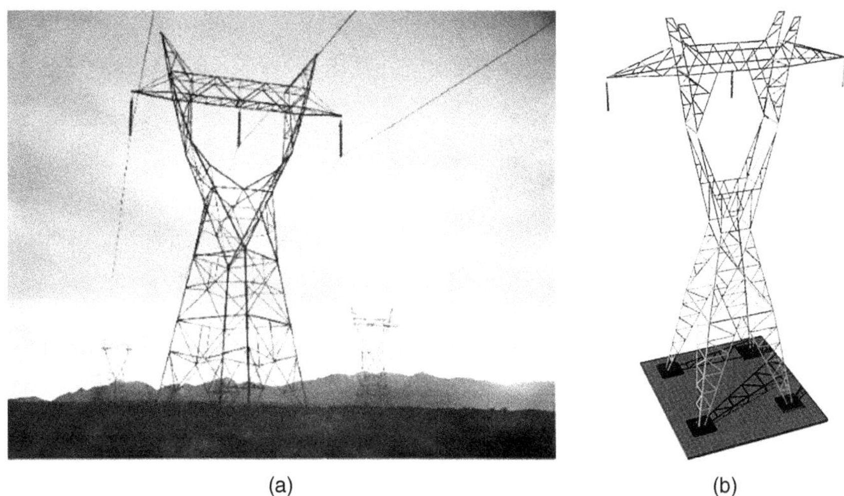

(a) (b)

FIGURE 2.4
(a), (b) Self-supporting structure.

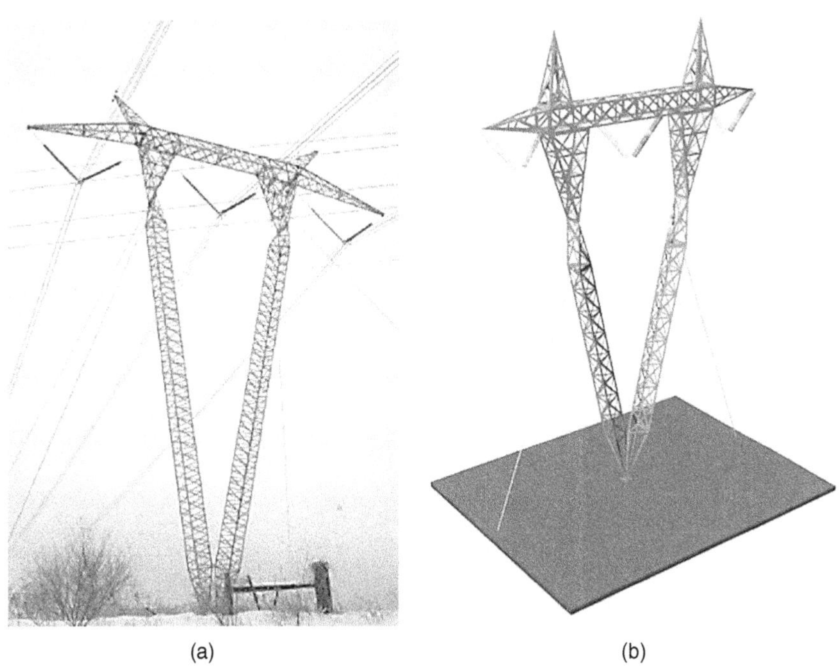

(a) (b)

FIGURE 2.5
(a), (b) Cable-stayed structure.

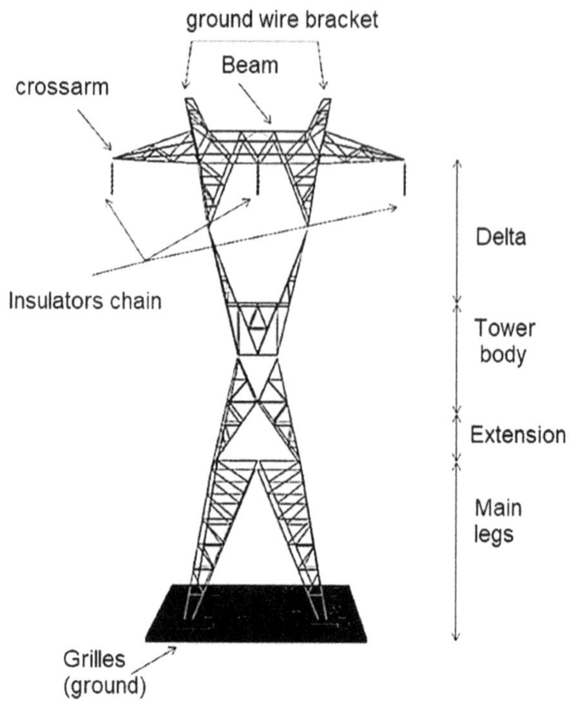

FIGURE 2.6
Suspension structure.

keeping them away from the ground and from each other. In this type of structure, the conductors are not mechanically sectioned but only stapled through the so-called suspension clamps, as shown in Figure 2.6.

2.3.1.2 Mooring or Anchoring Structure

Unlike suspension structures, it mechanically splits the TLs, serving as reinforcement point and eventual opening in specific situations, as shown in Figure 2.7.

2.3.1.3 Angle Structure

This is used if we require a lead at a point on the line, as shown in Figure 2.8.

2.3.1.4 Transposition Structure

It is intended to facilitate the execution of transposition on TLs, as shown in Figure 2.9.

FIGURE 2.7
Mooring structure.

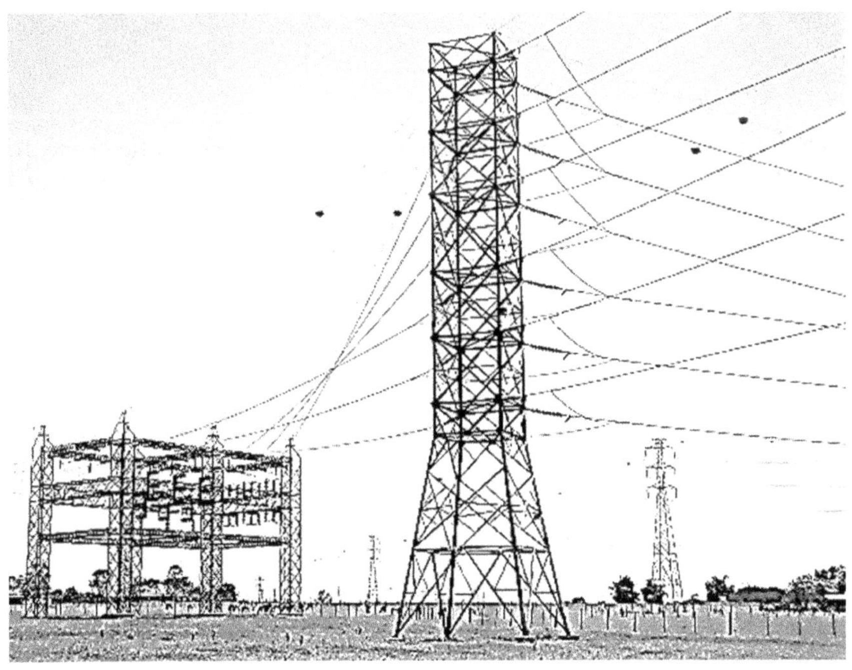

FIGURE 2.8
Angle structure.

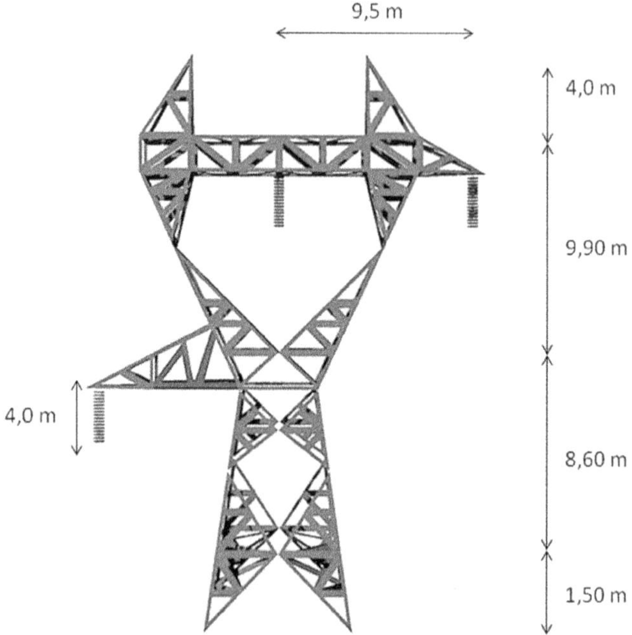

FIGURE 2.9
Transpose structure.

2.3.1.5 Wooden Structure

The tower structures, of which the most common are the lattice type, follow an almost standard architecture, being used to transport energy in one or two circuits. For TLs with large extensions and consequently high voltages (>69 kV), metal towers are the most economical solution. For lower voltages (<69 kV), other materials are also used, such as reinforced concrete and wood, as shown in Figure 2.10.

2.3.2 Conductors and Ground Wire Cables

In a TL, there are the conductors that belong to the phases of the three-phase system and the ground wires that serve as a protection for the TL.

2.3.2.1 Conductors

The American Wire Gauge (AWG) gauge scale is used worldwide as a function of the circular thousand sections up to number 4. Above this gauge are expressed in Circular Mils for copper, aluminum and aluminum-steel (MCM). A circular mil is a unit of area equal to the area of a circle with a diameter of 1,000 in (one thousandth of an inch). It corresponds to $5{,}067 \times 10^{-4}$ mm^2.

FIGURE 2.10
Wooden structure.

There are two types of conductors: solid wire and cable (Figure 2.11 (a), (b)). Massive wires were formerly employed, even with large gauges, and their employment is currently limited to the 4 AWG gauge above which cables are preferred because of their flexibility and ease of handling.

The cables are conductors formed by a series of thin wires, stranded in one or more layers, and may be composed of wires of the same material (homogeneous cables) and wires of different materials (heterogeneous cables) or copper steel wires (copperweld) or aluminum (alumoweld). Copper and aluminum are employed in their purest or alloyed electrolytic forms, while steel is employed to increase mechanical strength and as ground wires.

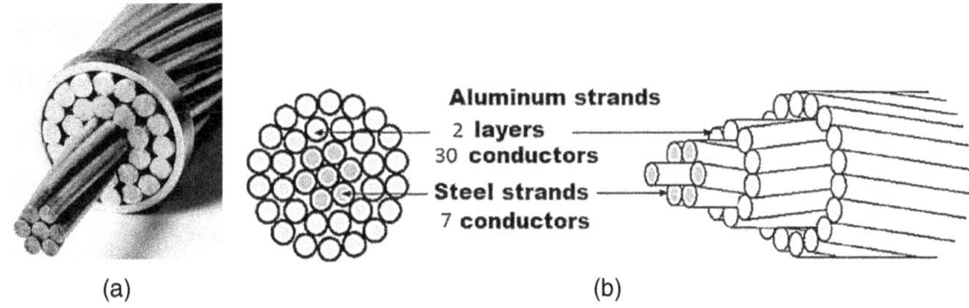

FIGURE 2.11
(a) Aluminum Cable Steel Reinforced (ACSR) 26/7. (b) Aluminum Cable Steel Reinforced (ACSR) 30/7.

TABLE 2.4

Aluminum Conductors Steel Reinforced

Composition	transversal section from the cable	Composition	transversal section from the cable
6/1 (1)+6		18/7 (1)+6+12	
12/7 (1+6)+12		26/7 (1+6)+10+16	
30/7 (1+6)+12+18		45/7 (1+6)+9+15+21	
54/7 (1+6)+12+18+24		54/19 (1+6+12)+12+18+24	

The cable types are:

a. Aluminum Cable Steel Reinforced (ACSR), which is an inner wire–reinforced aluminum cable (core) as shown in Figure 2.11, where 26/7 shows 26 aluminum conductors per 7 steel conductors. It has a larger diameter, which allows the corona effect to reduce, as shown in Table 2.4.

   For the calculation of TL parameters, for example, a 26/7 composite cable, the inner diameter is the sum of three steel wire diameters and the outer diameter is the inner diameter plus the sum of four diameters of aluminum wires. Thus, $D_{\text{inside}} = 3 \times 0.1054'' \times 0.0254$ (inches to m conversion factor) = 0.00803148 m and $D_{\text{outside}} = 0.00803148 + 4 \times 0.1355'' \times 0.0254 = 0.02179828$ m.

b. ALPAC cables are aluminum cables with flat steel core (wire section is trapezoidal), as shown in Figure 2.12.

c. Expanded copper and expanded ACSR cables have a non-metallic material filler layer between the steel wires and aluminum wires

FIGURE 2.12
ALPAC cables.

(Figure 2.13). Expanded copper cables are obtained by winding copper wires over the twisted fill in the opposite direction to the steel core for a better grip. They are used for very high voltage lines, getting a larger diameter with the use of less conductive material, which facilitates the transport of energy.

2.3.2.2 Ground Wire Cables

High-intensity electrical discharges usually damage the insulator surfaces, reducing their efficiency, and there may be arcing between the conductor and the structure. To intercept lightning discharges and discharge them to earth, we use ground wires as shown in Figure 2.14.

When lightning strikes the ground wire conductors, there will be a two-way current flow, as shown in Figure 2.14. These currents partially reach the

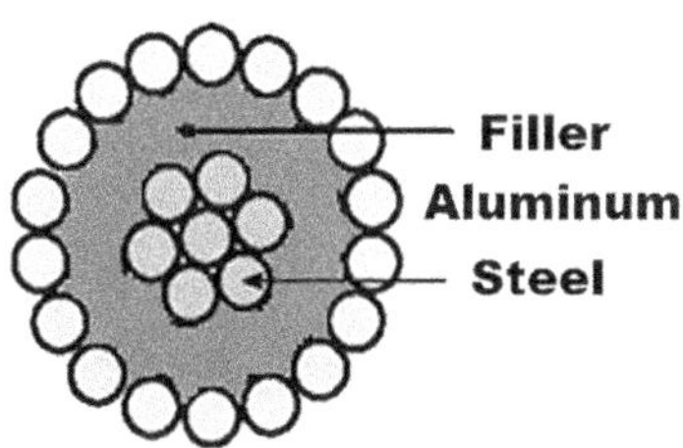

FIGURE 2.13
Expanded ACSR cable.

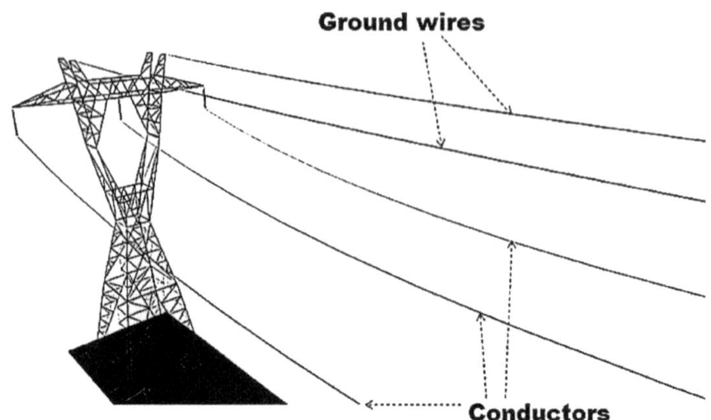

FIGURE 2.14
Ground wire cables.

structures and are discharged to the earth. Thus, the insulators are subjected to a lower voltage as the conductors are not directly struck and will only suffer a voltage induction because of the current in the ground wire cables. Ground wires are currently used for the following cable types:

a. 7-wire SM, HS or HSS type galvanized steel cables with 5/16", 3/8", ½", and 5/8" gauge.
b. Copperweld cable and alumoweld cable, with gauges equivalent to the cables of item a (Figure 2.15 (a), (b)). These cables are more expensive but have greater durability.
c. ACSR cables of high mechanical strength.

Electrical material manufacturers provide cables with the mechanical and electrical characteristics of ground wire cables and conductor cables.

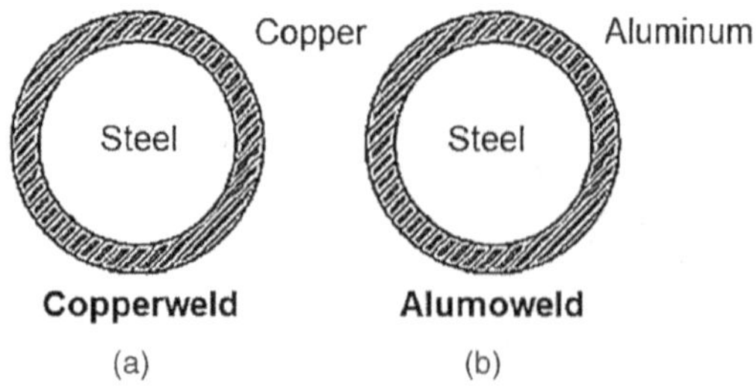

FIGURE 2.15
(a) Copperweld cable. (b) Alumoweld cable.

2.3.3 Insulators

The functions of insulators are to prevent conductor current from flowing into the support or bracket and mechanically supporting the cables.

The manufacturing materials used are glazed porcelain and tempered glass. Both have equivalent performance regarding mechanical strength and durability. Glass insulators have a high impact resistance on their backs, are fragile on the soffit or skirt, and break apart entirely because they are temperate, making it easier to find fault elements at a distance. This is not the case with porcelain insulators, which are difficult to locate when cracked.

As for form, three types are employed in TL.

a. Pin insulators—These are fixed to the structures through steel pins to which they are bolted (Figure 2.16). The steel pin head is finished with a conquered lead thread that joins that of the insulator. Depending on the number of parts that make up the chain, up to 66 kV are used.
b. Pillar insulators—These are single-body insulators of small diameter in relation to the length and have in their lower part a malleable iron body cemented to it. They have high flexural strength and low electrostatic capacity because of the distance between the conductor and the clamping system (Figure 2.17).
c. Disc insulators—These are the most commonly used isolators in voltages above 33 kV. They are used as a suspension or anchor chain (tensioned). The number of insulators in the chain depends on the line voltage. They have high mechanical strength and can withstand tensions of 8,000 kg. In Brazil, three suspension insulator diameters are more common: 10" or 254 mm, with a useful height of 145 mm;

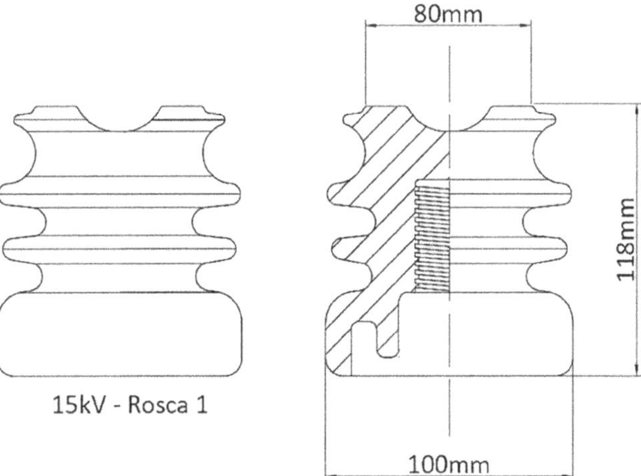

FIGURE 2.16
Pin Insulator.

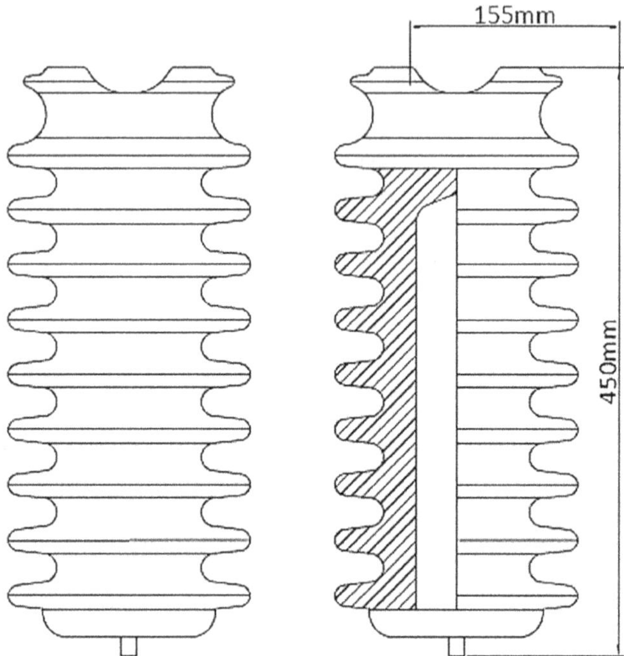

FIGURE 2.17
Pillar Insulator.

8" or 203.5 mm with a working height of 145 mm; and 6" or 152.4 mm with a working height of 145 mm. 6" and 8" insulators are used as TL anchor isolators up to 25 kV and are used as a suspension when heavier cables are employed. 10" isolators are used for any voltage, varying only the number of discs in the chain.

The types of disk insulators are as follows:

Ball-socket—This system offers greater freedom of movement between the chain elements, as the insulator is subjected to the axis direction, being preferred in high voltage lines (Figure 2.18).

Eye-fork—This system allows lateral movement only and is used in lines up to 25 kV or at higher voltages as anchor chains (Figure 2.19).

The insulator chain may be a suspension or passageway chain or an anchor or mooring chain. In the first case, the insulator chain is subjected only to the action of the conductor weight passing through the structure. The position of the chain is vertical as shown in Figure 2.20.

In the second case, the cable is attached to the frame by the intended working chain. The purpose of anchor chains is to tension the conductor cables,

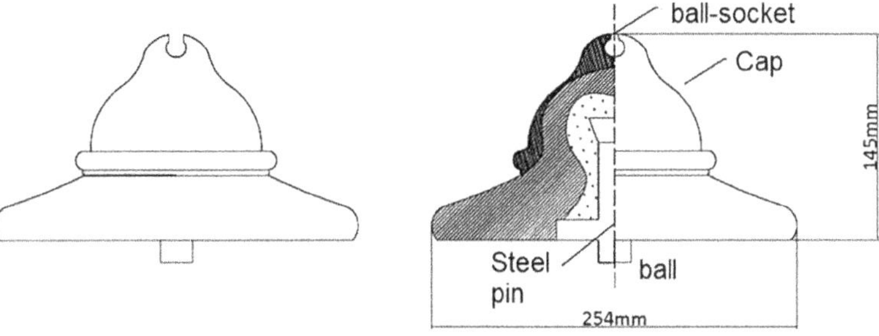

FIGURE 2.18
Insulator ball-shell.

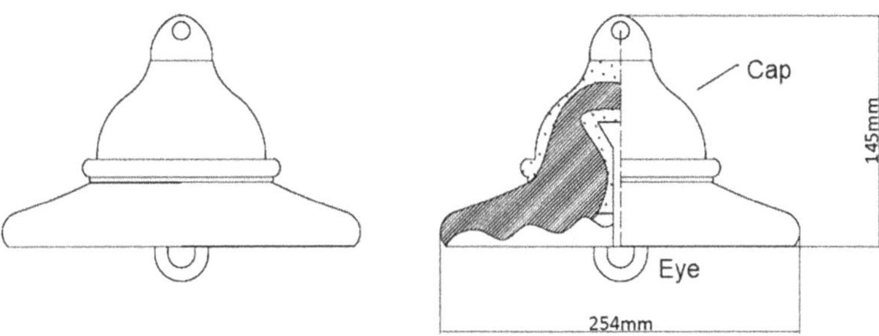

FIGURE 2.19
Eye-fork insulator.

FIGURE 2.20
Suspension insulator chain.

FIGURE 2.21
Chain of anchor insulators.

and we use them on crossings and very uneven terrain. The position of the chain is horizontal, as shown in Figure 2.21.

Consider a chain with n insulators, as shown in Figure 2.22. In this chain, there are the following capacitances:

- K—This is the element's own capacitance or the capacitance between the insulators;
- C—This is the capacitance of each insulator with respect to earth;
- $C1$—This is the capacitance of each insulator in relation to the line.

FIGURE 2.22
Chain with n Insulators.

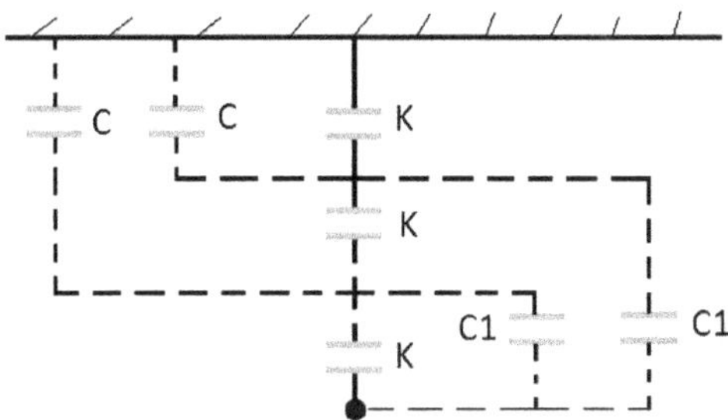

FIGURE 2.23
Capacitance of a corona ring insulator chain.

These capacitances are shown in Figure 2.23.

Generically, we can represent the capacitances shown in Figure 2.23, as shown in Figure 2.24.

Applying Kirchhoff's current law to the circuit of Figure 2.24, we have:

$$i_n + i''_{n-1} = i_{n-1} + i'_{n-1} \tag{2.1}$$

where:

$$I_n = jwK(E_n - E_{n-1}) \tag{2.2}$$

$$I''_{n-1} = jwC1(E_n - E_{n-1}) \tag{2.3}$$

$$I_{n-1} = jwK((E_{n-1} - E_{n-2}) \tag{2.4}$$

$$I'_{n-1} = jwC(E_{n-1}) \tag{2.5}$$

For a particular insulator q, we have:

$$I_q = jwK(E_q - E_{q-1}) \tag{2.6}$$

$$I''_{q-1} = jwC1(E_n - E_{q-1}) \tag{2.7}$$

$$I_{q-1} = jwK((E_{q-1} - E_{q-2}) \tag{2.8}$$

$$I'_{q-1} = jwC(E_{q-1}) \tag{2.9}$$

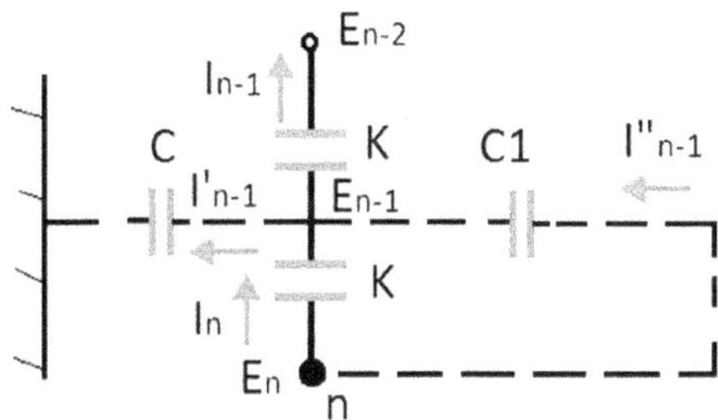

FIGURE 2.24
Capacitance representation in an insulator chain.

Capacitances C1 are always calculated with respect to n.

Soon,

$$I_q + I''_{q-1} = I_{q-1} + I'_{q-1} \tag{2.10}$$

$$jwK\left(E_q - E_{q-1}\right) + jwC1\left(E_n - E_{q-1}\right) = jwK\left(\left(E_{q-1} - E_{q-2}\right) + jwC(E_{q-1}\right) \tag{2.11}$$

Eliminating jw, dividing by K and making $\beta = C1/K$, $\alpha = C/K$,

$$E_q - E_{q-1} + \beta E_n - \beta E_{q-1} = E_{q-1} - E_{q-2} + \alpha E_{q-1} \tag{2.12}$$

$$E_q = (2+\beta+\alpha)E_{q-1} - \beta E_n - E_{q-2} \tag{2.13}$$

Let's consider the following parameters:

$$\rho = 2+\beta+\alpha \tag{2.14}$$

$$\tau = \beta E_n \tag{2.15}$$

$$E_1 = 1 \tag{2.16}$$

Voltages, as a function of τ, can be calculated.

If capacitance C1 is removed, we arrive at equation (2.17).

$$E_q = (2+\alpha)E_{q-1} - E_{q-2} \tag{2.17}$$

Example 2.1

Calculate the potential distribution of a chain of five pin-block insulators in a 69-kV line with and without corona ring, assuming that:

$$\alpha = 0.1,\ \beta = 0.05,\ \text{and}\ E_1 = 1$$

SOLUTION:

Figure 2.25 represents the problem insulator chain.

We have:

$$E_1 = 1$$

According to equation (2.13), it comes:

$$E_2 = \rho E_1 - \tau - E_0 = (2+0.01+0.05)1 - \tau - 0 = 2.15 - \tau$$

$$E_3 = \rho E_2 - E_1 - \tau = 2.15(2.15-\tau) - 1 - \tau = 3.6225 - 3.15\tau$$

$$\begin{aligned} E_4 &= \rho E_3 - E_2 - \tau = 2.15(3.6225 - 3.15\tau) - (2.15-\tau) - \tau \\ &= 5.6384 - 6.7725\tau \end{aligned}$$

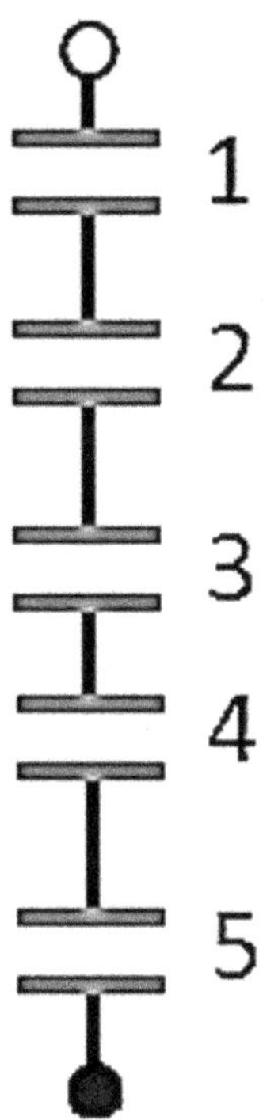

FIGURE 2.25
Insulator chain.

$$E_5 = \rho E_4 - E_3 - \tau = 2.15(5.6384 - 6.7725\tau) - 3.6225 - 3.15\tau - \tau$$
$$= 8.5 - 12.411\tau$$

However, $\tau = \beta E_n = \beta E_5 = 0.05(8.5 - 12.411\tau) \rightarrow \tau = 0.425 - 0.62055\tau \rightarrow$

$$\tau = \frac{0.425}{1.62055} = 0.2623$$

Distribution voltage calculation: substituting the value of τ to calculate E_2, E_3, E_4, and E_5, we have:

$$E_2 = 2.15 - \tau = 2.15 - 0.2623 = 1.8877$$

$$E_3 = 3.6225 - 3.15\tau = 2.7963$$

$$E_4 = 5.6384 - 6.7725\tau = 3.862$$

$$E_5 = 8.5 - 12.411\tau = 5.2446$$

The voltage distribution curve is obtained as follows:

Voltage per element in % of E_5

$$e_1 = \frac{1}{5.2446} \times 100 = 19.07\%$$

$$e_2 = \frac{1.8877}{5.2446} \times 100 = 35.99\%$$

$$e_3 = \frac{2.7963}{5.2446} \times 100 = 53.32\%$$

$$e_4 = \frac{3.862}{5.2446} \times 100 = 73.74\%$$

$$e_5 = \frac{5.2446}{5.2446} \times 100 = 100\%$$

The percentage voltage variations on the insulators are:

$$\Delta e_1 = e_1 - 0 = 19.07 - 0 = 19.07\%$$

$$\Delta e_2 = e_2 - e_1 = 35.99 - 19.07 = 16.92\%$$

$$\Delta e_3 = e_3 - e_2 = 53.32 - 35.99 = 17.33\%$$

$$\Delta e_4 = e_4 - e_3 = 73.74 - 53.32 = 20.32\%$$

$$\Delta e_5 = e_5 - e_4 = 100 - 73.74 = 26.36\%$$

Therefore, the voltage variations in kV are:

$$\Delta e_1 = (19.07\%)\frac{69000}{\sqrt{3}} = (19.07\%)39837.1686 = 7596.9\ V$$

$$\Delta e_2 = (16.92\%)39837.1686 = 6740.4\ V$$

$$\Delta e_3 = (17.33\%)39837.1686 = 6903.8\ V$$

$$\Delta e_4 = (20.32\%)39837.1686 = 8094.9\ V$$

$$\Delta e_5 = (26.36\%)39837.1686 = 10501\ V$$

Using equation (2.17), we arrive at:

$$\Delta e_1 = 5517.5\ V$$

$$\Delta e_2 = 6071.2\ V$$

$$\Delta e_3 = 7230,5\ V$$

$$\Delta e_4 = 9114.7\ V$$

$$\Delta e_5 = 11903.4\ V$$

The graph is shown in Figure 2.26.

Therefore, corona rings (Figure 2.27), which distribute the voltage between the chain insulators to decrease the voltage on the first ones, are widely used in the overhead lines.

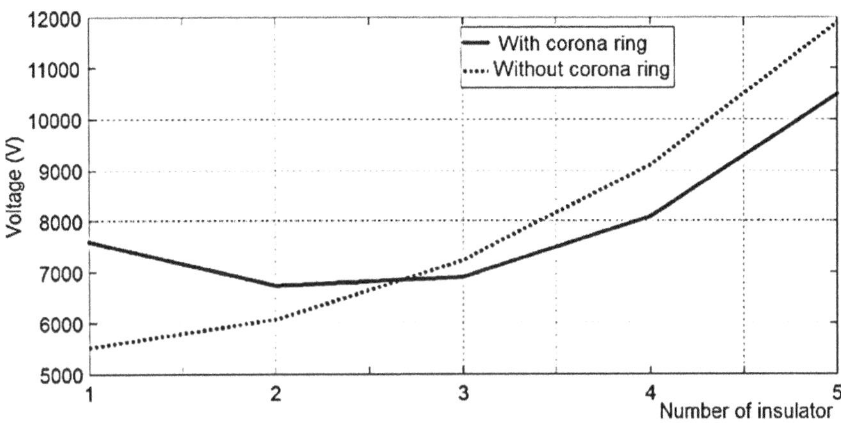

FIGURE 2.26
Insulator chain voltage distribution.

2.3.4 Fittings

The fittings are intended to form a pivotal connection with the conductors and to connect them with insulating columns and the latter with the structures.

2.3.4.1 Suspension Chain Fittings

U-clevis—This is used to attach the chain to the tower (Figure 2.28). The material is galvanized forged steel.

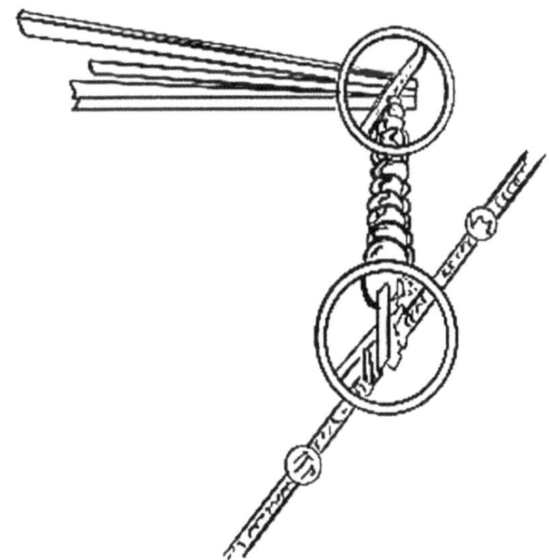

FIGURE 2.27
Corona rings.

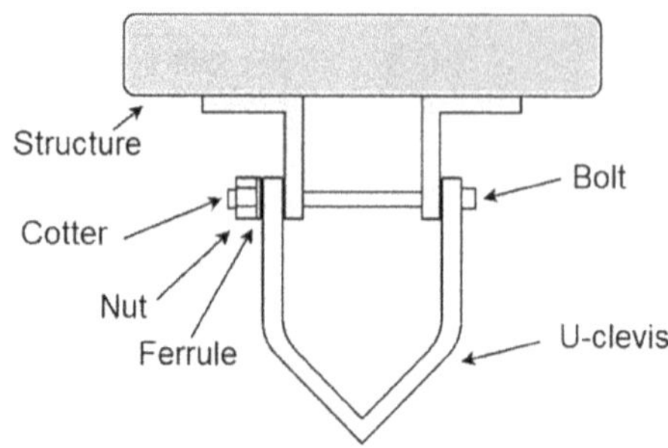

FIGURE 2.28
U-clevis.

Shackle – is used to attach the U-clevis to the other accessories (Figure 2.29). The material is galvanized forged steel.

Ball-eye—This is the part that joins the shackle to the insulator chain, as shown in Figure 2.30. The material, too, is galvanized forged steel.

Socket-Y-clevis—This is used to attach the insulator chain to the yoke plate (Figure 2.31). The material is galvanized malleable steel.

Yoke plate—This is used to attach, at each end, the clamps that hold the cables (Figure 2.32). The material is galvanized malleable steel.

Y Clevis-eye—This is used to attach the yoke plate to the cable clamp, as shown in Figure 2.33. The material is galvanized forged steel.

Suspension clamp—This is used to hold the cable and attach it to the insulators (Figure 2.34 (a), (b)). It is made of aluminum alloy. The clamp comprises two pieces: the clamp itself and another piece above the clamp, over the conductor, function of which is to hold the conductor and is called the clamp tile.

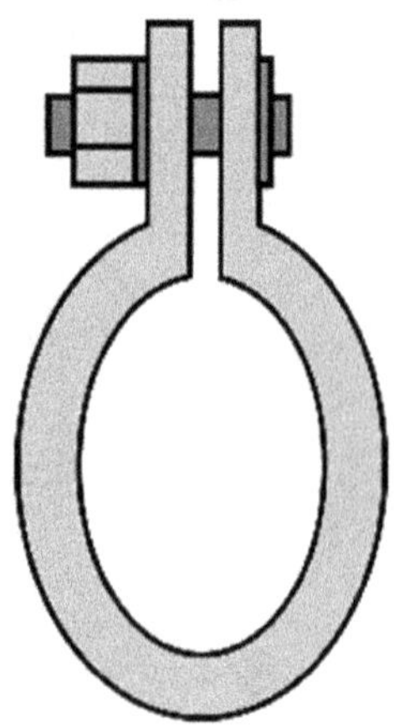

FIGURE 2.29
Shackle.

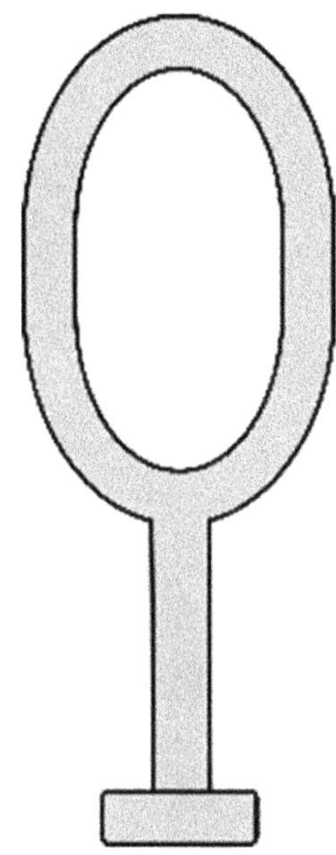

FIGURE 2.30
Ball-eye.

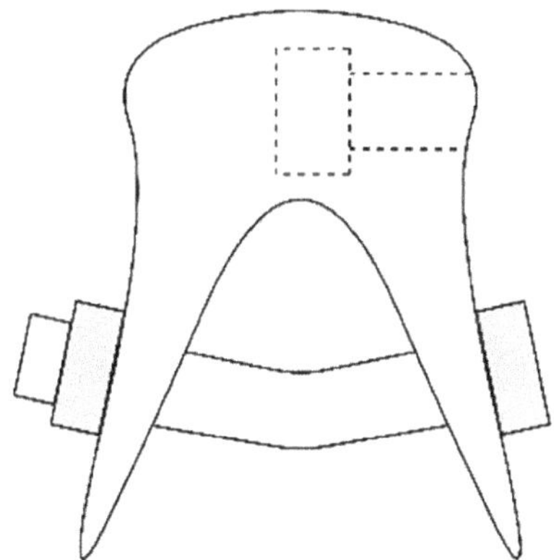

FIGURE 2.31
Socket-Y-clevis.

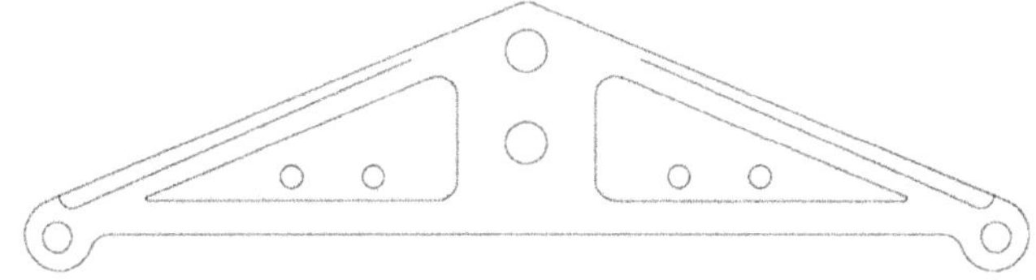

FIGURE 2.32
Yoke plate.

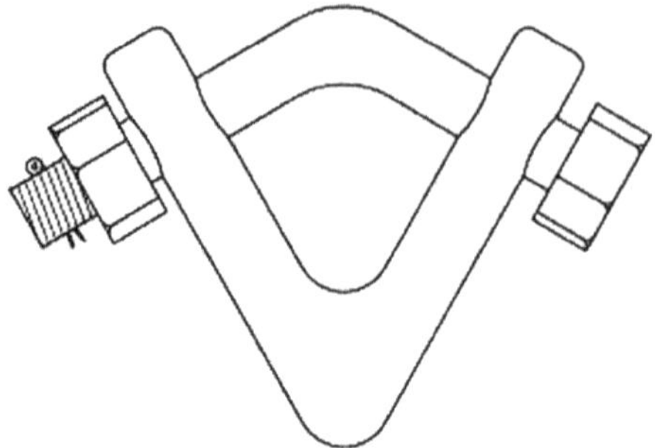

FIGURE 2.33
Y Clevis-eye.

Figures 2.35 and 2.36 show details of mounting an outer suspension insulator chain and an inner suspension insulator chain.

Other fittings are:

Spacer—This is used to prevent mechanical contact of conductors of the same phase (Figure 2.37). The spacers are made of malleable galvanized steel.

Preformed armor rods—These are aluminum rods that are wrapped around the cable to be stapled (Figure 2.38).

Stockbridge damper—These are used to absorb the vibration of the cables caused by wind action (Figure 2.39). Stockbridge damper may only be installed on one side of the tower or on each side of the tower.

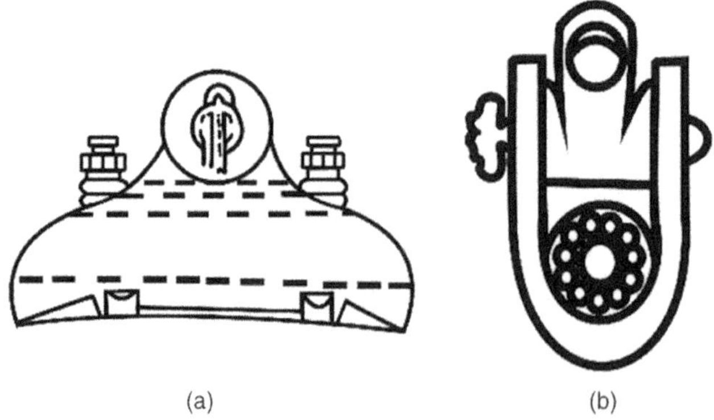

FIGURE 2.34
(a) Suspension clamp. (b) Tile.

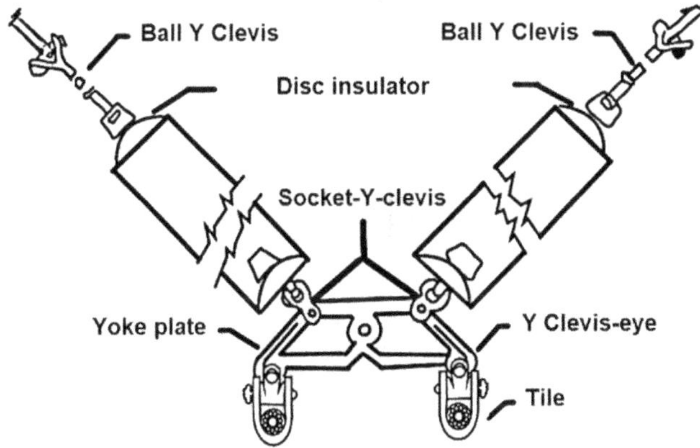

FIGURE 2.35
Inside-structure suspension insulator chain mounting details.

2.3.4.2 Anchor Chain Fittings

Shackle—This is used to attach the eye-fork, which holds the chain, as in Figure 2.40.

Clevis-eye—This is used to attach the shackle to the yoke plate (Figure 2.41).

Ball-clevis—This is used to attach the yoke plate to the insulators (Figure 2.42).

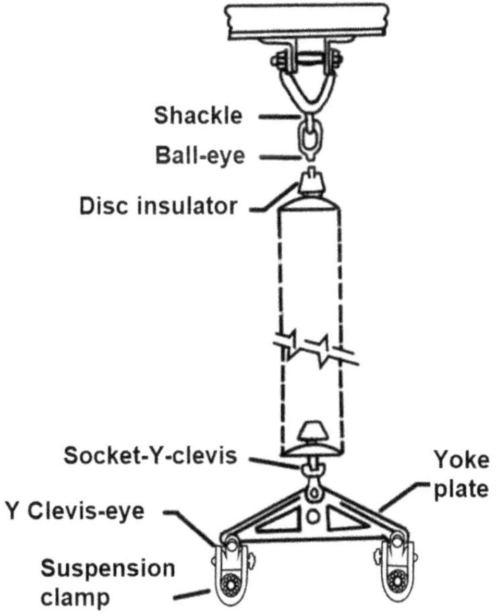

FIGURE 2.36
Suspension insulator chain assembly mounting details outside the structure.

FIGURE 2.37
Spacers for a two-conductor line per phase.

FIGURE 2.38
Preformed armor.

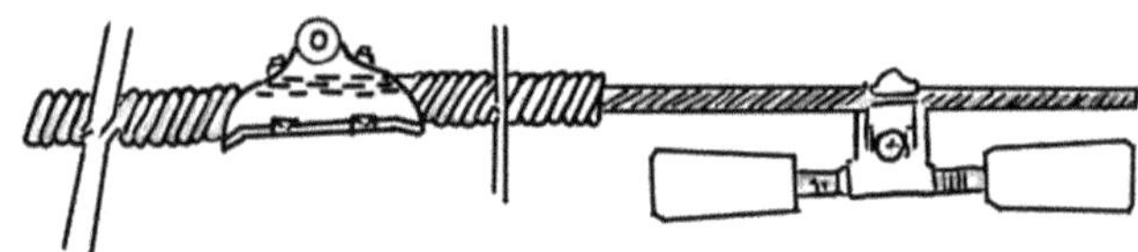

FIGURE 2.39
Stockbridge damper.

Ball-Y-clevis—This is used to attach the insulators to another yoke plate (Figure 2.43).

Yoke plate—This yoke plate differs from the previous ones because it has the purpose of tensioning the cables. In this yoke plate, the anti-corona rings are connected. The shape of the yoke plate is rectangular and with bent ends (Figure 2.44).

Oval grading ring—This is used for better voltage distribution in the insulator chain, to decrease energy loss and to protect against arcing on the insulators (Figure 2.45).

Turnbuckles—These are used to maintain a correct distance between the conductor jumper and the insulators (Figure 2.46).

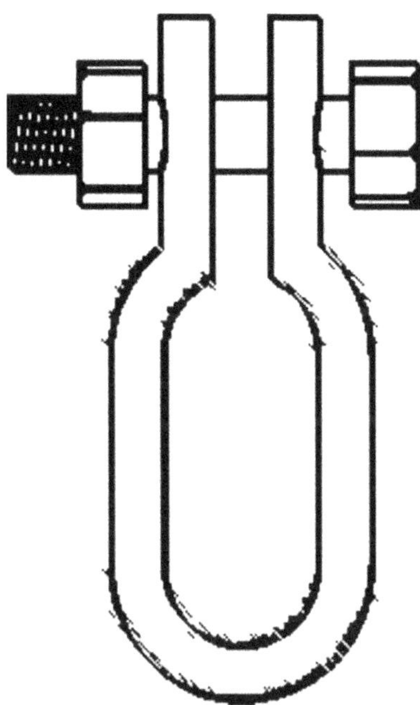

FIGURE 2.40
Anchor chain shackle.

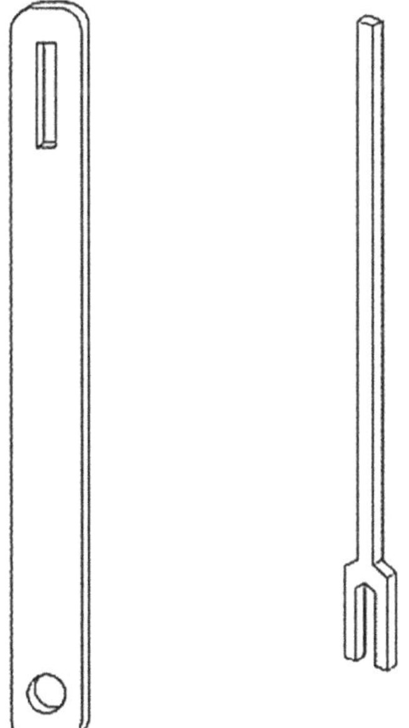

FIGURE 2.41
Eye-fork.

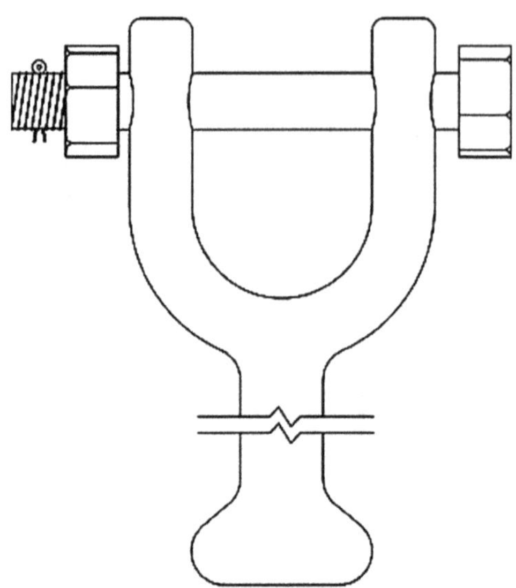

FIGURE 2.42
Ball-clevis.

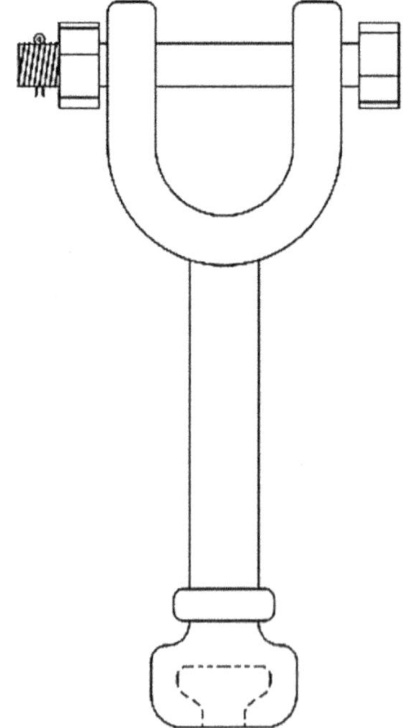

FIGURE 2.43
Ball-Y-clevis.

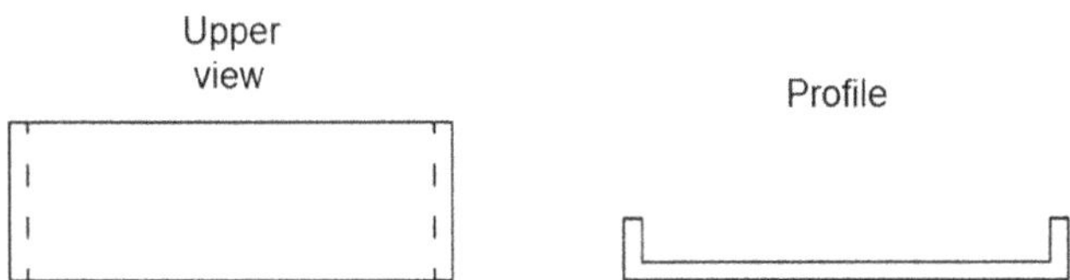

FIGURE 2.44
Yoke plate.

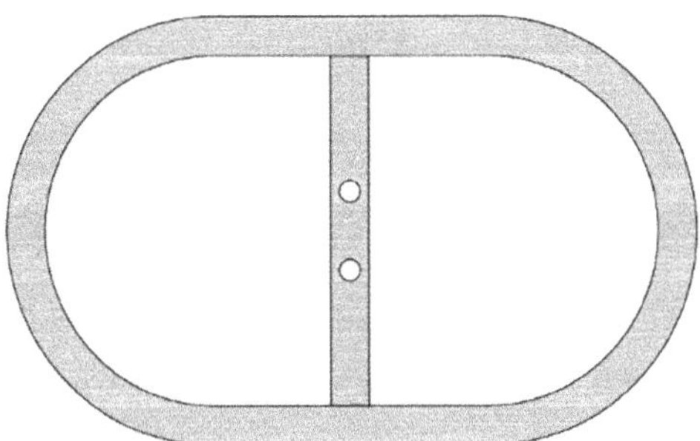

FIGURE 2.45
Corona ring.

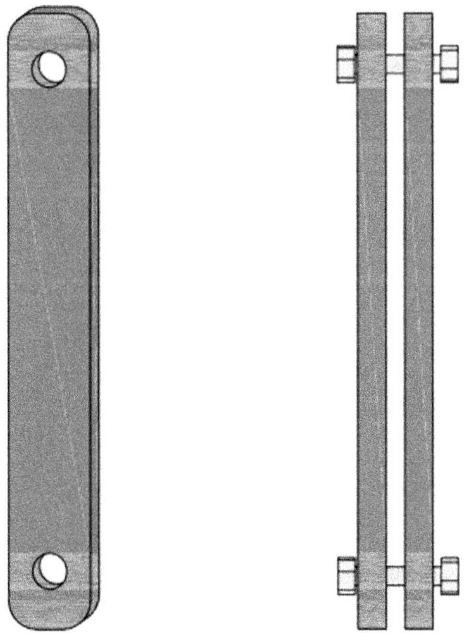

FIGURE 2.46
Turnbuckles.

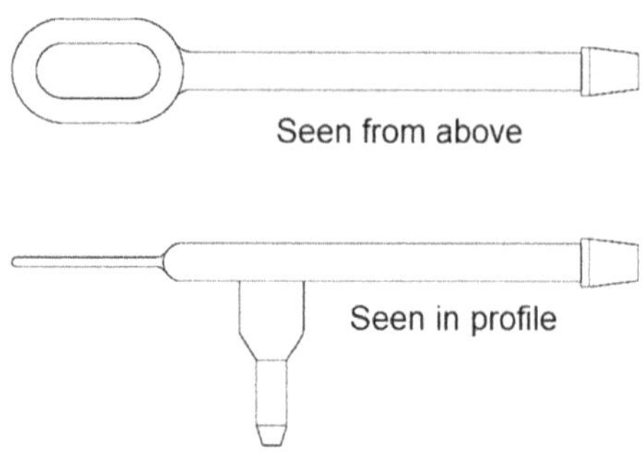

FIGURE 2.47
Compression eye clamp.

Compression eye clamp—This is used as a cable terminal (Figure 2.47). From this clamp comes the jumper.

Jumper—This is used to join cables that are broken by the use of anchor chains. The cable used to make the jumper is the same as the cable used for the energy transmission (Figure 2.48).

Figures 2.49 and 2.50 show details of the jumper chain and anchor chain assembly.

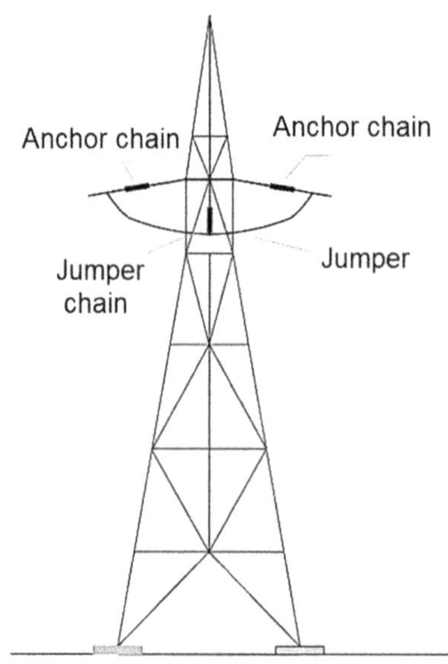

FIGURE 2.48
Jumper.

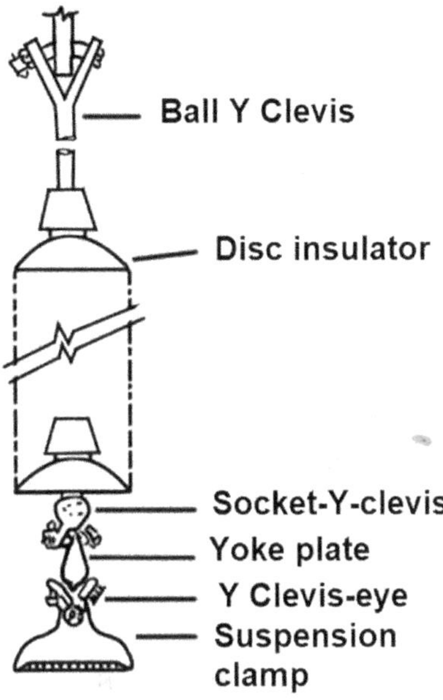

FIGURE 2.49
Jumper chain assembly details.

2.4 Construction of a Transmission Line

The elaboration of the basic documents for the implantation of a TL requires detailed field survey, to provide the knowledge of the physical environment, the living components of the environment, and the result of human intervention, where the TL will pass. It is also necessary to collect historical, meteorological, and geotechnical data of the region to choose the best alternative to be implanted.

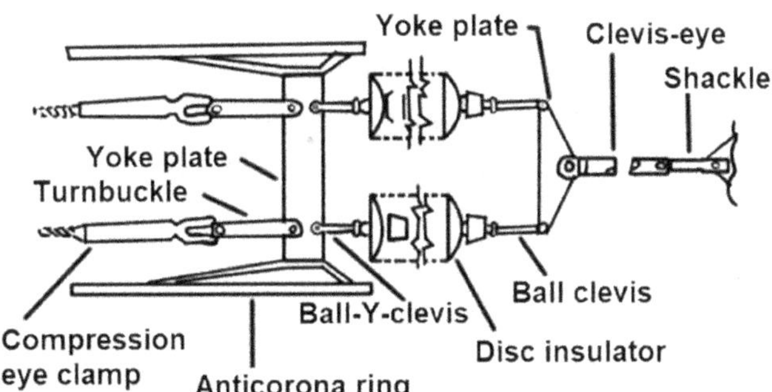

FIGURE 2.50
Anchor chain assembly details.

The steps for building a TL are:

1. Choose guidelines for the best layout and cost reduction of TL:
   - Defining the smallest possible total extent (this may not be possible as there are protection areas, and the easement strip should be removed from villages), thus reducing the number of towers, materials and services associated with the construction of the TL.
   - Avoid strong deflections, because the sharper the angles between two structures, the greater the stress on the towers and foundations, thus requiring the installation of more robust and economically expensive anchors.
   - Choose reliefs favorable to the allocation of structures, avoiding the use of towers with greater heights or spans of short length.
   - Choose soils suitable for the execution of normal foundations.
   - Avoid crossing the TL route (highways, rivers, other TL, railways, etc.).
   - If possible, maintain parallelism of the right of way with TLs already built in place.
   - If possible, deploy the line corridor at locations close to transportation to facilitate logistical support, the arrival of materials, equipment, and worker access to construction sites.
2. To make the topographic survey, to elaborate the plan and profile drawings, and to estimate the coordinates of the trace within the TL corridor with the future points of deflection of the stretch.
3. Elaboration of the basic project:
   - Technical standards used;
   - Climate weather data, wind speeds, and wind loads;
   - Technical documentation of existing structures;
   - Study of crossings;
   - Selected guidelines;
   - Wind vibration protection system;
   - Conductors and lightning conductors;
   - Mechanical study of conductors and lightning conductors;
   - Insulators and hardware;
   - Transmission towers and loading assumptions;
   - Program of loading tests;
   - Width of easement range;
   - Safety distances for structure allocation;
   - Typical foundations;

- Grounding system;
- TL electric project;
- Isolation coordination.

2.4.1 Project Execution in the Field

Executive designs are required for the different standard and special foundation types, construction and assembly drawings with specifics for various hardware, accessories, grounding and mounting items, as well as specifications, instructions, the arrow chart, and the construction list of the project, which gathers data such as tower numbers, types, heights, leg lengths and extensions, span lengths, foundation types, chain arrangements, and so on.

2.4.1.1 Execution of the Work

1. Cleaning the TL right-of-way—The TL right-of-way, which is a limit used by the company to maintain line safety, must be clean. The width of the line's safety range is determined based on three parameters: electrical effects, winding of the cables, and positioning of the support and snap foundations. In compliance with safety criteria, a clean and sufficiently wide strip must be provided to permit the deployment, operation, and maintenance of the line.

   In the case of a single TL, the following equation determines the minimum safety TL right-of-way width:

$$L = 2\left[b + d + (f + l) sen\alpha\right] \quad \text{(m)} \tag{2.18}$$

   where:

   L = the width of the right of way (m)
   b = the distance from the centerline of the structure to the phase attachment point m
   f = the typical conductor cable sag (m)
   l = the length of the insulator and fittings chain (m)
   α = the swing angle of the conductor and chain
   d = the distance in meters, equal to $\frac{V_{máx}}{150}$, where $V_{máx}$ is the voltage TL operating maximum (kV).

   The higher the line voltage, the larger the TL right-of-way. Figure 2.51 shows the TL right-of-way for a 345-kV line.

   In the TL right-of-way, all trees and shrubs should be cut as close to the ground as possible (around 20 cm).

2. Tower foundation—The holes are marked and excavated to the width corresponding to the tower foundation design, as shown in Figure 2.52.

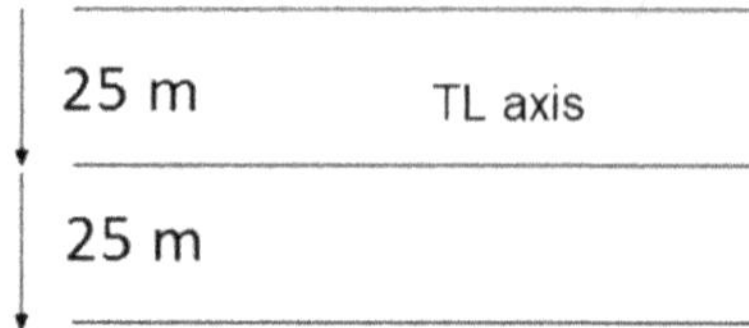

FIGURE 2.51
TL right-of-way limits for a 345-kV line.

3. Grilles placement—After excavation, the grilles (made of galvanized steel) are placed and the tower base assembled, without tightening the clamping screws to allow the correct placement of the base and its leveling (Figure 2.53 (a) and (b)).
4. Mounting the towers—Initially the preassembly, that is the assembly of parts of the structure on the ground. These parts are brought into place and screwed together following the design sequence of the project. An additional hardware, called a hawk (approximately 7 m length), is used for lifting all material, as shown in Figure 2.54.
5. Grounding the structures—Generally, the structures are grounded in two ways: (a) through ground rods; (b) through copper wires, steel wires, or metal tapes buried at a certain depth (counterweight wires), as shown in Figure 2.55.
6. TL parallel fence grounding—Fence grounding is required to enable TL-induced currents to flow into the ground.
7. Insulator installation—After the towers are assembled and overhauled, insulator chains are installed using pulleys (Figure 2.56 (a) and (b)).

   Then we have: Launching of the guide rope (made by a helicopter), as shown in Figure 2.57 (a) and (b), launching of the messenger (a 3/8″ steel cable designed to carry the pilot cable to the square

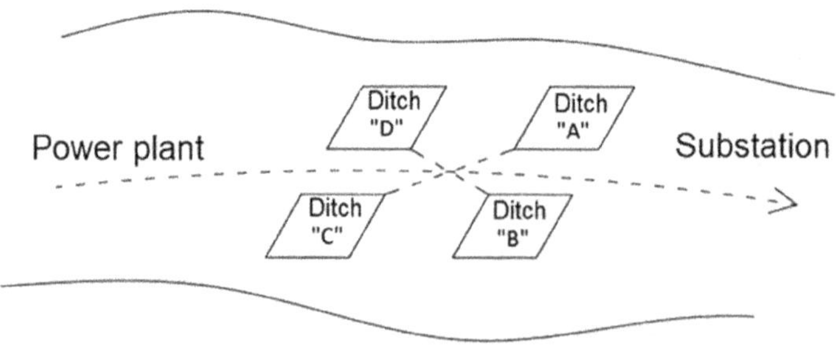

FIGURE 2.52
Foundation of a Tower.

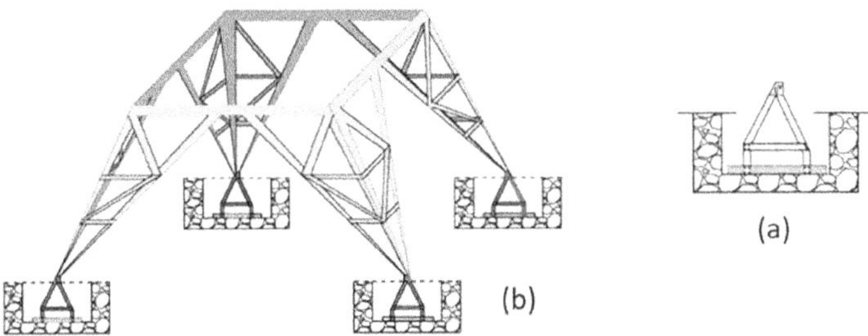

FIGURE 2.53
(a) Grilles. (b) Tower base.

where the conductor coils are located). One messenger cable for each ground wire cable. With a double circuit (two lightning conductor cables), two messenger cables are released simultaneously, as shown in Figure 2.58, cable launching ground wire, pilot cable launch (a 1″ steel cable designed to carry one-phase sub-conductors), and conductor cable launch.

8. Ground wires launch—The ground wires are initially launched analogously to the phase conductors.
9. Conductor launches—Cable installation should begin near an anchor tower to facilitate cable leveling. If the start is between suspension towers, it will be more difficult to level the cables.

The length of the gaps for the installation of equipment takes into account the length of the coils (2,200 m), the length of the pilot cable, the place of

FIGURE 2.54
Mounting a tower.

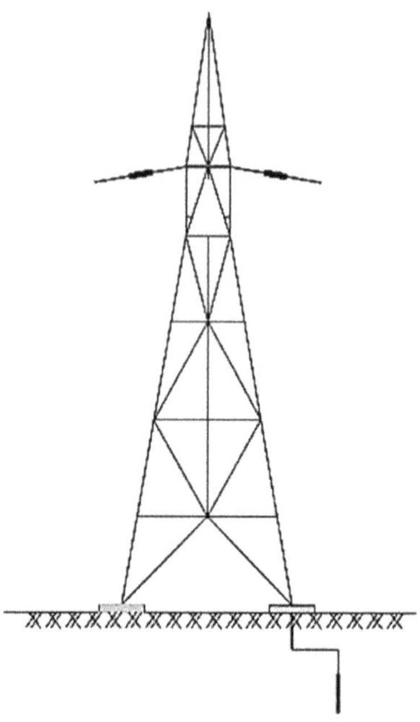

FIGURE 2.55
Grounded tower with rod.

installation of the equipment, obstacles (highways, railways), and gaps for splicing (Figure 2.59).

At crossings on railways, highways, and other lines, trestles made with eucalyptus sticks are mounted (Figure 2.60).

Initially, the pilot cable is launched. A coil of approximately 4,500 m is placed in the pilot cable reel, having at each end a device for placing the distortion. The winders are placed in phase alignment, grounded on the floor (Figure 2.61).

The process is repeated for the other towers. The pilot's tip is engaged with the hydraulic winch (Figure 2.62). Whenever possible following the TL axis, the winch will pull the pilot until it passes approximately 50 m in front of the first tower. They are taken from the winch and placed in their pulleys with the aid of a rope in advance placed in the central groove. The end of the rope is tied to the pilot and the other end is manually pulled by the escort to the other side of the tower, keeping the hydraulic winch. The process is repeated for the other towers.

Then we have the flexing (placement of the conductors in arrow every two launch sections), the stapling (comprises the removal of the launching pulleys and the final installation of the suspension accessories), and the installation of the warning sphere (Figure 2.63).

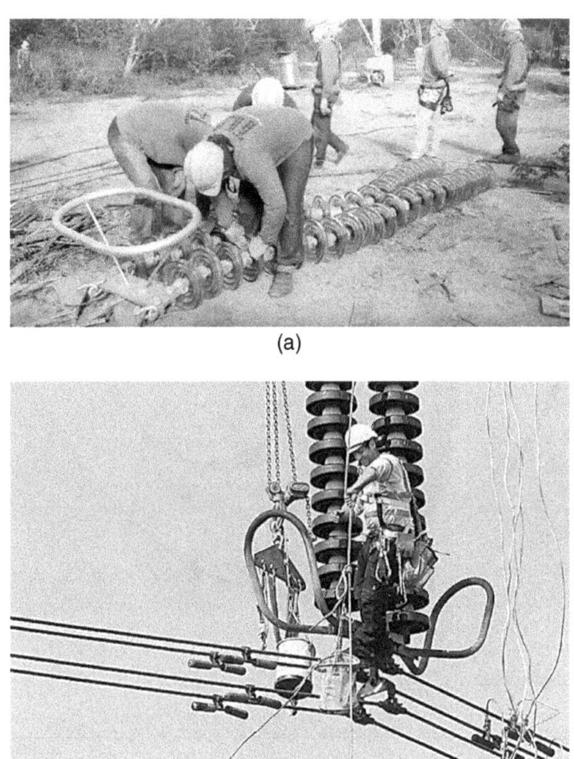

(a)

(b)

FIGURE 2.56
(a), (b) Installation of insulators.

(a) (b)

FIGURE 2.57
(a), (b) Putting in place the guide rope.

FIGURE 2.58
345-kV TL messenger cable.

2.5 Corona Effect

The leakage current in the overhead TLs is usually tiny and rises proportionally with the voltage to a certain limit. From that point onward, growth becomes very rapid, no longer negligible. Thus, for very high voltage values, dry air is no longer a perfect insulator and the leakage current becomes significant. Many tests show that dry air at normal temperature and pressure (25°C and 76 cm barometric pressure) is no longer insulating at 29.8 kV/cm (peak value). In points where there are edges or protrusions, because of the power of the points, the electric field becomes high and luminous effluvia appears, producing a slight

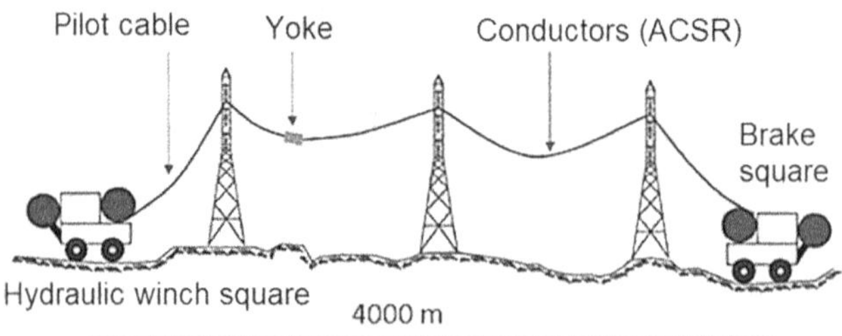

FIGURE 2.59
Location of cable launching equipment.

FIGURE 2.60
Road-mounted cable trestles.

crackle. These effluvia make up the beginning of the dielectric perforation. From a certain voltage value, and when observed in darkness, the entire driver appears surrounded by a bluish light halo, which produces a whistle-like noise, as shown in Figure 2.64. This phenomenon is the corona effect.

The main consequences of this phenomenon are:

- light emission;
- audible noise;
- Radio noise (interference in communication circuits);
- Conductor vibration;

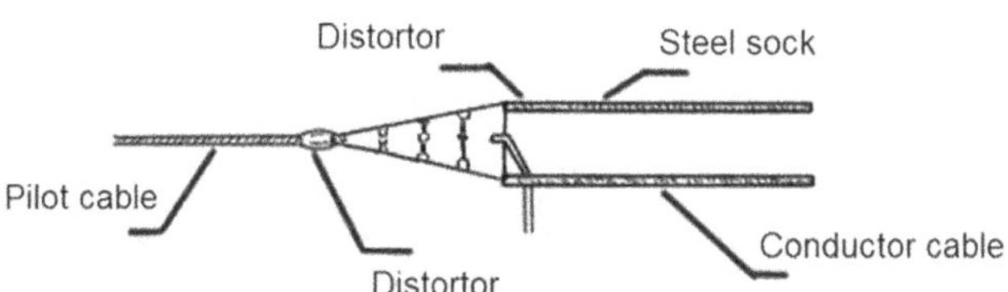

FIGURE 2.61
Conductors caught in pilot cable by distortors.

FIGURE 2.62
345-kV TL pilot cable puller.

- Ozone release;
- Increased power losses.

The critical disruptive voltage under normal temperature and pressure conditions (25°C and 76 cm Hg) can be calculated by:

$$V_0 = 21.2rln\left(\frac{D}{r}\right) kV \tag{2.19}$$

FIGURE 2.63
Aerial warning sphere installation.

FIGURE 2.64
Corona effect.

where:

V_0 = the disruptive critical phase voltage (mean square value) at kV
r = the radius of the conductor at cm
D = the distance between two conductors at cm

Critical visual voltage, according to Peek, is given by:

$$V_v = 21.1\delta m_v r\left(1+\frac{0.301}{\sqrt{\delta r}}\right)ln\frac{D}{r} \tag{2.20}$$

where:

V_v = the critical visual phase voltage (mean square value) at kV
r = the radius of the conductor at cm
D = the distance between two conductors at cm
m_v = the irregularity factor for visible corona $(0 < m_v \leq 1)$

$$\delta = \frac{3.9211p}{273+t} \tag{2.21}$$

where:

δ = the relative density factor of air
p = the barometric pressure in cm Hg
t = the temperature in degrees Celsius

Corona losses (Peek, 1929), considering good weather, can be calculated using equation (2.22).

$$P_c = \frac{221}{\delta}(f+25)\left(\frac{r}{D}\right)^{1/2}(V-V_0)^2.10^{-5}\ kW/km \tag{2.22}$$

where:

δ = the relative density factor of air
f = the frequency at Hz
r = the radius of the conductor at cm
D = the distance between two conductors at cm
V = the phase-neutral operating voltage in kV
V_0 = the disruptive critical phase voltage (mean square value) at kV

2.6 Problems

2.6.1 Explain what support structures are.
Answer: See item 2.3.1.

2.6.2 Explain what transposition structures are.
Answer: See item 2.3.1.

2.6.3 Explain what expanded ACSR cables are.
Answer: See item 2.3.2.1.

2.6.4 What are ground wire cables?
Answer: See item 2.3.2.2.

2.6.5 Describe the disk isolators.
Answer: See item 2.3.3.

2.6.6 Explain what the potential distributor rings are.
Answer: See item 2.3.3.

2.6.7 Describe the fittings used on the transmission lines.
Answer: See item 2.3.4.

2.6.8 What is a jumper chain?
Answer: See item 2.3.4.

2.6.9 Explain the steps for building a TL.
Answer: See item 2.4.1.

2.6.10 Judge the next items regarding transmission lines of electric power systems.

a. The function of insulators in a transmission line is only mechanical: to support the conductors in the transmission lines.
b. In high voltage power transmission lines, conductor cables obtained by stranding aluminum wires shall be used. The choice of conductors with these characteristics is justified because these conductors have lower inductive reactance than solid conductors of the same diameter and length, among other factors.

Answers: a) Wrong, b) Wrong.

2.6.11 Regarding the transmission and distribution of electricity in Brazil, the National Interconnected System Basic Network comprises transmission lines, buses, power transformers, and substation equipment and power transformers with equal primary voltage or greater than X and secondary and tertiary stresses less than X from July 1, 2004. This X value is

a. 115 kV
b. 230 kV
c. 345 kV
d. 460 kV

Answer: b.

2.6.12 Regarding the electrical requirements considered for the sizing of transmission lines that make up the basic network of the SIN, judge the following items.

a. For sizing of transmission line insulators, consideration shall be given, among other factors, to the maximum operating voltage and balance of the insulator chain under critical wind action, the minimum expected return period of which shall be based on relevant standard.
b. In long-term operation, a phase-to-ground conductor clearance of up to 50% of the minimum safety distance of the transmission line operating under normal conditions shall be permitted.

Answers: a) Right, b) Wrong.

2.6.13 What is the type of cable most commonly used in Brazil as a conductor of electricity in overhead transmission lines?

a. Copper Cable
b. Aluminum (AC) cable
c. Galvanized steel cable
d. Aluminum Cable Steel Reinforced (ACSR)
e. Copperweld Cable

Answer: d.

2.6.14 The choice of an appropriate voltage level for the transmission of electricity should take into account technical and economical design aspects. Regarding this matter, judge the following items.

a. With the transmission of a fixed value of power between two points of a transmission line, a higher-rated voltage level will require lower rated current and therefore under these conditions a smaller diameter conductor will be required.
b. Transmission system design shall include fixed and operating costs. Losses associated with the transmission system are part

of the fixed costs, while expenses related to conductors, isolators, and passageway are associated with operating costs.

Answers: a) Right, b) Wrong.

2.6.15 A positive aspect in the Brazilian electricity system is the possibility of integrating the various regions of the country with large transmission lines. These lines, because of their size and operational responsibility, must comply with technical guidelines regarding their various components. Regarding this matter, judge the following item.

a. The choice of tower type and cables is independent of the topographic characteristics of the transmission line layout.

Answer: Wrong.

3

Transmission Line Parameter Calculation

3.1 Introduction

The determination of the serial impedance and shunt admittance of an overhead transmission line (TL), with or without ground wire cables, is vitally important for the analysis of a TL.

In this chapter, we will deal with determining the calculation of TL parameters.

3.2 Resistance

The direct current (DC) resistance of a conductor at a given temperature is calculated as:

$$R_{DC} = \frac{\rho l}{A}\ (\Omega) \tag{3.1}$$

where:

ρ = the resistivity of the conductor at a given temperature
l = the conductor length
A = the cross-sectional area

The resistance of a conductor depends on the following factors:

- Spiraling;
- Temperature;
- Frequency;
- Current magnitude.

The DC resistance is higher than that calculated using equation (3.1), as placing the spiral conductors (as shown in Figure 3.1) makes them longer by 1% to 2% of the original length.

FIGURE 3.1
Stranded conductors.

The temperature resistance variation of metal conductors is approximately linear for normal operation.

$$R_2 = \frac{\rho_1 l_1}{A_1}\left(\frac{t_2 + |T|}{t_1 + |T|}\right) \tag{3.2}$$

where:

T = the constant temperature that depends on the conductive material
Index 1 = the conductor parameters at temperature t_1
R_2 = the DC resistance of the conductor at the new temperature t_2

Temperatures (T) for commonly used materials are:

- Annealed copper of 100% conductivity (234.5°C);
- Hard-drawn copper of 97% conductivity (241.5°C);
- Hard-drawn aluminum of 61% conductivity (228.1°C);
- Steel of 2–14% conductivity (180–980°C).

The DC resistance can be calculated directly using equation (3.3):

$$R_{DC} = \frac{\rho_{20} t_{20}}{A_{20}}\left[1 + \alpha_{20}(t - 20)\right] \tag{3.3}$$

Where R_{DC} is DC resistance at t°C, ρ is volume resistivity, and the subscript 20 indicates that ρ, t, A, and α are at 20°C.

Alternating current (AC) resistance or effective resistance is given by:

$$R_{AC} = \frac{\textit{Conductor real power loss}}{I_{rms}^2} \tag{3.4}$$

For DC, the current distribution is uniform across the conductor cross section and the DC resistance can be calculated from equation (3.1). But for AC, the current distribution is uneven across the conductor cross section and as the frequency increases, the current in a solid cylindrical conductor increases toward the conductor surface, with a lower current density in the conductor center. This phenomenon is called the skin effect, as shown in Figure 3.2.

In Figure 3.2 (a), we have a tubular conductor excited with DC or zero frequency. In Figure 3.2 (b), the tubular conductor is excited by AC. As the

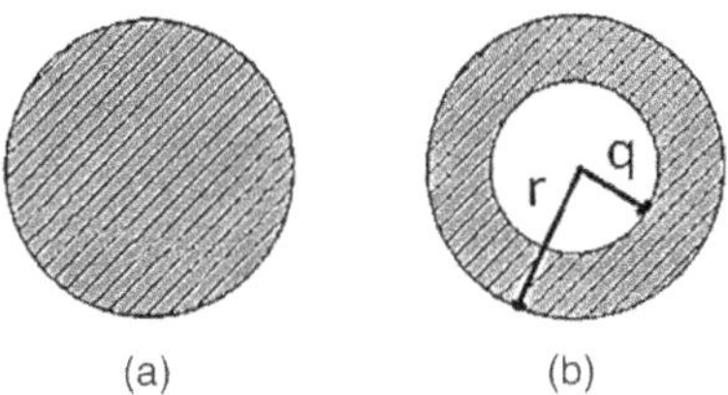

FIGURE 3.2
(a), (b) Current distribution in a tubular conductor.

frequency of the AC increases, the current concentration shifts to the periphery of the conductor, as shown in Figure 3.3. The skin effect causes increased resistance and decreased internal inductance as the effective conduction area decreases.

At a frequency of 60 Hz, the conductor resistance in AC increases slightly compared to the resistance in DC.

Table 3.1 presents the skin effect on a tubular conductor up to 1 MHz within the range of an electromagnetic transient.

We see that in Table 3.1, up to 60 Hz, the influence of the skin effect is small, but in the electromagnetic transient frequencies, the skin effect is significant.

For magnetic conductors, such as steel conductors used for shielded wires, the resistance depends on the current module.

3.3 Inductance

The inductance of a TL is calculated as flux linkages per ampere. Considering the constant permeability μ, the resulting flux linkages produced by and in phase with the sinusoidal current can be expressed as the phasor λ. Thus:

$$L = \frac{\lambda}{I} \tag{3.5}$$

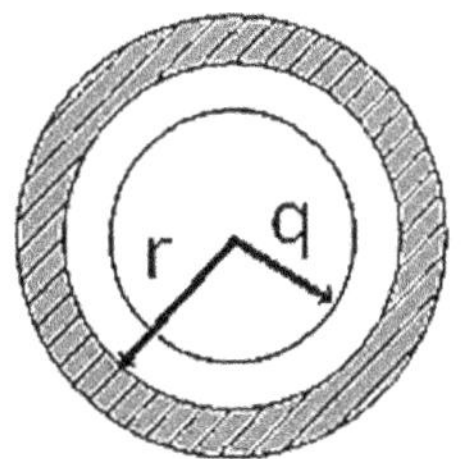

FIGURE 3.3
Skin effect.

TABLE 3.1

Skin Effect on a Tubular Conductor

f (Hz)	R_{AC}/R_{DC}	$L_{AC\ internal}/L_{DC\ internal}$
2	1.0002	0.909662
4	1.0007	0.909456
6	1.0015	0.909114
8	1.0025	0.908634
10	1.0041	0.908019
20	1.0164	0.902948
40	1.0532	0.883578
60	1.1347	0.854219
80	1.2233	0.818272
100	1.3213	0.779083
200	1.7983	0.602534
400	2.4554	0.427612
600	2.9421	0.350272
800	3.3559	0.304017
1000	3.7213	0.272232
2000	5,1961	0.192896
4000	7.1876	0.136534
6000	8.7471	0.111515
8000	10.0622	0.096590
10000	11.2289	0.086401
20000	15.7678	0.061106
40000	22.1958	0.043212
60000	27.1337	0.035284
80000	31.2942	0.030557
1000000	34.9597	0.027331

To calculate the approximate inductance of a TL, it is necessary to consider the flux inside and outside the conductor. Consider the long cylindrical conductor "i" shown in Figure 3.4, which has a concentric magnetic field.

The ampere law states that:

$$fmm = \oint H_{tangent} d_S = I_{closed} \ (Ampere - turns) \tag{3.6}$$

where:

$H_{tangent}$ = the component of the intensity of the magnetic field tangent to d_S (Ampere-turns/metro)

S = the distance along the path (meter)

I = the current in the closed path (Amps)

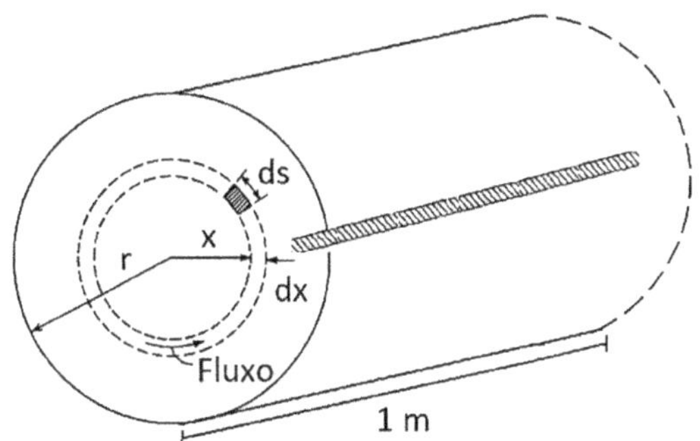

FIGURE 3.4
Cross section of a cylindrical conductor "i."

Performing the integration stated in equation (3.6) around the concentric circular path with the conductor at x meters from the center where H_x is a constant over the path and tangent to it, we have:

$$H_x(2\pi x) = I_x \ \ para \ \ x < r \tag{3.7}$$

Taking the value of H_x:

$$H_x = \frac{I_x}{2\pi x} \tag{3.8}$$

We assume an even distribution of current within the conductor, thus:

$$I_x = \frac{\pi x^2}{\pi r^2} I \tag{3.9}$$

Replacing equation (3.9) in (3.8), we get:

$$H_x = \frac{x^2}{2\pi x r^2} I = \frac{x}{2\pi r^2} I \ \ At/m \tag{3.10}$$

The flux density x meters from the center of the conductor "i" is:

$$B_x = \mu H_x = \frac{\mu x}{2\pi r^2} I \ \ Wb/m^2 \tag{3.11}$$

The differential flow $d\phi$ per unit length of conductor in the dashed rectangle of Figure 3.2 is:

$$d\phi = B_x dx = \frac{\mu x}{2\pi r^2} I dx \ \ Wb/m \tag{3.12}$$

The flux linkages $d\lambda$ per meter of length, which are originated by the flux in the tubular element, are the product of the flux per meter in length and the fraction of the current linked.

$$d\lambda_i = d\phi\,\frac{\pi x^2}{\pi r^2} = \frac{x^2}{r^2}\,d\phi = \frac{\mu x^3 I}{2\pi r^4}\,dx \tag{3.13}$$

$$\lambda_{int(i)} = \int_0^r \frac{\mu x^3 I}{2\pi r^4}\,dx = \frac{\mu I}{8\pi} \tag{3.14}$$

Knowing that $\mu = \mu_r \mu_0 = 1x\ 4\pi x 10^{-7}\ H/m$

$$\lambda_{int(i)} = \frac{I}{2}10^{-7}\ Wb/m \tag{3.15}$$

Therefore, the internal inductance of the conductor "i" is:

$$L_{int(i)} = \frac{1}{2}10^{-7}\ H/m \tag{3.16}$$

Let us now consider the external flux linkages to conductor "i," as shown in Figure 3.5.

From equation (3.12) the flux density external to conductor "i," that is, for $x > r$, with a non-conductor permeability $\mu = \mu_0$, is calculated as:

$$B_x = \mu H_x = 4\pi x 10^{-7}\,\frac{I}{2\pi x} = 2x10^{-7}\,\frac{I}{x}\ Wb/m^2 \tag{3.17}$$

The current I is connected to the flux outside the conductor:

$$d\phi = d\lambda = 2x10^{-7}\,\frac{I}{x}\,dx\ (Wb/m) \tag{3.18}$$

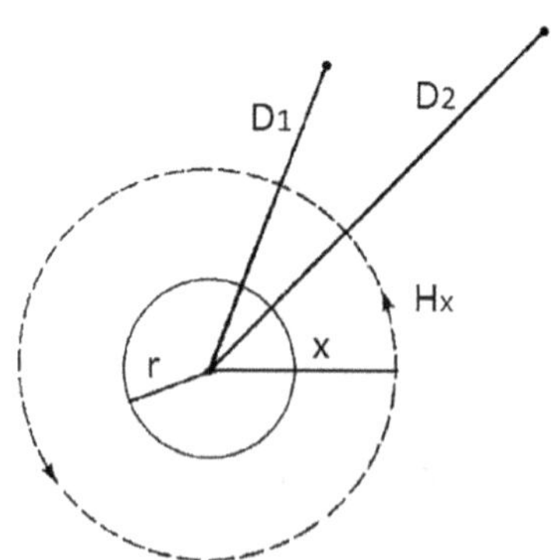

FIGURE 3.5
Magnetic field outside conductor "i."

Integrating equation (3.18) between two external points at distances D_1 and D_2, we have:

$$\lambda_{12} = \int_{D1}^{D2} 2x10^{-7}\frac{I}{x}dx = 2x10^{-7}I\int_{D1}^{D2}\frac{dx}{x} = 2x10^{-7}\,Iln\frac{D2}{D1}\ (Wb-t/m) \quad (3.19)$$

The external inductance to conductor "i," due to the flux linkages λ_{12}, is:

$$L_{12} = \frac{\lambda_{12}}{I} = 2x10^{-7}ln\frac{D2}{D1}\ H/m \quad (3.20)$$

Considering the full flux linkages λ_T, which connects conductor "i" to external point P at distance D, causes $D1 = r$ and $D2 = D$, therefore:

$$\lambda_T = \frac{I}{2}10^{-7} + 2x10^{-7}\,Iln\frac{D}{r} = 10^{-7}I\left(\frac{1}{2} + 2ln\frac{D}{r}\right) = 10^{-7}I\left(2lne^{1/4} + 2ln\frac{D}{r}\right)$$
$$= 2x10^{-7}I\left(ln\frac{De^{1/4}}{r}\right) = 2x10^{-7}I\left(ln\frac{D}{re^{-1/4}}\right) \quad (3.21)$$

The term

$$GMR = r' = re^{-1/4} = 0.7788r \quad (3.22)$$

is known as geometric mean radius (GMR) and applies to solid cylindrical conductors. Therefore, λ_T and the total inductance of conductor "i" are respectively:

$$\lambda_T = 2x10^{-7}I\left(ln\frac{D}{RMG}\right) \quad (3.23)$$

and

$$L_T = \frac{\lambda_T}{I} = 2x10^{-7}\left(ln\frac{D}{RMG}\right) \quad (3.24)$$

3.4 Flux Linkages of One Conductor in a Group

A more general problem is that of a conductor in a group of "N" conductors, as shown in Figure 3.6, where the sum of the currents that circulates in all of them is zero. This is:

$$I_1 + I_2 + \cdots + I_i + \cdots + I_N = \sum_{n=1}^{n=N} I_N = 0 \quad (3.25)$$

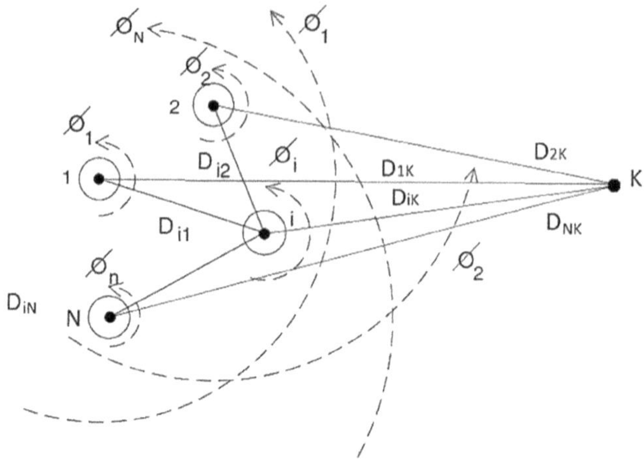

FIGURE 3.6
Group of "N" cylindrical conductors.

According to equation (3.23), the full flux linkages connecting each conductor "i" to the distance of a point K can be calculated using the superposition theorem. Therefore:

$$\lambda_{iK} = \lambda_{i1} + \lambda_{i2} + \cdots + \lambda_{iN} = 2x10^{-7} \sum_{n=1}^{n=N} I_n ln\left(\frac{D_{Kn}}{D_{in}}\right) \tag{3.26}$$

We should highlight that for $n = i \rightarrow D_{ii} = GMR$.

Breaking down equation (3.26) into two parts, we have:

$$\lambda_{iK} = 2x10^{-7} \sum_{n=1}^{n=N} I_n ln\left(\frac{1}{D_{in}}\right) + 2x10^{-7} \sum_{n=1}^{n=N} I_n ln D_{Kn} \tag{3.27}$$

Separating the last term from the second sum, we have:

$$\lambda_{iK} = 2x10^{-7} \sum_{n=1}^{n=N} I_n ln\left(\frac{1}{D_{in}}\right) + 2x10^{-7} \sum_{n=1}^{n=N-1} I_n ln D_{Kn} + I_n ln D_{KN} \tag{3.28}$$

From equation (3.25) we have

$$I_n = -(I_1 + I_2 + \cdots I_i + \cdots + I_{N-1}) = -\sum_{n=1}^{n=N-1} I_n \tag{3.29}$$

Substituting equation (3.28) into equation (3.27), we have:

$$\begin{aligned}\lambda_{iK} &= 2x10^{-7}\sum_{n=1}^{n=N} I_n ln\left(\frac{1}{D_{in}}\right) + 2x10^{-7}\sum_{n=1}^{n=N-1} I_n ln D_{Kn} - \sum_{n=1}^{n=N-1} I_n ln D_{KN} \\ &= 2x10^{-7}\sum_{n=1}^{n=N} I_n ln\left(\frac{1}{D_{in}}\right) + 2x10^{-7}\sum_{n=1}^{n=N-1} I_n ln\left(\frac{D_{Kn}}{D_{KN}}\right)\end{aligned} \tag{3.30}$$

Making the point $K \to \infty$, all distances become equal; therefore, $\frac{D_{Kn}}{D_{KN}} \to 1$ and $ln\left(\frac{D_{Kn}}{D_{KN}}\right) \to 0$. Thus, the total flux linkage is equal to:

$$\lambda_{iT} = 2x10^{-7}\sum_{n=1}^{n=N} I_n ln\left(\frac{1}{D_{in}}\right) \tag{3.31}$$

To calculate the inductance of a line of composite conductors, which comprise two or more solid cylindrical filaments in parallel, we will use equation (3.31).

3.5 Inductance of Composite-Conductor Lines

In Figure 3.7, we have a single-phase TL composed of two conductors: conductor X has N identical parallel filaments and conductor Y has M identical parallel filaments. The only restriction is that each parallel filament is cylindrical and conducts the same current. Therefore, each X parallel filament conducts an I/N current and each Y parallel filament conducts an $-I/M$ current.

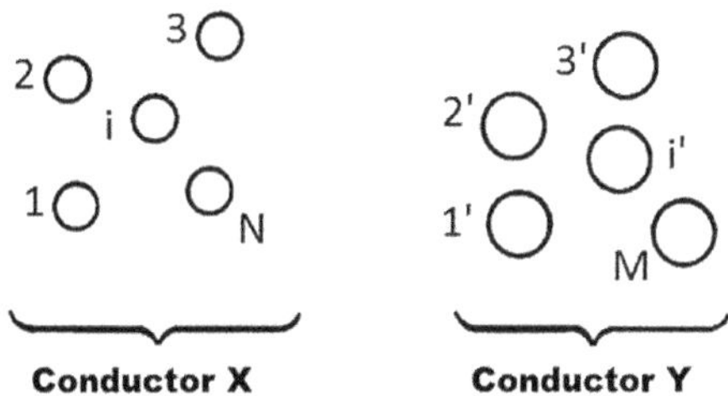

FIGURE 3.7
Single-phase two-conductor line with composite conductors.

By applying equation (3.30) to conductor X parallel filament 1, we get the flux linkages:

$$\lambda_1 = 2\times10^{-7}\frac{I}{N}\left[ln\left(\frac{1}{GMR_1}\right)+ln\left(\frac{1}{D_{12}}\right)+ln\left(\frac{1}{D_{13}}\right)+\cdots+ln\left(\frac{1}{D_{1N}}\right)\right]$$
$$-2\times10^{-7}\frac{I}{M}\left[ln\left(\frac{1}{D_{11'}}\right)+ln\left(\frac{1}{D_{12'}}\right)+ln\left(\frac{1}{D_{13'}}\right)+\cdots+ln\left(\frac{1}{D_{1M}}\right)\right] \quad (3.32)$$

$$\lambda_1 = 2\times10^{-7}\, I ln\left(\frac{\sqrt[M]{D_{11'}D_{12'}D_{13'}\cdots D_{1M}}}{\sqrt[N]{GMR_1 D_{12}D_{13}\cdots D_{1N}}}\right) Wb-t/m \quad (3.33)$$

The inductance of filament 1 is calculated as:

$$L_1 = \frac{\lambda_1}{\frac{I}{N}} = 2N\times10^{-7} ln\left(\frac{\sqrt[M]{D_{11'}D_{12'}D_{13'}\cdots D_{1M}}}{\sqrt[N]{GMR_1 D_{12}D_{13}\cdots D_{1N}}}\right) H/m \quad (3.34)$$

Similarly, we can calculate the inductances of parallel filaments 2, 3, and N. As, for example, for parallel filament 2, we have:

$$L_2 = 2N\times10^{-7} ln\left(\frac{\sqrt[M]{D_{21'}D_{22'}D_{23'}\cdots D_{2M}}}{\sqrt[N]{D_{21}GMR_2 D_{23}\cdots D_{2N}}}\right) H/m \quad (3.35)$$

The average inductance of the parallel filaments of conductor X is:

$$L_{médio} = \frac{L_1+L_2+L_3+\cdots+L_N}{N} \quad (3.36)$$

Conductor X consists of N parallel filaments that are electrically in parallel.

$$L_X = \frac{L_{médio}}{N} = \frac{\frac{L_1+L_2+L_3+\ldots+L_N}{N}}{N} = \frac{L_1+L_2+L_3+\cdots+L_N}{N^2} \quad (3.37)$$

Substituting the values of $L_1, L_2, L_3 \ldots L_N$, we have:

$$L_X = 2\times10^{-7} ln$$
$$\times\left(\frac{\sqrt[MN]{(D_{11'}D_{12'}D_{13'}\cdots D_{1M})(D_{21'}D_{22'}D_{23'}\cdots D_{2M})\ldots(D_{M1'}D_{M2'}D_{M3'}\cdots D_{MN})}}{\sqrt[N^2]{(GMR_1 D_{12}D_{13}\cdots D_{1N})(D_{21}GMR_2 D_{23}\ldots D_{2N})\ldots(D_{N1}D_{N2}D_{N3}\ldots GMR_N)}}\right) H/m$$
$$(3.38)$$

The numerator of equation (3.38) is known as the mutual geometric mean distance (D_m) between the X and Y conductors and the denominator is known as the proper geometric mean distance (GMD), which is represented by D_s. Thus,

$$L_X = 2 \times 10^{-7} \ln\left(\frac{D_m}{D_s}\right) H/m \tag{3.39}$$

The inductance of conductor Y is similarly determined and the line inductance is given by:

$$L_{line} = L_X + L_Y \tag{3.40}$$

3.6 GMD Proper Bundled Conductor Calculation

In voltage TLs above 230 kV, with a single conductor per phase, corona loss and communications interference can be very high. Thus, when we increase the grouping of conductors within the group, the high voltage gradient is reduced if two or more conductors are placed per phase at a distance that, compared to the distance between phases, is small. Another effect of conductor grouping is to reduce reactance.

Figure 3.8 show the composed of bundled used.

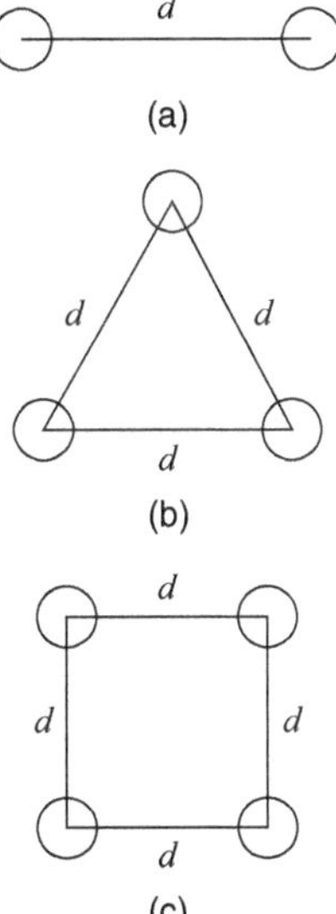

FIGURE 3.8
Bundle arrangements: (a) Two-strand bundle. (b) Three-strand bundle. (c) Four-strand bundle.

GMD calculation is done using the expression of the denominator of equation (3.38):

For a two-strand bundle:

$$D_{s2} = \sqrt[2^2]{GMR_1(d)GMR_2(d)} = \sqrt[4]{[(GMR)(d)]^2} = \sqrt{(GMR)(d)} \tag{3.41}$$

With $GMR_1 = GMR_2 = GMR$

For a three-strand bundle:

$$D_{s3} = \sqrt[3^3]{GMR_1(d)(d)GMR_2(d)(d)GMR_3(d)(d)} = \sqrt[9]{[(GMR)(d)(d)]^3} = \sqrt[3]{GMR(d)^2} \tag{3.42}$$

With $GMR_1 = GMR_2 = GMR_3 = GMR$

For a four-strand bundle:

$$D_{s4} = \sqrt[4^4]{GMR_1(d)(d)\left(d\sqrt{2}\right)GMR_2(d)(d)\left(d\sqrt{2}\right)GMR_3(d)(d)\left(d\sqrt{2}\right)GMR_4(d)(d)\left(d\sqrt{2}\right)}$$
$$= \sqrt[16]{[(GMR)(d)(d)(d\sqrt{2})]^4} = 1.0905\sqrt[4]{GMR(d)^3} \tag{3.43}$$

With $GMR_1 = GMR_2 = GMR_3 = GMR_4 = GMR$

Example 3.1

To determine the proper geometric mean distance for each of the unconventional cables shown in Figure 3.9, assuming that each of the filaments has a diameter $2r = 2.3$ mm and the same current density. For cable (d) in Figure 3.9, the center wire shall be considered with nonzero conductivity and zero conductivity.

SOLUTION:

Cable (a)

The distances between each conductor are:

$$D_{11} = D_{22} = D_{33} = D_{44} = D_{55} = D_{66} = RMG$$
$$DA = D_{12} = D_{13} = D_{24} = D_{34} = D_{35} = D_{46} = D_{56} = 2r$$
$$DB = D_{14} = D_{23} = D_{36} = D_{45} = 2r\sqrt{2}$$
$$DC = D_{15} = D_{26} = 4r$$
$$DD = D_{16} = D_{25} = 2r\sqrt{5}$$

Using these distances and using the expression of the denominator of equation (3.38):

$$D_s = \sqrt[6^2]{GMR^6 * \left((D_A)^7\right)^2 * \left((D_B)^4\right)^2 * \left((D_C)^2\right)^2 * \left((D_D)^2\right)^2}$$

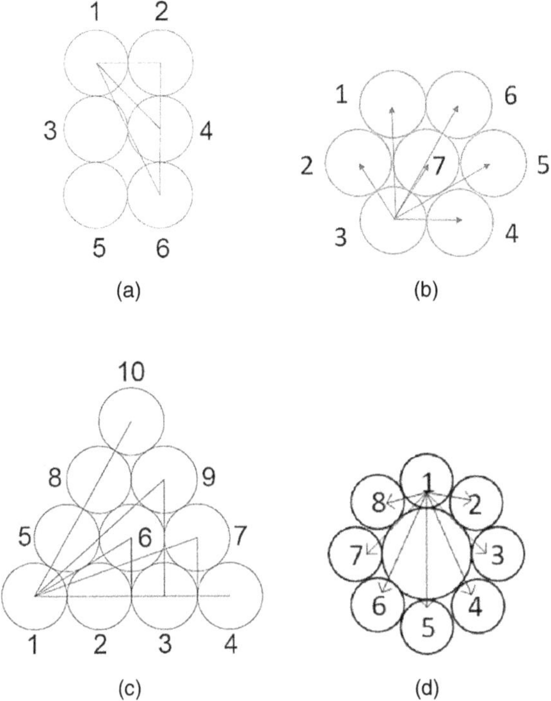

FIGURE 3.9
(a), (b), (c), (d) Non-conventional conductor cables.

$$D_s = \sqrt[36]{(0.7788r)^6 * \left((2r)^7\right)^2 * \left(\left(2r\sqrt{2}\right)^4\right)^2 * \left((4r)^2\right)^2 * \left(\left(2r\sqrt{5}\right)^2\right)^2}$$

$$D_s = \sqrt[36]{0.223128814r^6 * 16384r^{14} * 4096r^8 * 256r^4 * 400r^4}$$

$$D_s = \sqrt[36]{(0.223128814 * 16384 * 4096 * 256 * 400) * r^{36}}$$

$$D_s = 2.180167569r$$

Cable (b)
The distances between each conductor are:

$$D_{11} = D_{22} = D_{33} = D_{44} = D_{55} = D_{66} = D_{77} = RMG$$

For conductor 3:

$$D_{32} = D_{34} = D_{37} = 2r$$

$$D_{36} = 4r$$

$$D_{35} = D_{31} = \frac{sin(120°) .2r}{sin(30°)} = sin(60°).4r$$

From Figure 3.9 (b), for conductors 1, 2, 4, 5, and 6 we have the same distances.

For cable 7:

$$D_{71} = D_{72} = D_{73} = D_{74} = D_{75} = D_{76} = 2r$$

Using these distances and using the expression of the denominator of equation (3.38):

$$Ds = \sqrt[7^2]{(RMG)^7 * \left((D_{32})^3\right)^6 * (D_{36})^6 * \left((D_{35})^2\right)^6 * (D_{71})^6}$$

$$Ds = \sqrt[49]{(RMG)^7 * \left((2r)^3\right)^6 * (4r)^6 * \left((sen(60°).4r)^2\right)^6 * (2r)^6}$$

$$Ds = \sqrt[49]{(0,7788r)^7 * (2r)^{24} * (4r)^6 * \left((sen(60°).4r)^2\right)^6}$$

$$Ds = \sqrt[49]{(0,7788)^7 * (2)^{24} * (4)^6 * (sen(60°).4)^{12} * r^{49}}$$

$$Ds = 2,176701906r$$

Cable (c)

The distances between each conductor are:

$$D_{11} = D_{22} = D_{33} = D_{44} = D_{55} = D_{66} = D_{77} = D_{88} = D_{99} = D_{1010} = GMR$$

$$DA = D_{12} = D_{15} = D_{23} = D_{25} = D_{26} = D_{34} = D_{36} = D_{37} = D_{47} = D_{56} = D_{58} = D_{67} = D_{68} = D_{69} = D_{79} = D_{89} = D_{810} = D_{910} = 2r$$

$$DB = D_{13} = D_{18} = D_{24} = D_{29} = D_{38} = D_{49} = D_{57} = D_{510} = D_{710} = 4r$$

$$DC = D_{16} = D_{27} = D_{35} = D_{46} = D_{59} = D_{78} = 2r\sqrt{2}$$

$$DD = D_{14} = D_{110} = D_{410} = 6r$$

$$DE = D_{17} = D_{19} = D_{210} = D_{310} = D_{45} = D_{48} = 2r\sqrt{7}$$

$$DF = D_{28} = D_{39} = D_{610} = 2r\sqrt{3}$$

Using these distances and using the expression of the denominator of equation (3.38):

$$D_s = \sqrt[10^2]{GMR^{10} * \left((D_A)^{18}\right)^2 * \left((D_B)^9\right)^2 * \left((D_C)^6\right)^2 * \left((D_D)^3\right)^2 * \left((D_E)^6\right)^2 * \left((D_F)^3\right)^2}$$

$$D_s = \sqrt[100]{(0.7788r)^{10} * \left((2r)^{18}\right)^2 * \left((4r)^9\right)^2 * \left((2r\sqrt{2})^6\right)^2 * \left((6r)^3\right)^2 * \left((2r\sqrt{7})^6\right)^2 * \left((2r\sqrt{3})^3\right)^2}$$

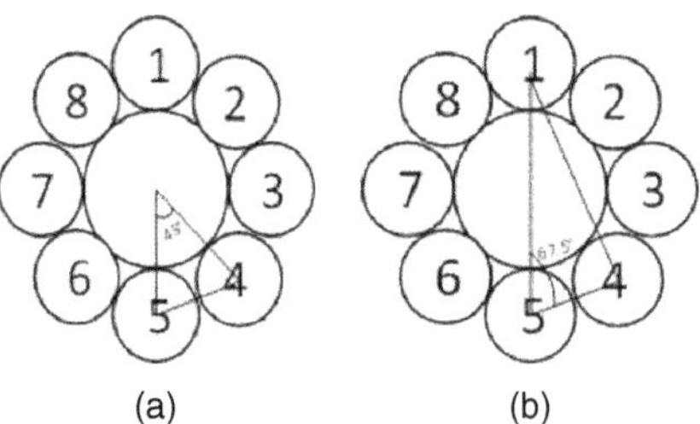

FIGURE 3.10
(a), (b) Angles.

$$D_s = \sqrt[100]{0.082084173r^{10} * 2^{36}r^{36} * 2^{36}r^{18} * 2^{18}r^{12} * 46656r^6 * 481890304r^{12} * 1728r^6}$$

$$D_s = \sqrt[100]{(0.082084173 * 2^{36} * 2^{36} * 2^{18} * 46656 * 481890304 * 1728) * r^{100}}$$

$$D_s = 2.666635059r$$

Cable (d)

For the calculation of angles, we redraw Figure 3.9 (d) and obtain Figure 3.10:

$$D_{\text{steel core}} = 11.9 - 2 * 2.377 = 7.146$$

$$D = 2.377 \text{ mm}$$

$$D_{12} = D$$

$$D_{13} = \sqrt{\left(\frac{D_{\text{steel core}}}{2} + \frac{D}{2}\right)^2 x2} = 6{,}73377 \text{ mm}$$

$$D_{14} = \sqrt{(D_{\text{steel core}} + D)^2 + 2D^2 - 2(D_{\text{alma}} + D\ )2D\cos(67.5)} = 10.99 \text{ mm}$$

$$D_{15} = D_{\text{steel core}} + D = 11.9 \text{ mm}$$

$$GMR = 0.7778D/2$$

$$D_s = \sqrt[8^2]{(GMR)^8 \times (D_{12}{}^2)^8 \times (D_{13}{}^2)^8 \times (D_{14}{}^2)^8 \times (D_{15})^8} = 4.91 \text{ mm}$$

Considering the steel core (Figure 3.11):

$$D_{19} = \frac{D_{\text{steel core}}}{2} + \frac{D}{2} = 4.7615,\ GMR1, GMR2 = \{0.7788D/2,\ 0.7788D_{\text{steel core}}/2\}$$

$$D_s = \sqrt[9^2]{(GMR1)^8 \times GMR2 \times (D_{12}{}^2)^9 \times (D_{13}{}^2)^9 \times (D_{14}{}^2)^9 \times (D_{15})^9 \times (D_{19})^9} = 4.965 \text{ mm}$$

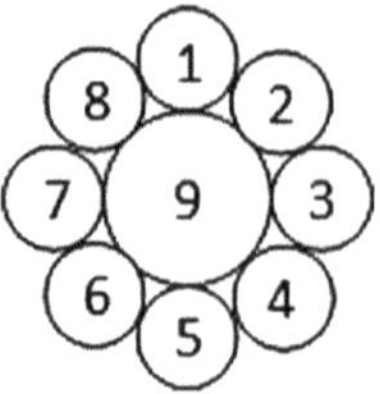

FIGURE 3.11
Steel core item (d) cable.

3.7 TL Parameter Matrix Calculation

The longitudinal impedance of a TL can be divided into three components:

- $Z_{int} \rightarrow$ the internal impedance of the conductor;
- $Z_{ext\rightarrow}$ the external impedance of the conductor;
- $Z_{solo\rightarrow}$ the impedance because of the earth effect.

3.7.1 Internal Impedance Calculation

Commonly used conductors are generally tubular conductors (Figure 3.12), presenting two types of materials in their constitution. Thus, two radii are defined for these conductors: an inside radius q, which delimits a particular material, usually steel, function of which is to support the weight of the cable; and an outside radius r, which delimits the other material, which has the desired conduction properties, usually aluminum or copper.

The expression for calculating the internal impedance of the conductor is given by:

$$Z_{int} = R_{int} + jwL_{int} = R_{DC}j\frac{1}{2}mr(1-s^2)\frac{[ber(mr)+jbei(mr)+\varphi(ker(mr)+jkei(mr))]}{[ber'(mr)+jbei'(mr)+\varphi(ker'(mr)+jkei'(mr))]} \tag{3.44}$$

where:

R_{int} = the AC resistance with skin effect included in Ω/km
L_{int} = the internal inductance with skin effect included in Ω/km
R_{DC} = the DC resistance at Ω/km

$$\varphi = -\frac{[ber'(mq)+jbei'(mq)]}{ker'(mq)+jkei'(mq)} \tag{3.45}$$

$$s = q/r \tag{3.46}$$

$$(mr)^2 = k\frac{1}{1-s^2} \tag{3.47}$$

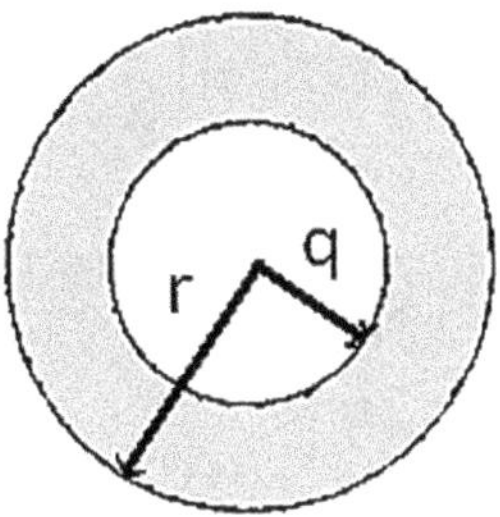

FIGURE 3.12
Tubular conductor model.

$$(mq)^2 = k\frac{s^2}{1-s^2} \tag{3.48}$$

$$k = \frac{8\pi 10^{-4} f}{R_{DC}}\mu_r \tag{3.49}$$

where:

f = the frequency (Hz)

μ_r = the relative permeability of the medium

ber, bei, ker, and kei are Kelvin functions, which belong to the Bessel family of functions, and ber′, bei′, ker′, and kei′ are their derivatives, respectively

The terms ber and bei are abbreviations of "real Bessel" and "imaginary Bessel," respectively.

Bessel's functions were first defined by mathematician Daniel Bernoulli and generalized by German mathematician Friedrich Wilhelm Bessel. The polynomial approximations of Bessel functions are defined from equations (3.50 to 3.61) and used in the ATPdraw program.

For $-8 \leq x \leq 8$

$$\begin{aligned} ber(x) = 1 - 64\left(\left(\frac{x}{8}\right)^4\right) + 113.77777774\left(\left(\frac{x}{8}\right)^8\right) - 32.36345652\left(\left(\frac{x}{8}\right)^{12}\right) \\ + 2.64191397\left(\left(\frac{x}{8}\right)^{16}\right) - 0.08349609\left(\left(\frac{x}{8}\right)^{20}\right) + 0.00122552\left(\left(\frac{x}{8}\right)^{24}\right) \\ - 0.00000901\left(\left(\frac{x}{8}\right)^{28}\right) + \varepsilon \end{aligned} \tag{3.50}$$

$$|\varepsilon| < 1x10^{-9}$$

$$bei(x) = 16\left(\left(\frac{x}{8}\right)^2\right) - 113.77777774\left(\left(\frac{x}{8}\right)^6\right) + 72.81777742\left(\left(\frac{x}{8}\right)^{10}\right)$$
$$- 10.56765779\left(\left(\frac{x}{8}\right)^{14}\right) + 0.52185615\left(\left(\frac{x}{8}\right)^{18}\right) \quad (3.51)$$
$$- 0.01103667\left(\left(\frac{x}{8}\right)^{22}\right) + 0.00011346\left(\left(\frac{x}{8}\right)^{26}\right) + \varepsilon$$

$$|\varepsilon| < 6x10^{-9}$$

For $0 < x \leq 8$

$$ker(x) = -ln\left(\frac{1}{2}x\right)ber(x) + \frac{1}{4}\pi bei(x) - 0.57721566 - 59.05819744\left(\left(\frac{x}{8}\right)^4\right)$$
$$+ 171.36272133\left(\left(\frac{x}{8}\right)^8\right) - 60.60977451\left(\left(\frac{x}{8}\right)^{12}\right) + 5.65539121\left(\left(\frac{x}{8}\right)^{16}\right)$$
$$- 0.19636347\left(\left(\frac{x}{8}\right)^{20}\right) + 0.00309699\left(\left(\frac{x}{8}\right)^{24}\right)$$
$$- 0.00002458\left(\left(\frac{x}{8}\right)^{28}\right) + \varepsilon$$

(3.52)

$$|\varepsilon| < 1x10^{-8}$$

$$kei(x) = -ln\left(\frac{1}{2}x\right)bei(x) - \frac{1}{4}\pi ber(x) + 6.76454936\left(\left(\frac{x}{8}\right)^2\right)$$
$$- 142.91827687\left(\left(\frac{x}{8}\right)^6\right) + 124.23569650\left(\left(\frac{x}{8}\right)^{10}\right)$$
$$- 21.30060904\left(\left(\frac{x}{8}\right)^{14}\right) + 1.17509064\left(\left(\frac{x}{8}\right)^{18}\right) \quad (3.53)$$
$$- 0.02695875\left(\left(\frac{x}{8}\right)^{22}\right) + 0.00029532\left(\left(\frac{x}{8}\right)^{26}\right) + \varepsilon$$

$$|\varepsilon| < 3x10^{-9}$$

Derivatives of these functions are given by:

For $-8 \le x \le 8$

$$ber'(x) = x\left[-4\left(\left(\frac{x}{8}\right)^{2}\right) + 14.22222222\left(\left(\frac{x}{8}\right)^{6}\right) - 6.06814810\left(\left(\frac{x}{8}\right)^{10}\right) + 0.66047849\left(\left(\frac{x}{8}\right)^{14}\right) - 0.02609253\left(\left(\frac{x}{8}\right)^{18}\right) + 0.00045957\left(\left(\frac{x}{8}\right)^{22}\right) - 0.00000394\left(\left(\frac{x}{8}\right)^{26}\right)\right] + \varepsilon \tag{3.54}$$

$$|\varepsilon| < 2.1x10^{-8}$$

$$bei'(x) = x\left[\frac{1}{2} - 10.66666666\left(\left(\frac{x}{8}\right)^{4}\right) + 11.37777772\left(\left(\frac{x}{8}\right)^{8}\right) - 2.31167514\left(\left(\frac{x}{8}\right)^{12}\right) + 0.14677204\left(\left(\frac{x}{8}\right)^{16}\right) - 0.00379386\left(\left(\frac{x}{8}\right)^{20}\right) + 0.00004609\left(\left(\frac{x}{8}\right)^{24}\right)\right] + \varepsilon \tag{3.55}$$

$$|\varepsilon| < 7x10^{-8}$$

For $0 < x \le 8$

$$\ker'(x) = -ln\left(\frac{1}{2}x\right)\text{ber}'(x) - x^{-1}ber(x) + \frac{1}{4}\pi\text{bei}'(x) + x\left[-3.69113734\left(\left(\frac{x}{8}\right)^{2}\right) + 21.42034017\left(\left(\frac{x}{8}\right)^{6}\right) - 11.36433272\left(\left(\frac{x}{8}\right)^{10}\right) + 1.41384780\left(\left(\frac{x}{8}\right)^{14}\right) - 0.06136358\left(\left(\frac{x}{8}\right)^{18}\right) + 0.00116137\left(\left(\frac{x}{8}\right)^{22}\right) - 0.00001075\left(\left(\frac{x}{8}\right)^{26}\right)\right] + \varepsilon \tag{3.56}$$

$$|\varepsilon| < 8x10^{-8}$$

$$\text{kei}'(x) = -ln\left(\frac{1}{2}x\right)\text{bei}'(x) - x^{-1}bei(x) - \frac{1}{4}\pi\text{ber}'(x) + x\left[0.21139217 - 13.39858846\left(\left(\frac{x}{8}\right)^4\right)\right.$$
$$+19.41182758\left(\left(\frac{x}{8}\right)^8\right) - 4.65950823\left(\left(\frac{x}{8}\right)^{12}\right) + 0.33049424\left(\left(\frac{x}{8}\right)^{16}\right)$$
$$\left.-0.00926707\left(\left(\frac{x}{8}\right)^{20}\right) + 0.00011997\left(\left(\frac{x}{8}\right)^{24}\right)\right] + \varepsilon \tag{3.57}$$

$$|\varepsilon| < 7x10^{-8}$$

For $8 \le x < \infty$

$$\ker(x) + jkei(x) = f(x)(1 + \epsilon_1) \tag{3.58}$$

where:

$$f(x) = \sqrt{\left(\frac{\pi}{2x}\right)}exp\left[-\frac{1+j}{\sqrt{2}}x + \theta(-x)\right]$$

$$|\epsilon_1| < 1x10^{-7}$$

$$ber(\text{x}) + jbei(\text{x}) - \frac{j}{\pi}(\ker(x) + jkei(x) = g(x)(1 + \epsilon_2) \tag{3.59}$$

where:

$$g(x) = \frac{1}{\sqrt{2\pi x}}exp\left[\frac{1+j}{\sqrt{2}}x + \theta(x)\right]$$

$$|\epsilon_2| < 3x10^{-7}$$

$$\theta(x) = (0.0 - 0.3926991j) + (0.0110486 - 0.0110485j)\left(\frac{8}{x}\right)$$
$$+(0.0 - 0.0009765j)\left(\left(\frac{8}{x}\right)^2\right) + (-0.0000906 - 0.0000901j)\left(\left(\frac{8}{x}\right)^3\right)$$
$$+(-0.000252 + 0.0j)\left(\left(\frac{8}{x}\right)^4\right) + (-0.0000034 + 0.0000051j)\left(\left(\frac{8}{x}\right)^5\right)$$
$$+(0.0000006 + 0.0000019j)\left(\left(\frac{8}{x}\right)^6\right)$$

$$\text{ker}'(x) + j\text{kei}'(x) = -f(x)\varnothing(-x)(1+\epsilon_3) \tag{3.60}$$

$$|\epsilon_3| < 2x10^{-7}$$

$$\text{ber}'(x) + j\text{bei}'(x) - \frac{j}{\pi}\left(\text{ker}'(x) + j\text{kei}'(x)\right) = g(x)\varnothing(x)(1+\epsilon_4) \tag{3.61}$$

$$|\epsilon_4| < 3x10^{-7}$$

where:

$$\begin{aligned}\varnothing(x) = &(0.7071068 + 0.7071068j) + (-0.0625001 - 0.0000001j)\left(\frac{8}{x}\right)\\ &+(-0.0013813 + 0.0013811j)\left(\left(\frac{8}{x}\right)^2\right) + (0.000005 + 0.0002452j)\left(\left(\frac{8}{x}\right)^3\right)\\ &+(0.000346 + 0.0000338j)\left(\left(\frac{8}{x}\right)^4\right) + (0.0000117 - 0.0000024j)\left(\left(\frac{8}{x}\right)^5\right)\\ &+(0.0000016 - 0.0000032j)\left(\left(\frac{8}{x}\right)^6\right)\end{aligned}$$

For arguments $x \geq 8$, all Bessel and Kelvin functions are multiplied by $exp\left[-\frac{1+j}{\sqrt{2}}x\right]$ to avoid very large numbers.

Example 3.2

To determine the values of the $\frac{R_{CA}}{R_{DC}}$ e $\frac{L_{CA}}{L_{DC}}$ ratios of a conductor, at 60 Hz, which has a DC resistance of 0.02474 Ω/km and a ratio $\frac{q}{r} = 0.2258$. The internal inductance of the cable in DC is given by:

$$L_{DC} = 2\times10^{-4}\left[\frac{q^4}{\left(r^2-q^2\right)^2}ln\left(\frac{r}{q}\right) - \frac{(3q^2-r^2)}{4\left(r^2-q^2\right)}\right]\Omega/km$$

SOLUTION:

L_{DC} calculation: using the given equation, we have:

$$L_{DC} = 4.5487x10^{-5}\ \Omega/km$$

$$s = \frac{q}{r} = 0.2258$$

$$k = \frac{8\pi10^{-4}f}{R_{DC}}\mu_r = \frac{8\pi10^{-4}60}{0.02474}1 = 6.0976$$

$$mr = \sqrt{\frac{k}{1-s^2}} = \sqrt{\frac{6.0976}{1-0.2258^2}} = 2.5348$$

$$mq = \sqrt{\frac{ks^2}{1-s^2}} = \sqrt{\frac{6.0976(0.2258)^2}{1-0.2258^2}} = 0.5724$$

Calculation of Bessel functions: using equations (3.50) to (3.57), we have:

$$ber(mr) = 0.3665$$

$$bei(mr) = 1.4919$$

$$ker(mr) = -0.0702$$

$$kei(mr) = -0.1056$$

$$ber'(mr) = -0.9816$$

$$bei'(mr) = 0.9978$$

$$\text{ker}'(mr) = -0.0130$$

$$\text{kei}'(mr) = 0.1444$$

To calculate φ using equation (3.45), we need to calculate the derivatives as a function of mq.

$$\text{ber}'(mq) = -0.0117$$

$$bei'(mq) = 0.286000$$

$$\text{ker}'(mq) = -1.0955$$

$$\text{kei}'(mq) = -0.4665$$

φ calculation:

$$\varphi = 0.0851 + j0.2249$$

Calculation of internal impedance: using equation (3.44), we have:

$$Z_{\text{int}} = R_{\text{int}} + jwL_{\text{int}} = 0.0273 + j0.0158\ \Omega/km$$

Ratio calculation:

$$\frac{R_{\text{CA}}}{R_{\text{DC}}} = 1.1036$$

$$\frac{L_{\text{CA}}}{L_{\text{DC}}} = 0.9225$$

3.7.2 Carson Method

Carson's corrections terms introduce the earth return effects. The impedance matrix can be calculated using the equations developed by Carson (1928) and changed by Clarke (1943) and Calabrese (1959), and the method is preferably for computational execution. Figure 3.13 shows the main data required for the method.

Carson's method can be applied to:

- Inherently unbalanced circuits;
- Distances between unequal conductors;
- Lines without transposition;
- Lines with arbitrary number of conductors.

The method requires data from conductors, bundled conductors, ground wires, height and distances between conductors and between conductors and their images.

Carson's method considers the following assumptions:

- Infinite ground plane;
- Uniform solid surface;
- Constant ground resistivity;
- Neutral grounding effects do not interfere with power frequency;
- It uses an image conductor for each conductor. The image conductor is at a distance below ground equal to the distance above ground, as shown in Figure 3.14.

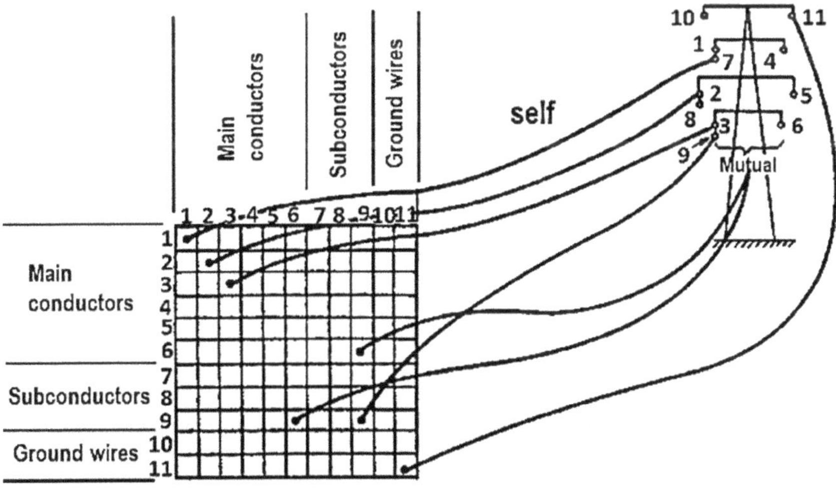

FIGURE 3.13
Data placement to process Carson's method.

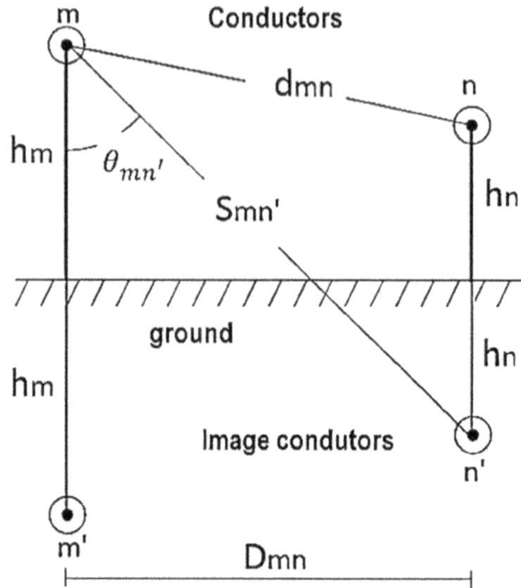

FIGURE 3.14
Conductors and images.

3.7.2.1 External Impedance Calculation

The external impedance is because of the magnetic field present in the air surrounding the conductors, as shown in Figure 3.15.

Therefore, from equation (3.39) and Figure 3.8, the conductor external impedance m, considering the ground resistivity (ρ) equal to zero, we have:

$$Z_{\text{extmm}} = R_{\text{extmm}} + jX_{\text{extmm}} = Z_{\text{mm}} = R_{\text{m}} + j4\pi 10^{-4} f ln\left(\frac{2h_m}{GMR_m}\right) \Omega/km \quad (3.62)$$

If m is a bundled conductor, GMR_{m} is replaced by D_S.

And the mutual external impedance is given by:

$$Z_{\text{extmn}} = R_{\text{extmn}} + jX_{\text{extmm}} = Z_{\text{mn}} = j4\pi 10^{-4} f ln\left(\frac{S_{mn'}}{d_{mn}}\right) \Omega/km \quad (3.63)$$

In the external impedance equations, the terms of the internal reactances are already included.

3.7.2.2 Ground Impedance Calculation (Carson's Corrections)

External impedance correction because of earth effect is expressed by:

$$Z_{\text{ground}} = \Delta R' + j\Delta X' \quad (3.64)$$

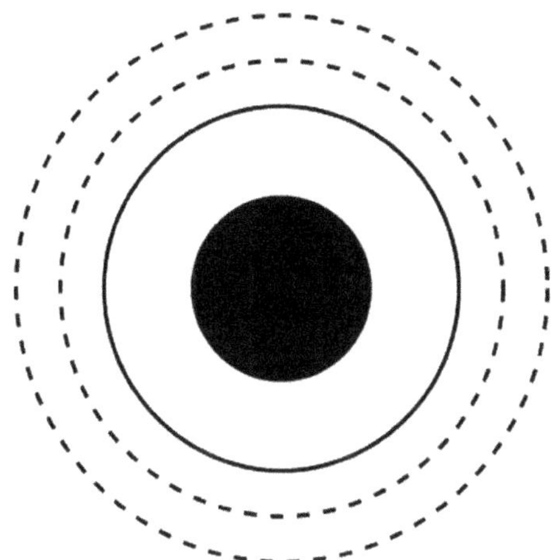

FIGURE 3.15
Magnetic field surrounding the conductors.

Carson's correction terms $\Delta R'$ and $\Delta X'$ are functions of the angle $\theta_{mn'}$ and the parameter k_i calculated by equation (3.65):

$$k_i = 4\sqrt{5}\pi D 10^{-4}\sqrt{\frac{f}{\rho}} \tag{3.65}$$

where, according to Figure 3.8:

$D = 2h_i$ (m) for self-impedances and $S_{mn'}$ for mutual impedances
f = the frequency (Hz)
ρ = the resistivity of the ground at $\Omega - m$

Carson's correction terms, for their self and mutual impedances, are given as an infinite integral for $\Delta R'$ and $\Delta X'$, which he developed in a sum of four infinite series for $k_i \leq 5$. For $k_i \rightarrow \infty$ (this happens for very high frequencies or very low earth resistivities), $\Delta R'$ and $\Delta X'$ are null.

For $k_i \leq 5$, the series placed in a way to facilitate computational programming are given by:

$$\begin{aligned}\Delta R' = 8\pi f 10^{-4} \Bigg\{ & \frac{\pi}{8} - b_1 k_i cos\theta_{mn'} + b_2\left[(c_2 - lnk_i)k_i^2 cos2\theta_{mn'} + \theta_{mn'}k_i^2 sin2\theta_{mn'}\right] \\ & + b_3 k_i^3 cos3\theta_{mn'} - d_4 k_i^4 cos4\theta_{mn'} - b_5 k_i^5 cos5\theta_{mn'} + b_6\left[(c_6 - lnk_i)k_i^6 cos6\theta_{mn'}\right. \\ & \left. + \theta_{mn'}k_i^6 sin6\theta_{mn'}\right] + b_7 k_i^7 cos7\theta_{mn'} - b_8 k_i^8 cos8\theta_{mn'} - \cdots \Bigg\} \Omega/km\end{aligned} \tag{3.66}$$

$$\Delta X' = 8\pi f 10^{-4}\left\{\frac{1}{2}(0.6159315 - lnk_i) + b_1 k_i cos\theta_{mn'} - d_2 k_i^2 cos2\theta_{mn'} + b_3 k_i^3 cos3\theta_{mn'}\right.$$

$$- b_4\left[(c_4 - lnk_i)k_i^4 cos4\theta_{mn'} + \theta_{mn'} k_i^4 sin4\theta_{mn'}\right] + b_5 k_i^5 cos5\theta_{mn'} - d_6 k_i^6 cos6\theta_{mn'}$$

$$\left. + b_7 k_i^7 cos7\theta_{mn'} - b_8\left[(c_8 - lnk_i)k_i^8 cos8\theta_{mn'} + \theta_{mn'} k_i^8 sin8\theta_{mn'}\right] - \cdots\right\}\Omega/km$$

(3.67)

In equations (3.66) and (3.67), each four successive terms form a repeating pattern. The coefficients b_i, c_i, and d_i are constant and can be calculated by the following recursive equations.

$$b_i = b_{i-2}\frac{signal}{i(i+2)} \tag{3.68}$$

These terms start with $b_1 = \sqrt{2}/6$ for odd subscribers and $b_2 = 1/16$ for even subscribers.

$$c_i = c_{i-2} + \frac{1}{i} + \frac{1}{i+2} \tag{3.69}$$

These terms start with $c_2 = 1.3659315$

$$d_i = \frac{\pi}{4} b_i \tag{3.70}$$

The signal changes after four consecutive terms. That is, for $i = 1, 2, 3, 4 \rightarrow signal = +1$; for $i = 5, 6, 7, 8 \rightarrow signal = -1$; and so on.

For $k_i > 5$, the following finite series are used:

$$\Delta R' = \frac{8\pi f 10^{-4}}{\sqrt{2}}\left(\frac{cos\theta_{mn'}}{k_i} - \frac{\sqrt{2}cos2\theta_{mn'}}{k_i^2} + \frac{cos3\theta_{mn'}}{k_i^3} + \frac{5cos5\theta_{mn'}}{k_i^5} - \frac{45cos7\theta_{mn'}}{k_i^7}\right)$$

(3.71)

$$\Delta X' = \frac{8\pi f 10^{-4}}{\sqrt{2}}\left(\frac{cos\theta_{mn'}}{k_i} - \frac{cos3\theta_{mn'}}{k_i^3} + \frac{3cos5\theta_{mn'}}{k_i^5} - \frac{45cos7\theta_{mn'}}{k_i^7}\right) \tag{3.72}$$

With the following recursive equations:

$$k_i^{(p)}\cos(p\theta_{mn'}) = k_i\left[k_i^{(p-1)}\cos(p-1)\theta_{mn'}).\ cos\theta_{mn'} - k_i^{(p-1)}\sin(p-1)\theta_{mn'}).\ sin\theta_{mn'}\right]$$

(3.73)

$$k_i^{(p)}\sin(p\theta_{mn'}) = k_i\left[k_i^{(p-1)}\cos(p-1)\theta_{mn'}).\ sin\theta_{mn'} + k_i^{(p-1)}\sin(p-1)\theta_{mn'}).\ cos\theta_{mn'}\right] \tag{3.74}$$

Considering a TL, the matrix form, Carson's complete method equations are:

$$\begin{bmatrix} V_1 \\ V_2 \\ V_3 \\ \vdots \\ V_m \end{bmatrix} = \begin{bmatrix} Z_{11} & Z_{11} & \cdots & Z_{1n} \\ Z_{21} & Z_{22} & \ddots & Z_{2n} \\ \vdots & & \cdots & \vdots \\ Z_{m1} & Z_{m2} & \dots & Z_{mn} \end{bmatrix} \begin{bmatrix} I_1 \\ I_2 \\ I_3 \\ \vdots \\ I_m \end{bmatrix} \tag{3.75}$$

where:

For $m = n$,

$$Z_{mm} = R_{mm} + jX_{mm} = (R_{int} + \Delta R') + j(4\pi 10^{-4}\ fln\left(\frac{2h_m}{GMR_m}\right) + \Delta X')\ \Omega/km \tag{3.76}$$

In the case of bundled-conductor TL GMR_m is replaced by D_s given by equations (3.41), (3.42), and (3.43).

For $m \neq n$,

$$Z_{mn} = R_{mm} + jX_{mm} = \Delta R' + j(4\pi 10^{-4}\ fln\left(\frac{S_{mn'}}{d_{mn}}\right) + \Delta X')\ \Omega/km \tag{3.77}$$

where:

f = the frequency (Hz)
GMR_m = the geometric mean radius of the conductor m
h_m = the height of the conductor m to the ground
d_{mn} = the distance between conductors m and n
$S_{mn'}$ = the distance between conductor m and the image of conductor n
R_{mm}, R_{mn} = the AC resistances at Ω/km
X_{mm}, X_{mn} = the AC reactances at Ω/km

In case $m = n$:

$$k_i = 4\sqrt{5}\pi 2h_i 10^{-4}\sqrt{\frac{f}{\rho}} \tag{3.78}$$

$$\theta_{mn'} = 0 \tag{3.79}$$

In case $m \neq n$:

$$k_i = 4\sqrt{5}\pi S_{mn'} 10^{-4}\sqrt{\frac{f}{\rho}} \tag{3.80}$$

$$\theta_{mn'} = arcsin\frac{D_{mn}}{S_{mn'}} \tag{3.81}$$

$$S_{mn'} = \sqrt{4h_m h_n + d_{mn}^2} \tag{3.82}$$

where:

f = the frequency (Hz)
ρ = the resistivity of the ground $(\Omega - m)$
D_{mn} = the distance between conductors m and n (meters)
d_{mn} = the distance between conductors m and n (meters)
$S_{mn'}$ = the distance between conductor m and the image of conductor n

The conductor height value takes into account the conductor sag as shown in Figure 3.16 and is calculated by equation (3.69).

$$h_m = H_m - 2sag/3 \tag{3.83}$$

3.7.3 Approximate Carson Method

The approximate method uses a single return conductor, which has a unitary GMR (1 ft, 1 cm, 1 m), as shown in Figure 3.17.

The value of the unit conductor distance is calculated by:

$$D_e = 658.68\sqrt{\frac{\rho}{f}}\ (m) \tag{3.84}$$

Table 3.2 shows the values of D_e and ρ for various land type conditions.

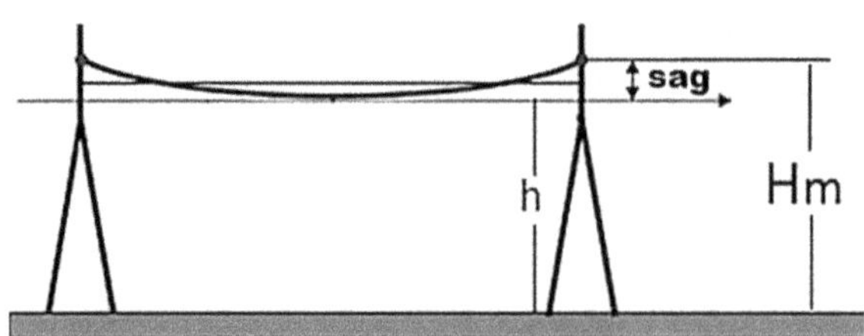

FIGURE 3.16
Sag conductor.

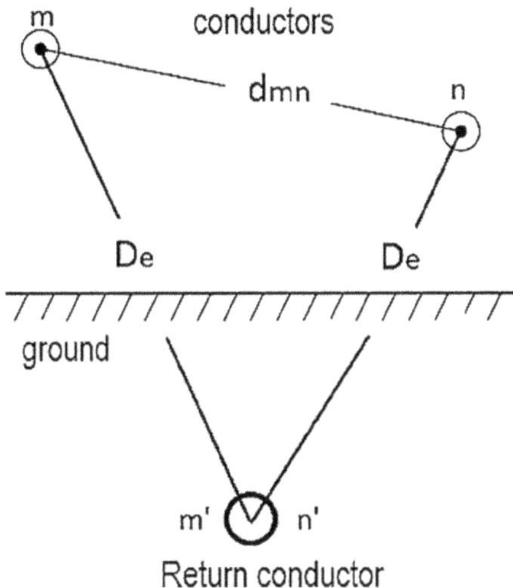

FIGURE 3.17
Single return conductor.

The following approximations are made for Carson's modified method: it uses only one term in $\Delta R'$ and two terms in $\Delta X'$.

$$\Delta R' = 8\pi f 10^{-4}\frac{\pi}{8} = 0.0009869604401 f \tag{3.85}$$

$$\Delta X' = -0.03860 + \frac{1}{2}ln\left(\frac{2}{k_i}\right) \tag{3.86}$$

The equations for the calculation of self and mutual impedances are given by:

$$Z_{mm} = R_{int} + 0.000988 f + j0.00125664 f ln\left(\frac{D_e}{RMG_m}\right)\ \Omega/km \tag{3.87}$$

$$Z_{mn} = 0.000988 f + j0.00125664 f ln\left(\frac{D_e}{d_{mn}}\right)\ \Omega/km \tag{3.88}$$

TABLE 3.2
D_e for Various Resistivities at 60 Hz

Return conductor condition	**$\rho\ (\Omega - m)$**	**$D_e\ (m)$**
Sea water	0.01 – 1.0	8.5 – 85.0
Marshland	10 – 100	268.8 – 850.0
Dry land	1000	2688
Boulder ground	10^7	268800
Sandstone Terrain	10^9	2688000
Average value of large number of measurements	100	850

The equations, for the calculation of self and mutual impedances, with the frequency of 60 Hz and $\rho = 100\ \Omega - m$, are given by:

$$Z_{mm} = R_{int} + 0.05928 + j0.0753984ln\left(\frac{D_e}{RMG_m}\right)\Omega/km \quad (3.89)$$

$$Z_{mn} = 0.05928 + j0.0753984ln\left(\frac{D_e}{d_{mn}}\right)\Omega/km \quad (3.90)$$

For bundled conductors, the GMR_m, also, should be replaced by D_s.

3.7.4 Carson's Method Operationalization

Let's consider six cases: TL without ground wire cables, TL with one ground wire, TL with two ground wires, TL with two parallel circuits and two ground wires, TL with three three-phase circuits in same tower, and TL with four three-phase circuits in the same tower. With extra-high-voltage TL, where the ground wire is isolated through low-voltage disruptive isolators equipped with spark arrestors, as shown in Figure 3.18, there are no meshes for current circulation and the current in the ground wire is equal to zero. Thus, in the calculation of inductive impedances, this ground wire is an open circuit and its presence can be ignored, coming back to the case of TL without ground wire.

Case 1: TL without ground wire cables

Carson's impedance matrix has the following form:

$$[Z_{imp}] = \begin{bmatrix} Z_{aa} & Z_{ab} & Z_{ac} \\ Z_{ba} & Z_{bb} & Z_{bc} \\ Z_{ca} & Z_{cb} & Z_{cc} \end{bmatrix} \quad (3.91)$$

Where the elements of the impedance matrix are calculated by expressions (3.60) and (3.61) if Carson's full method is used or by expressions (3.75) and (3.76) if Carson's approximate method is used.

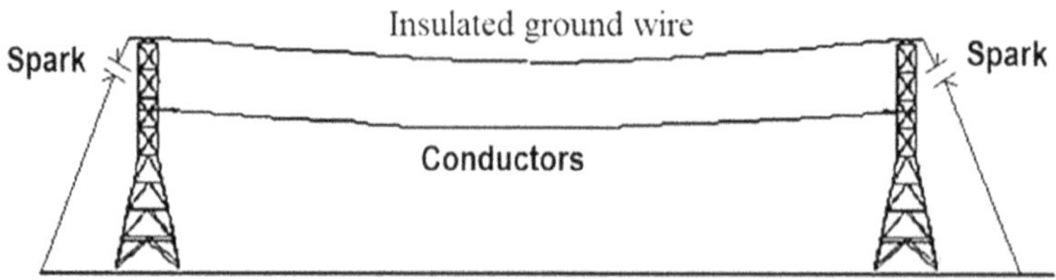

FIGURE 3.18
Insulated ground wire.

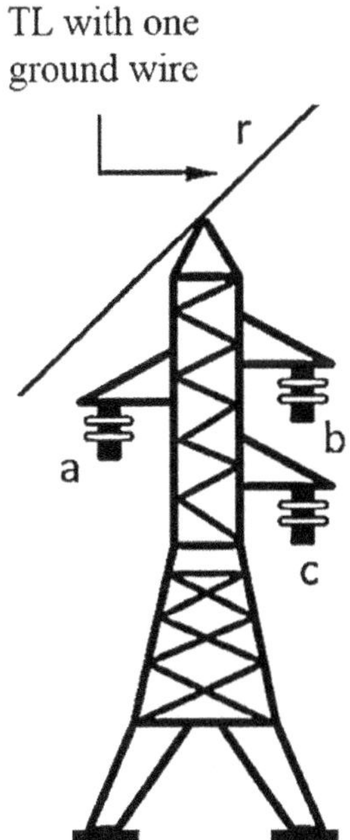

FIGURE 3.19
TL with one ground wire.

Case 2: TL with one ground wire

Figure 3.19 shows a TL with one ground wire.

Carson's impedance matrix has the following form:

$$\left[Z_{imp}\right]=\begin{bmatrix} Z_{aa} & Z_{ab} & Z_{ac} & | & Z_{ar} \\ Z_{ba} & Z_{bb} & Z_{bc} & | & Z_{br} \\ Z_{ca} & Z_{cb} & Z_{cc} & | & Z_{cr} \\ -- & -- & -- & & -- \\ Z_{ra} & Z_{rb} & Z_{rc} & | & Z_{rr} \end{bmatrix} \tag{3.92}$$

In order for the matrix to be 3 × 3 in size, the impedance matrix must be reduced using the Kron reduction and consequently the line and column referring to the ground wire are eliminated.

Kron reduction of a matrix is operationalized as follows:

Let $[Z]$ be a matrix with four sub-matrices:

$$[Z]=\begin{bmatrix} [Z_1] & [Z_2] \\ [Z_3] & [Z_4] \end{bmatrix} \tag{3.93}$$

The new matrix $[Z_1]$ is equal to:

$$[Z_{1nova}]=[Z_1]-[Z_2][Z_4]^{-1}[Z_3] \tag{3.94}$$

Demonstration:

Let the following system of equations, where the elements are submatrices:

$$\begin{bmatrix} [A_1] & [A_2] \\ [A_3] & [A_4] \end{bmatrix} \begin{bmatrix} [X_1] \\ [X_2] \end{bmatrix} = \begin{bmatrix} [B_1] \\ [B_2] \end{bmatrix} \tag{3.95}$$

Developing:

$$\left\{ \begin{matrix} [A_1][X_1]+[A_2][X_2]=[B_1] \\ [A_3][X_1]+[A_4][X_2]=[B_2] \end{matrix} \right\} \rightarrow \left\{ \begin{matrix} [A_1][X_1]+[A_2][X_2]=[B_1] \\ [A_4][X_2]=[B_2]-[A_3][X_1] \end{matrix} \right\} \tag{3.96}$$

Multiplying the second equation of (3.96) by $[A_4]^{-1}$

$$[X_2]=[A_4]^{-1}[B_2]-[A_4]^{-1}[A_3][X_1] \tag{3.97}$$

Substituting in the first equation of (3.92), we have:

$$[A_1][X_1]+[A_2]\left([A_4]^{-1}[B_2]-[A_4]^{-1}[A_3][X_1]\right)=[B_1] \tag{3.98}$$

Putting $[X_1]$ in evidence, we have:

$$([A_1]-[A_2][A_4]^{-1}[A_3])[X_1]=[B_1]-[A_2][A_4]^{-1}[B_2] \tag{3.99}$$

Therefore,

$$[A_{1novo}]=[A_1]-[A_2][A_4]^{-1}[A_3] \tag{3.100}$$

Case 3: TL with two ground wires

Figure 3.20 shows the TL with two ground wires.

Carson's impedance matrix has the following form:

$$[Z_{imp}]=\begin{bmatrix} Z_{aa} & Z_{ab} & Z_{ac} & | & Z_{ar1} & Z_{ar2} \\ Z_{ba} & Z_{bb} & Z_{bc} & | & Z_{br1} & Z_{br2} \\ Z_{ca} & Z_{cb} & Z_{cc} & | & Z_{cr1} & Z_{cr2} \\ -- & -- & -- & & -- & -- \\ Z_{r1a} & Z_{r1b} & Z_{r1c} & | & Z_{r1r1} & Z_{r1r2} \\ Z_{r2a} & Z_{r2b} & Z_{r2c} & | & Z_{r2r1} & Z_{r2r2} \end{bmatrix} \tag{3.101}$$

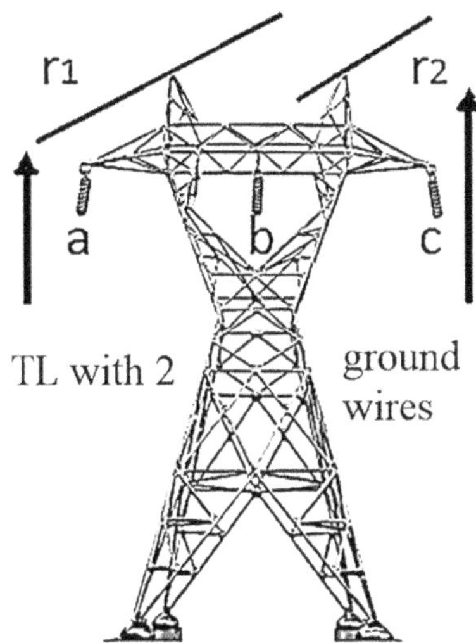

FIGURE 3.20
TL with two ground wires.

For the impedance matrix with two ground wires to be 3 × 3 in size, it must also be reduced using the Kron reduction.

Case 4: TL with two parallel circuits and two ground wires

Phase A consists of conductors a and a′ in parallel, and phases B and C are similarly constructed as shown in Figure 3.21.

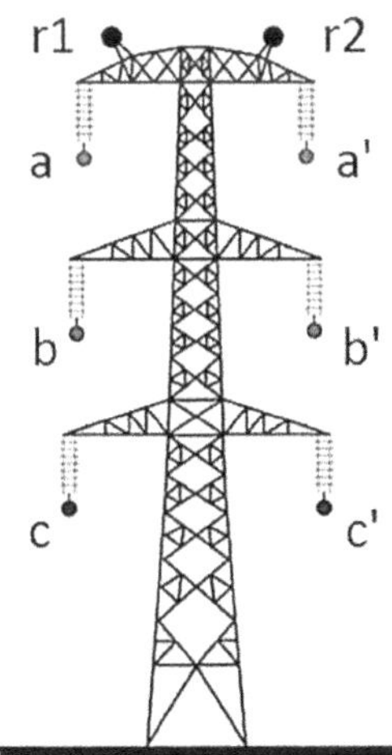

FIGURE 3.21
TL with parallel circuits and two ground wires.

Carson's impedance matrix has the following form:

$$\left[Z_{imp}\right]=\begin{bmatrix} Z_{aa} & Z_{ab} & Z_{ac} & Z_{aa'} & Z_{ab'} & Z_{ac'} & | & Z_{ar1} & Z_{ar2} \\ Z_{ba} & Z_{bb} & Z_{bc} & Z_{ba'} & Z_{bb'} & Z_{bc'} & | & Z_{br1} & Z_{br2} \\ Z_{ca} & Z_{cb} & Z_{cc} & Z_{ca'} & Z_{cb'} & Z_{cc'} & | & Z_{cr1} & Z_{cr2} \\ Z_{a'a} & Z_{a'b} & Z_{a'c} & Z_{a'a'} & Z_{a'b'} & Z_{a'c'} & | & Z_{a'r1} & Z_{a'r2} \\ Z_{b'a} & Z_{b'b} & Z_{b'c} & Z_{b'a'} & Z_{b'b'} & Z_{b'c'} & | & Z_{b'r1} & Z_{b'r2} \\ Z_{c'a} & Z_{c'b} & Z_{c'c} & Z_{c'a'} & Z_{c'b'} & Z_{c'c'} & | & Z_{c'r1} & Z_{c'r2} \\ - & - & - & - & - & - & | & - & - \\ Z_{r1a} & Z_{r1b} & Z_{r1c} & Z_{r1a'} & Z_{r1b'} & Z_{r1c'} & | & Z_{r1r1} & Z_{r1r2} \\ Z_{r2a} & Z_{r2b} & Z_{r2c} & Z_{r2a'} & Z_{r2b'} & Z_{r2c'} & | & Z_{r2r1} & Z_{r2r2} \end{bmatrix} \tag{3.102}$$

In order for the impedance matrix with two ground wires to be 6×6 in size, it must also be reduced using the Kron reduction.

When the conductors of a three-phase TL are not evenly spaced, the impedances of each phase are not equal and the TL becomes unbalanced. Equilibrium can be re-established if the position of the conductors is changed along the TL using special structures so that each conductor occupies the position of the other two at equal distances. This change of position of conductors is known as transposition. Figures 3.22 and 3.23 show a complete transposition cycle for a single TL of length l.

Each conductor A, B, and C occupies a position that corresponds to the length of the TL divided by 3.

If the line is completely transposed, the diagonal and non-diagonal impedances are calculated as an average value.

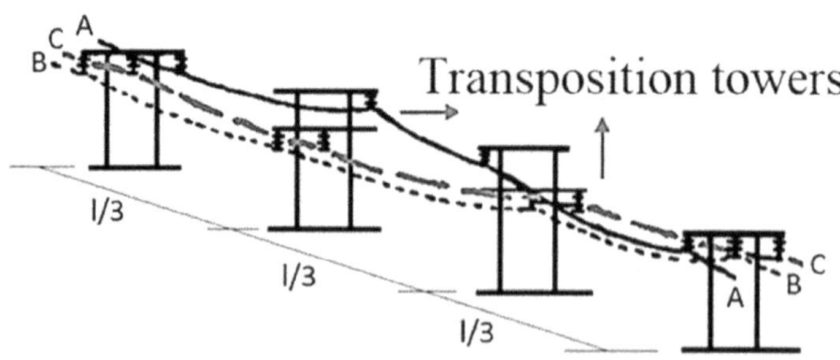

FIGURE 3.22
Conductor transposition.

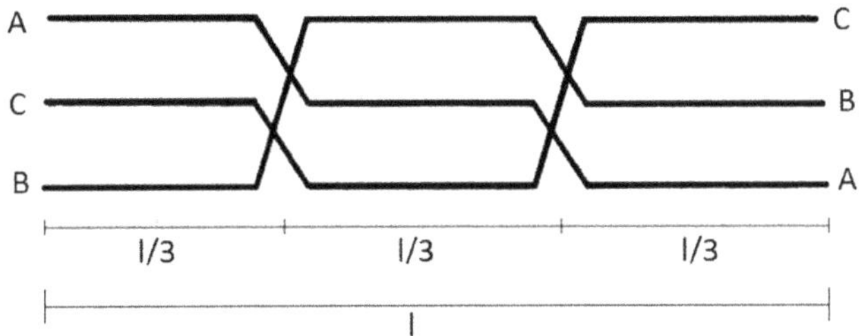

FIGURE 3.23
Complete transpose cycle.

Let the impedance matrix be 3 × 3, after Kron reduction:

$$[Z_{ABC}] = \begin{bmatrix} Z_{AA} & Z_{AB} & Z_{AC} \\ Z_{BA} & Z_{BB} & Z_{BC} \\ Z_{CA} & Z_{CB} & Z_{CC} \end{bmatrix} \tag{3.103}$$

The new diagonal and off-diagonal impedances after transposition are respectively:

$$Z_D = \frac{Z_{AA} + Z_{BB} + Z_{CC}}{3} \tag{3.104}$$

$$Z_M = \frac{Z_{AB} + Z_{BC} + Z_{CA}}{3} \tag{3.105}$$

The new impedance matrix with the transposed conductors is equal to:

$$[Z_{ABCT}] = \begin{bmatrix} Z_D & Z_M & Z_M \\ Z_M & Z_D & Z_M \\ Z_M & Z_M & Z_D \end{bmatrix} \tag{3.106}$$

In the case of double circuit TL, the conductors of phases A, B, and C would occupy the spatial positions shown in Table 3.3 in the three transposition sections.

TABLE 3.3
Double-Circuit Transposed Conductor Positions

Patch	*a*	*b*	*c*	*a′*	*b′*	*c′*	*r*1	*r*2
1º	1	2	3	4	5	6	7	8
2º	2	3	1	6	4	5	7	8
3º	3	1	2	5	6	4	7	8

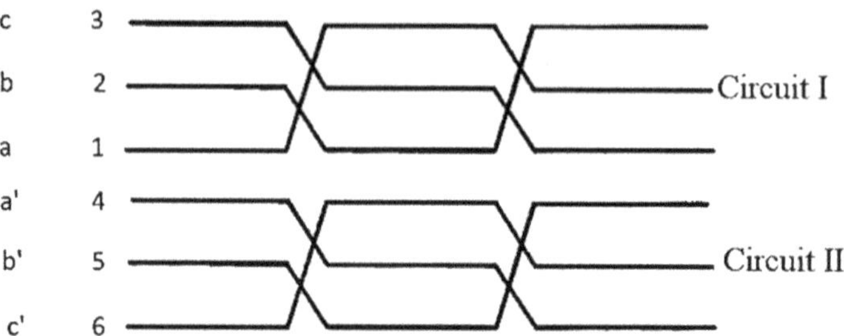

FIGURE 3.24
Double-circuit TL transposition scheme.

The transposition scheme of Table 3.3, shown in Figure 3.24, is called phase rotation in the opposite direction.

A transposition scheme that allows perfect decoupling between sequences except for zero sequence is shown in Figure 3.25.

This nine-section transposition scheme allows for the best decoupling between sequences. With eight transposition towers, it is more expensive than the previous scheme which uses two towers.

Initially, the impedances of the nine sections should be calculated. Thus, the reduced matrix is given by:

$$\left[Z'_{ABC}\right]=\frac{1}{9}\left\{\left[Z'_{seção1}\right]+\left[Z'_{seção2}\right]+\ldots+\left[Z'_{seção9}\right]\right\}$$

$$=\frac{1}{9}\begin{bmatrix}\begin{bmatrix}3Z_{D1} & 3Z_{M1} & 3Z_{M1}\\ 3Z_{M1} & 3Z_{D1} & 3Z_{M1}\\ 3Z_{M1} & 3Z_{M1} & 3Z_{D1}\end{bmatrix} & \begin{bmatrix}Z_E & Z_E & Z_E\\ Z_E & Z_E & Z_E\\ Z_E & Z_E & Z_E\end{bmatrix}\\ \begin{bmatrix}Z_E & Z_E & Z_E\\ Z_E & Z_E & Z_E\\ Z_E & Z_E & Z_E\end{bmatrix} & \begin{bmatrix}3Z_{D2} & 3Z_{M2} & 3Z_{M2}\\ 3Z_{M2} & 3Z_{D2} & 3Z_{M2}\\ 3Z_{M2} & 3Z_{M2} & 3Z_{D2}\end{bmatrix}\end{bmatrix} \quad (3.107)$$

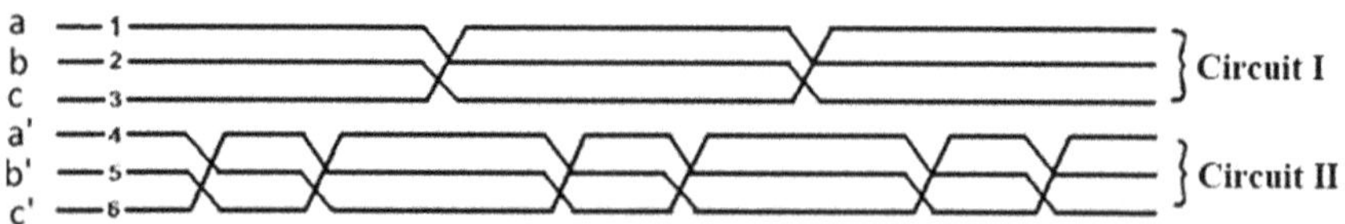

FIGURE 3.25
Nine-section transposition scheme.

where:

$$Z_{D1} = Z'_{11} + Z'_{22} + Z'_{33} \tag{3.108}$$

$$Z_{M1} = Z'_{12} + Z'_{23} + Z'_{31} \tag{3.109}$$

$$Z_{D2} = Z'_{44} + Z'_{55} + Z'_{66} \tag{3.110}$$

$$Z_{M2} = Z'_{45} + Z'_{56} + Z'_{64} \tag{3.111}$$

$$Z_E = Z'_{14} + Z'_{15} + Z'_{16} + Z'_{24} + Z'_{25} + Z'_{26} + Z'_{34} + Z'_{35} + Z'_{36} \tag{3.112}$$

And remembering that the matrix is symmetrical, that is, $Z_{ik} = Z_{ki}$.

Three or more three-phase circuits in the same tower or running in proximity in a corridor are sometimes used in power grids. Similar to dual circuit lines, there is a mutual inductive and capacitive coupling between all conductors forming a complex multi-conductor system.

The following two cases illustrate the cases of three and four three-phase circuits.

Case 5: TL with three three-phase circuits in same tower

Phase A consists of three parallel conductors, and phases B and C are similarly constructed as shown in Figure 3.26.

Case 6: TL with four three-phase circuits in the same tower

Phase A consists of four parallel conductors, and phases B and C are similarly constructed as shown in Figure 3.27.

Consider a line with three circuits 1, 2, and 3 mutually coupled in the same tower, as shown in Figure 3.22. The 9 × 9 series transposed and balanced phase impedance matrix keeps mutual impedances between the positive,

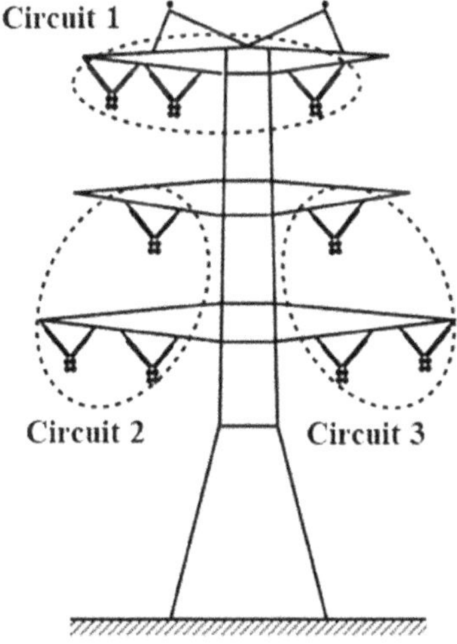

FIGURE 3.26
TL with three parallel circuits and two ground wires.

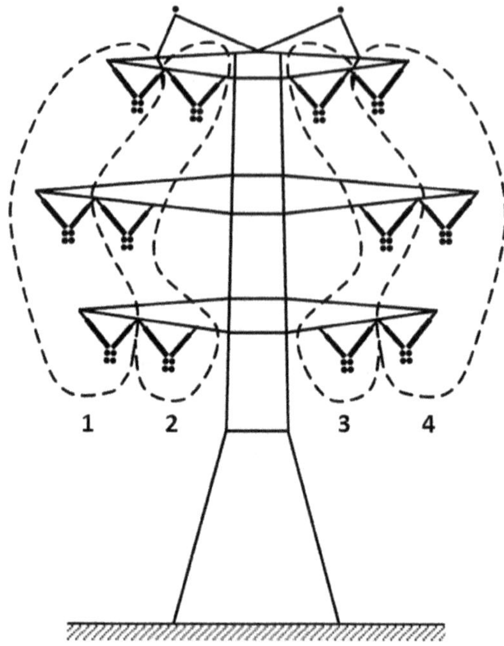

FIGURE 3.27
TL with four parallel circuits and two ground wires.

negative, and zero sequence circuits. Each circuit is perfectly transposed and the mutual impedance between two circuits results in equal impedances between them. However, the mutual coupling impedances between circuits 1 and 2, circuits 1 and 3, and circuits 2 and 3 are considered unequal to maintain generality. It can be shown that the resulting phase balanced impedance matrix is given by:

$$[Z_{ABCT}] = \begin{bmatrix} Z_{D1} & Z_{M1} & Z_{M1} & Z_{D12} & Z_{M12} & Z_{M12} & Z_{D13} & Z_{M13} & Z_{M13} \\ Z_{M1} & Z_{D1} & Z_{M1} & Z_{M12} & Z_{D12} & Z_{M12} & Z_{M13} & Z_{D13} & Z_{M13} \\ Z_{M1} & Z_{M1} & Z_{D1} & Z_{M12} & Z_{M12} & Z_{D12} & Z_{M13} & Z_{M13} & Z_{D13} \\ Z_{D12} & Z_{M12} & Z_{M12} & Z_{D2} & Z_{M2} & Z_{M2} & Z_{D23} & Z_{M23} & Z_{M23} \\ Z_{M12} & Z_{D12} & Z_{M12} & Z_{M2} & Z_{D2} & Z_{M2} & Z_{M23} & Z_{D23} & Z_{M23} \\ Z_{M12} & Z_{M12} & Z_{D12} & Z_{M2} & Z_{M2} & Z_{D2} & Z_{M23} & Z_{M23} & Z_{D23} \\ Z_{D13} & Z_{M13} & Z_{M13} & Z_{D23} & Z_{M23} & Z_{M23} & Z_{D3} & Z_{M3} & Z_{M3} \\ Z_{M13} & Z_{D13} & Z_{M13} & Z_{M23} & Z_{D23} & Z_{M23} & Z_{M3} & Z_{D3} & Z_{M3} \\ Z_{M13} & Z_{M13} & Z_{D13} & Z_{M23} & Z_{M23} & Z_{D23} & Z_{M3} & Z_{M3} & Z_{D3} \end{bmatrix} \quad (3.113)$$

For optimal transposition, the impedance arrays between circuits have all nine elements equal, thus keeping only the mutual coupling between sequence circuits. The assumption in optimal transposition results in $Z_{D12} = Z_{M12}$ for circuits 1 and 2, $Z_{D13} = Z_{M13}$ for circuits 1 and 3 and $Z_{D23} = Z_{M23}$ for circuits 2 and 3. The corresponding sequence impedance matrix can be calculated using:

$$[Z_{012}] = \begin{bmatrix} [A]^{-1} & [0] & [0] \\ [0] & [A]^{-1} & [0] \\ [0] & [0] & [A]^{-1} \end{bmatrix} [Z_{ABCT}] \begin{bmatrix} [A] & [0] & [0] \\ [0] & [A] & [0] \\ [0] & [0] & [A] \end{bmatrix}$$

$$= \begin{bmatrix} Z_{1-0} & 0 & 0 & Z_{12-0} & 0 & 0 & Z_{13-0} & 0 & 0 \\ 0 & Z_{1-1} & 0 & 0 & Z_{12-1} & 0 & 0 & Z_{13-1} & 0 \\ 0 & 0 & Z_{1-2} & 0 & 0 & Z_{12-2} & 0 & 0 & Z_{13-2} \\ Z_{12-0} & 0 & 0 & Z_{2-0} & 0 & 0 & Z_{23-0} & 0 & 0 \\ 0 & Z_{12-1} & 0 & 0 & Z_{2-1} & 0 & 0 & Z_{23-1} & 0 \\ 0 & 0 & Z_{12-2} & 0 & 0 & Z_{2-2} & 0 & 0 & Z_{23-2} \\ Z_{13-0} & 0 & 0 & Z_{23-0} & 0 & 0 & Z_{3-0} & 0 & 0 \\ 0 & Z_{13-1} & 0 & 0 & Z_{23-1} & 0 & 0 & Z_{3-1} & 0 \\ 0 & 0 & Z_{13-2} & 0 & 0 & Z_{23-2} & 0 & 0 & Z_{3-2} \end{bmatrix} \quad (3.114)$$

where:

$$Z_{i-1} = Z_{i-2} = Z_{Di} - Z_{Mi} \text{ para } i = 1,3 \quad (3.115)$$

$$Z_{i-0} = Z_{Di} + 2Z_{Mi} \text{ para } i = 1,3 \quad (3.116)$$

$$Z_{12-1} = Z_{12-2} = Z_{D12} - Z_{M12} \quad (3.117)$$

$$Z_{13-1} = Z_{13-2} = Z_{D13} - Z_{M13} \quad (3.118)$$

$$Z_{23-1} = Z_{23-2} = Z_{D23} - Z_{M23} \quad (3.119)$$

$$Z_{12-0} = Z_{D12} + 2Z_{M12} \quad (3.120)$$

$$Z_{13-0} = Z_{D13} + 2Z_{M13} \quad (3.121)$$

$$Z_{23-0} = Z_{D23} + 2Z_{M23} \quad (3.122)$$

Briefly, the impedance matrix is presented for an unusual case of four identical circuits or two dual circuits running near each other. In deriving such a matrix, it is assumed that each circuit is perfectly transposed and therefore represented by its own impedance and mutual impedance. Mutual impedance arrays between the circuits, any two circuits, are also considered balanced, reflecting a transposition assumption similar to that of dual circuit lines. In addition, the mutual coupling between the two circuit pair arrays is considered uneven to maintain generality. It can be shown that the phase balanced impedance matrix is given by:

$$[Z_{ABCT}] = \begin{bmatrix} \begin{matrix} D_1 & M_1 & M_1 \\ M_1 & D_1 & M_1 \\ M_1 & M_1 & D_1 \end{matrix} & \begin{matrix} D_{12} & M_{12} & M_{12} \\ M_{12} & D_{12} & M_{12} \\ M_{12} & M_{12} & D_{12} \end{matrix} & \begin{matrix} D_{13} & M_{13} & M_{13} \\ M_{13} & D_{13} & M_{13} \\ M_{13} & M_{13} & D_{13} \end{matrix} & \begin{matrix} D_{14} & M_{14} & M_{14} \\ M_{14} & D_{14} & M_{14} \\ M_{14} & M_{14} & D_{14} \end{matrix} \\ \begin{matrix} D_{12} & M_{12} & M_{12} \\ M_{12} & D_{12} & M_{12} \\ M_{12} & M_{12} & D_{12} \end{matrix} & \begin{matrix} D_2 & M_2 & M_2 \\ M_2 & D_2 & M_2 \\ M_2 & M_2 & D_2 \end{matrix} & \begin{matrix} D_{23} & M_{23} & M_{23} \\ M_{23} & D_{23} & M_{23} \\ M_{23} & M_{23} & D_{23} \end{matrix} & \begin{matrix} D_{24} & M_{24} & M_{24} \\ M_{24} & D_{24} & M_{24} \\ M_{24} & M_{24} & D_{24} \end{matrix} \\ \begin{matrix} D_{13} & M_{13} & M_{13} \\ M_{13} & D_{13} & M_{13} \\ M_{13} & M_{13} & D_{13} \end{matrix} & \begin{matrix} D_{23} & M_{23} & M_{23} \\ M_{23} & D_{23} & M_{23} \\ M_{23} & M_{23} & D_{23} \end{matrix} & \begin{matrix} D_3 & M_3 & M_3 \\ M_3 & D_3 & M_3 \\ M_3 & M_3 & D_3 \end{matrix} & \begin{matrix} D_{34} & M_{34} & M_{34} \\ M_{34} & D_{34} & M_{34} \\ M_{34} & M_{34} & D_{34} \end{matrix} \\ \begin{matrix} D_{14} & M_{14} & M_{14} \\ M_{14} & D_{14} & M_{14} \\ M_{14} & M_{14} & D_{14} \end{matrix} & \begin{matrix} D_{24} & M_{24} & M_{24} \\ M_{24} & D_{24} & M_{24} \\ M_{24} & M_{24} & D_{24} \end{matrix} & \begin{matrix} D_{34} & M_{34} & M_{34} \\ M_{34} & D_{34} & M_{34} \\ M_{34} & M_{34} & D_{34} \end{matrix} & \begin{matrix} D_4 & M_4 & M_4 \\ M_4 & D_4 & M_4 \\ M_4 & M_4 & D_4 \end{matrix} \end{bmatrix} \tag{3.123}$$

The sequence impedance matrix can be calculated by:

$$[Z_{012}] = \begin{bmatrix} [A]^{-1} & [0] & [0] & [0] \\ [0] & [A]^{-1} & [0] & [0] \\ [0] & [0] & [A]^{-1} & [0] \\ [0] & [0] & [0] & [A]^{-1} \end{bmatrix} [Z_{ABCT}] \begin{bmatrix} [A] & [0] & [0] & [0] \\ [0] & [A] & [0] & [0] \\ [0] & [0] & [A] & [0] \\ [0] & [0] & [0] & [A] \end{bmatrix} \tag{3.124}$$

The result of multiplication is given by:

$$[Z_{012}] =$$

$$\begin{bmatrix} \begin{matrix} Z_{1-0} & 0 & 0 \\ 0 & Z_{1-1} & 0 \\ 0 & 0 & Z_{1-2} \end{matrix} & \begin{matrix} Z_{12-0} & 0 & 0 \\ 0 & Z_{12-1} & 0 \\ 0 & 0 & Z_{12-2} \end{matrix} & \begin{matrix} Z_{13-0} & 0 & 0 \\ 0 & Z_{13-1} & 0 \\ 0 & 0 & Z_{13-2} \end{matrix} & \begin{matrix} Z_{14-0} & 0 & 0 \\ 0 & Z_{14-1} & 0 \\ 0 & 0 & Z_{14-2} \end{matrix} \\ \begin{matrix} Z_{12-0} & 0 & 0 \\ 0 & Z_{12-1} & 0 \\ 0 & 0 & Z_{12-2} \end{matrix} & \begin{matrix} Z_{2-0} & 0 & 0 \\ 0 & Z_{2-1} & 0 \\ 0 & 0 & Z_{2-2} \end{matrix} & \begin{matrix} Z_{23-0} & 0 & 0 \\ 0 & Z_{23-1} & 0 \\ 0 & 0 & Z_{23-2} \end{matrix} & \begin{matrix} Z_{24-0} & 0 & 0 \\ 0 & Z_{24-1} & 0 \\ 0 & 0 & Z_{24-2} \end{matrix} \\ \begin{matrix} Z_{13-0} & 0 & 0 \\ 0 & Z_{13-1} & 0 \\ 0 & 0 & Z_{13-2} \end{matrix} & \begin{matrix} Z_{23-0} & 0 & 0 \\ 0 & Z_{23-1} & 0 \\ 0 & 0 & Z_{23-2} \end{matrix} & \begin{matrix} Z_{3-0} & 0 & 0 \\ 0 & Z_{3-1} & 0 \\ 0 & 0 & Z_{3-2} \end{matrix} & \begin{matrix} Z_{34-0} & 0 & 0 \\ 0 & Z_{34-1} & 0 \\ 0 & 0 & Z_{34-2} \end{matrix} \\ \begin{matrix} Z_{14-0} & 0 & 0 \\ 0 & Z_{14-1} & 0 \\ 0 & 0 & Z_{14-2} \end{matrix} & \begin{matrix} Z_{24-0} & 0 & 0 \\ 0 & Z_{24-1} & 0 \\ 0 & 0 & Z_{24-2} \end{matrix} & \begin{matrix} Z_{34-0} & 0 & 0 \\ 0 & Z_{34-1} & 0 \\ 0 & 0 & Z_{34-2} \end{matrix} & \begin{matrix} Z_{4-0} & 0 & 0 \\ 0 & Z_{4-1} & 0 \\ 0 & 0 & Z_{4-2} \end{matrix} \end{bmatrix} \tag{3.125}$$

where:

$$Z_{i\text{-}1} = Z_{i\text{-}2} = Z_{Di} - Z_{Mi} \text{ para } i = 1,4 \tag{3.126}$$

$$Z_{i\text{-}0} = Z_{Di} + 2Z_{Mi} \text{ para } i = 1,4 \tag{3.127}$$

$$Z_{12-1} = Z_{12-2} = Z_{D12} - Z_{M12} \tag{3.128}$$

$$Z_{13-1} = Z_{13-2} = Z_{D13} - Z_{M13} \tag{3.129}$$

$$Z_{14-1} = Z_{14-2} = Z_{D14} - Z_{M14} \tag{3.130}$$

$$Z_{23-1} = Z_{23-2} = Z_{D23} - Z_{M23} \tag{3.131}$$

$$Z_{24-1} = Z_{24-2} = Z_{D24} - Z_{M24} \tag{3.132}$$

$$Z_{34-1} = Z_{34-2} = Z_{D34} - Z_{M34} \tag{3.133}$$

$$Z_{12-0} = Z_{D12} + 2Z_{M12} \tag{3.134}$$

$$Z_{13-0} = Z_{D13} + 2Z_{M13} \tag{3.135}$$

$$Z_{14-0} = Z_{D14} + 2Z_{M14} \tag{3.136}$$

$$Z_{23-0} = Z_{D23} + 2Z_{M23} \tag{3.137}$$

$$Z_{24-0} = Z_{D24} + 2Z_{M24} \tag{3.138}$$

$$Z_{34-0} = Z_{D34} + 2Z_{M34} \tag{3.139}$$

3.8 Sequence Impedances

To calculate the impedances of positive, negative, and zero sequences, the traditional transformation matrices of symmetrical components are used.

$$[A] = \begin{bmatrix} 1 & 1 & 1 \\ 1 & a^2 & a \\ 1 & a & a^2 \end{bmatrix} \tag{3.140}$$

$$[A]^{-1} = \frac{1}{3}\begin{bmatrix} 1 & 1 & 1 \\ 1 & a & a^2 \\ 1 & a^2 & a \end{bmatrix} \tag{3.141}$$

where:

$$a = 1\angle 120^\circ \tag{3.142}$$

For a simple and transposed TL, the impedance matrix in symmetrical components is calculated as:

$$[Z_{012}] = \frac{1}{3}\begin{bmatrix} 1 & 1 & 1 \\ 1 & a & a^2 \\ 1 & a^2 & a \end{bmatrix}\begin{bmatrix} Z_D & Z_M & Z_M \\ Z_M & Z_D & Z_M \\ Z_M & Z_M & Z_D \end{bmatrix}\begin{bmatrix} 1 & 1 & 1 \\ 1 & a^2 & a \\ 1 & a & a^2 \end{bmatrix} \tag{3.143}$$

$$[Z_{012}] = \begin{bmatrix} Z_0 & 0 & 0 \\ 0 & Z_1 & 0 \\ 0 & 0 & Z_2 \end{bmatrix} = \begin{bmatrix} Z_D + 2Z_M & 0 & 0 \\ 0 & Z_D - Z_M & 0 \\ 0 & 0 & Z_D - Z_M \end{bmatrix} \tag{3.144}$$

For the case of double-circuit TL, the matrix of equation (3.102) is reduced to a 6×6 matrix.

$$[Z_{ABC}] = \begin{bmatrix} Z_{11} & Z_{12} & Z_{13} & Z_{14} & Z_{15} & Z_{16} \\ Z_{21} & Z_{22} & Z_{23} & Z_{24} & Z_{25} & Z_{26} \\ Z_{31} & Z_{32} & Z_{33} & Z_{34} & Z_{35} & Z_{36} \\ Z_{41} & Z_{42} & Z_{43} & Z_{44} & Z_{45} & Z_{46} \\ Z_{51} & Z_{52} & Z_{53} & Z_{54} & Z_{55} & Z_{56} \\ Z_{61} & Z_{62} & Z_{63} & Z_{64} & Z_{65} & Z_{66} \end{bmatrix} \tag{3.145}$$

The new impedances after transposition are respectively:

$$[Z_{ABCT}] = \begin{bmatrix} \begin{bmatrix} Z_{D1} & Z_{M1} & Z_{M1} \\ Z_{M1} & Z_{D1} & Z_{M1} \\ Z_{M1} & Z_{M1} & Z_{D1} \end{bmatrix} & \begin{bmatrix} Z_{D12} & Z_{M12} & Z_{M12} \\ Z_{M12} & Z_{D12} & Z_{M12} \\ Z_{M12} & Z_{M12} & Z_{D12} \end{bmatrix} \\ \begin{bmatrix} Z_{D12} & Z_{M12} & Z_{M12} \\ Z_{M12} & Z_{D12} & Z_{M12} \\ Z_{M12} & Z_{M12} & Z_{D12} \end{bmatrix} & \begin{bmatrix} Z_{D2} & Z_{M2} & Z_{M2} \\ Z_{M2} & Z_{D2} & Z_{M2} \\ Z_{M2} & Z_{M2} & Z_{D2} \end{bmatrix} \end{bmatrix} \tag{3.146}$$

where:

$$Z_{D1} = \frac{Z_{11} + Z_{22} + Z_{33}}{3} \tag{3.147}$$

$$Z_{D2} = \frac{Z_{44} + Z_{55} + Z_{66}}{3} \tag{3.148}$$

$$Z_{M1} = \frac{Z_{12} + Z_{13} + Z_{23}}{3} \tag{3.149}$$

$$Z_{M2} = \frac{Z_{45} + Z_{46} + Z_{56}}{3} \tag{3.150}$$

$$Z_{D12} = \frac{Z_{14} + Z_{25} + Z_{36}}{3} \tag{3.151}$$

$$Z_{M12} = \frac{Z_{15} + Z_{16} + Z_{26}}{3} \tag{3.152}$$

Obtaining the sequence impedance matrix is done by equation (3.153).

$$[Z_{012}] = \begin{bmatrix} [A]^{-1} & [0] \\ [0] & [A]^{-1} \end{bmatrix} \begin{bmatrix} [Z_{11}] & [Z_{12}] \\ [Z_{21}] & [Z_{22}] \end{bmatrix} \begin{bmatrix} [A] & [0] \\ [0] & [A] \end{bmatrix} \tag{3.153}$$

Where $[Z_{11}]$, $[Z_{12}]$, $[Z_{21}]$, and $[Z_{22}]$ are the submatrices of equation (3.145).

Example 3.3

A 60 kHz three-phase 69 kV TL uses wooden structures as shown in Figure 3.28, where the distances between the conductors and the corresponding heights are indicated. The single conductor cables are type 477 CAA - 26/7 MCM and the lightning conductor cable is 7-wire galvanized steel with a 3/8″ nominal diameter. The conductor sag is 2.0 m and the

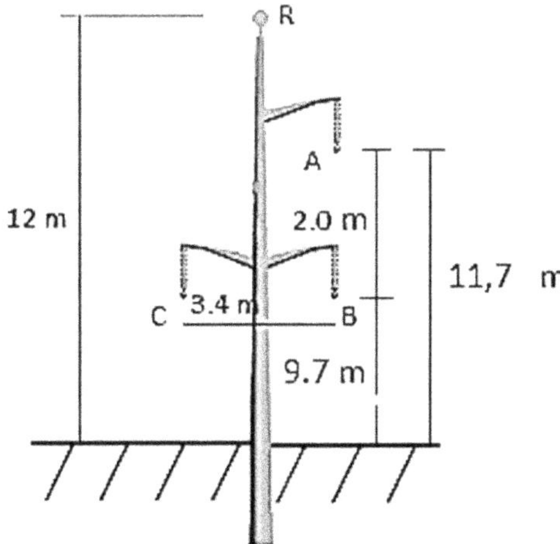

FIGURE 3.28
69 kV tower.

ground wire sag is 1.5 m. The soil resistivity is $\rho = 100\ \Omega - m$, and the temperature is 75°C.

a. Calculate for ideal soil $\rho = 0\ \Omega - m$, and considering TL without lightning conductor:
   1. The impedance matrix.
   2. The positive, negative, and null sequence impedances.
b. Calculate for ideal soil $\rho = 0\ \Omega - m$, and considering the TL with ground wire:
   1. The impedance matrix before Kron reduction.
   2. The impedance matrix after Kron reduction.
   3. The positive, negative, and null sequence impedances.

Repeat the procedure for items A and B, considering real soil $\rho = 100\ \Omega$-m. Use Carson's complete method and ATPdraw's LCC routine.

SOLUTION:

Considering the ideal soil $\rho = 0\ \Omega - m$, and TL without ground wire. Using equations (3.62) and (3.63), we have:

Self-impedances:

$$Z_{mm} = R_m + j4\pi 10^{-4} f ln\left(\frac{2h_m}{RMG_m}\right)\ \Omega/km$$

Therefore:

$$Z_{11} = R_{1cor} + j4\pi 10^{-4} 60 ln\left(\frac{2xh_1}{RMG_1}\right) = 0.1459 + j240\pi 10^{-4} ln\left(\frac{2x10.36666667}{0.00883}\right)$$

$$= 0.1459 + j0.585191473791685\ \Omega/km$$

$$Z_{22} = Z_{33} = R_{1cor} + j4\pi 10^{-4} 60 ln\left(\frac{2xh_2}{RMG_2}\right) = 0.1459 + j240\pi 10^{-4} ln\left(\frac{2x8.36666667}{0.00883}\right)$$

$$= 0.1459 + j0.569030620564765\ \Omega/km$$

where:

$RMG_1 = 0.00883\ (m)$, got from the conductor table.
R_1 is obtained from the cable table 477 (26/7), with a temperature of 25°C and a frequency of 60 Hz, is 0.1218 Ω/km. Using equation (3.2).

$$R_{1cor} = R_1 \frac{|T| + t_2}{|T| + t_1} = 0.1218 \frac{(228 + 75)}{(228 + 25)} = 0.1459\ \Omega/km$$

$$|T| = 228 \text{Hard-drawn aluminum}$$

From equation (3.83),

$$h_A = H_A - \frac{2fl}{3} = 11.70 - \frac{2x2.0}{3} = 10.36666667\ m$$

$$h_B = h_C = H_B - \frac{2fl}{3} = 9.70 - \frac{2x2.0}{3} = 8.36666667\ m$$

Mutual impedances:

$$Z_{mn} = j4\pi 10^{-4} fln\left(\frac{S_{mn'}}{d_{mn}}\right)$$

From equation (3.20):

$$d_{AB} = 10.36666667 - 8.36666667 = 2.0\ m$$

$$d_{BC} = 3.4\ m$$

$$d_{AC} = \sqrt{2^2 + 3.4^2} = 3.944616584663204\ m$$

Using equation (3.82):

$$S_{AC'} = \sqrt{4h_A h_C + d_{AC}^2} = \sqrt{4x10.36666667x8.36666667 + 3.944616584663204^2}$$

$$= 19.039374412715237\ m$$

$$S_{BC'} = \sqrt{4x8.36666667x8.36666667 + 3.40^2} = 17.075258260640030\ m$$

$$S_{AB'} = \sqrt{4x10.36666667x8.36666667 + 2.0^2} = 18.733333340000001\ m$$

Therefore:

$$Z_{13} = j4\pi 10^{-4} x60ln\left(\frac{19.039374412715237}{3.944616584663204}\right) = j0.118688672740578\ \Omega/km$$

$$Z_{12} = j4\pi 10^{-4} x60ln\left(\frac{18.733333340000001}{2.0}\right) = j0.168677685623807\ \Omega/km$$

$$Z_{23} = j4\pi 10^{-4} x60ln\left(\frac{17.075258260640030}{3.4}\right) = j0.121681807735945\ \Omega/km$$

Impedance matrix:

$$\left[Z_{imp}\right] = \begin{bmatrix} 0.1459 + j0.585191473791685 & j0.168677685623807 & j0.118688672740578 \\ j0.168677685623807 & 0.1459 + j0.569030620564765 & j0.121681807735945 \\ j0.118688672740578 & j0.121681807735945 & 0.1459 + j0.569030620564765 \end{bmatrix}$$

Impedance matrix with elements got with the mean:

Using equations (3.104) and (3.105), we have:

$$Z_D = \frac{0.1459 + j0.585191473791685 + 0.1459 + j0.569030620564765 + 0.1459 + j0.569030620564765}{3}$$

$$= 0.1459 + j0.574417571640405\ \Omega/km$$

$$Z_M = \frac{j0.168677685623807 + j0.118688672740578 + j0.121681807735945}{3}$$

$$= j0.136349388700110\ \Omega/km$$

$$\left[Z_{imp}\right] = \begin{bmatrix} 0.1459 + j0.574417571640405 & j0.136349388700110 & j0.136349388700110 \\ j0.136349388700110 & 0.1459 + j0.574417571640405 & j0.136349388700110 \\ j0.136349388700110 & j0.136349388700110 & 0.1459 + j0.57441757164040565 \end{bmatrix}$$

Sequence matrix:

Using equation (3.143), we have:

$$[Z_{012}] = \frac{1}{3}\begin{bmatrix} 1 & 1 & 1 \\ 1 & a & a^2 \\ 1 & a^2 & a \end{bmatrix} x$$

$$\begin{bmatrix} 0.1459 + j0.574417571640405 & j0.136349388700110 & j0.136349388700110 \\ j0.136349388700110 & 0.1459 + j0.574417571640405 & j0.136349388700110 \\ j0.136349388700110 & j0.136349388700110 & 0.1459 + j0.57441757164040565 \end{bmatrix}$$

$$x\begin{bmatrix} 1 & 1 & 1 \\ 1 & a^2 & a \\ 1 & a & a^2 \end{bmatrix}$$

$$[Z_{012}] = \begin{bmatrix} 0.1459 + j0.8471 & 0 & 0 \\ 0 & 0.1459 + j0.4381 & 0 \\ 0 & 0 & j0.1459 + 0.4381 \end{bmatrix} \Omega/km$$

Considering the ideal ground $\rho = 0\ \Omega - m$, and the TL with ground wire.

Ground wire self-impedance:

$$Z_{rr} = R_r + j4\pi 10^{-4} fln\left(\frac{2h_r}{RMG_r}\right) = 4.909 + j4\pi 10^{-4} 60ln\left(\frac{2x11}{0.00355911}\right)$$

$$= 4.909 + j0.658172750338629\ \Omega/km$$

$$h_r = H_r - \frac{2fl}{3} = 12.0 - \frac{2x1.5}{3} = 11.0\ m$$

$$RMG_r = 0.00355911\ m$$

Ground wire mutual impedance:

$$Z_{mn} = j4\pi 10^{-4} fln\left(\frac{S_{mn'}}{d_{mn}}\right)$$

Using equation (3.82):

$$d_{AR} = \sqrt{1.70^2 + (11 - 10.36666667)^2} = 1.814141975394674 \ m$$

$$d_{BR} = d_{CR} = \sqrt{1.70^2 + (11 - 8.36666667)^2} = 3.134396979785567 \ m$$

$$S_{AR'} = \sqrt{4h_A h_R + d_{AR}^2} = \sqrt{4x10.36666667x11.0 + 1.814141975394674^2}$$
$$= 21.434188685063145 \ m$$

$$S_{BR'} = S_{CR'} = \sqrt{4x8.36666667 \ x11.0 + 3.134396979785567^2} = 19.441136229832065 \ m$$

$$Z_{AR} = j4\pi 10^{-4} 60 ln\left(\frac{21.434188685063145}{1.814141975394674}\right) = j0.186186461035432$$

$$Z_{BR} = Z_{CR} = j4\pi 10^{-4} 60 ln\left(\frac{19.441136229832065}{3.134396979785567}\right) = j0.137598323138707$$

$$\left[Z_{imp}\right] = \begin{bmatrix} 0.1459 + j0.5852 & j0.1687 & j0.1187 & j0.1862 \\ j0.1687 & 0.1459 + j0.5690 & j0.1217 & j0.1376 \\ j0.1187 & j0.1217 & 0.1459 + j0.5690 & j0.1376 \\ j0.1862 & j0.1376 & j0.1376 & 4.909 + j0.6582 \end{bmatrix}$$

Kron reduction:

$$\left[Z_{impred}\right] = \begin{bmatrix} 0.1528 + j0.5843 & 0.0051 + j0.1680 & 0.0051 + j0.1180 \\ 0.0051 + j0.1680 & 0.1497 + j0.5685 & 0.0038 + j0.1212 \\ 0.0051 + j0.1180 & 0.0038 + j0.1212 & 0.1497 + j0.5685 \end{bmatrix}$$

Impedance matrix with elements got with the mean:
Using equations (3.104) and (3.105), we have:

$$Z_D = \frac{0.1528 + j0.5843 + 0.1497 + j0.5685 + 0.1497 + j0.5685}{3} = 0.1507 + j0.5738$$

$$Z_M = \frac{0.0051 + J0.1680 + 0.0051 + j0.1180 + 0.0038 + j0.1212}{3} = 0.0047 + j0.1357$$

$$\left[Z_{imp}\right] = \begin{bmatrix} 0.1507 + j0.5738 & 0.0047 + j0.1357 & 0.0047 + j0.1357 \\ 0.0047 + j0.1357 & 0.1507 + j0.5738 & 0.0047 + j0.1357 \\ 0.0047 + j0.1357 & 0.0047 + j0.1357 & 0.1507 + j0.5738 \end{bmatrix}$$

Sequence matrix:

Using equation (3.143), we have:

$$[Z_{012}]=\frac{1}{3}\begin{bmatrix}1 & 1 & 1\\ 1 & a & a^2\\ 1 & a^2 & a\end{bmatrix} x$$

$$\begin{bmatrix}0.1507+j0.5738 & 0.0047+j0.1357 & 0.0047+j0.1357\\ 0.0047+j0.1357 & 0.1507+j0.5738 & 0.0047+j0.1357\\ 0.0047+j0.1357 & 0.0047+j0.1357 & 0.1507+j0.5738\end{bmatrix} x \begin{bmatrix}1 & 1 & 1\\ 1 & a^2 & a\\ 1 & a & a^2\end{bmatrix}$$

$$[Z_{012}]=\begin{bmatrix}0.1601+j0.8452 & 0 & 0\\ 0 & 0.1461+j0.4380 & 0\\ 0 & 0 & 0.1461+j0.4380\end{bmatrix}$$

Considering the ideal ground $\rho = 100\ \Omega - m$, and the TL without ground wire.

Now, let's consider the Carson's correction effects given by equations (3.66) and (3.67).

Self-impedances:

$$Z_{mm}=(R_{int}+\Delta R')+j\left(4\pi 10^{-4} fln\left(\frac{2h_m}{RMG_m}\right)+\Delta X'\right)\ \Omega/km$$

The internal impedance is calculated by equation (3.44).

$$Z_{int} = R_{int} + jwL_{int}$$

$$= R_{CC} j\frac{1}{2}mr(1-s^2)\frac{[ber(mr)+jbei(mr)+\varphi(ker(mr)+jkei(mr))]}{[ber'(mr)+jbei'(mr)+\varphi(ker'(mr)+jkei'(mr))]}$$

For the conductor:

$$s=\frac{q}{r}=\frac{\frac{0.0080313}{2}}{\frac{0.0217932}{2}}=0.3685$$

$$k=\frac{8\pi 10^{-4} f}{R_{CC}}\mu_r=\frac{8\pi 10^{-4}60}{0.1459}=1.0336$$

$$mr=\sqrt{k\frac{1}{1-s^2}}=\sqrt{\frac{1.0336}{1-0.3685^2}}=1.0936$$

$$mq=\sqrt{\frac{1.0336x0.3685^2}{1-0.3685^2}}=0.4030$$

Using the equations from Bessel (3.50) to (3.57), it comes:

$$ber(mr) = 0.9777$$

$$bei(mr) = 0.2983$$

$$ker(mr) = 0.2266$$

$$kei(mr) = -0.4623$$

$$ber'(mr) = -0.0816$$

$$bei'(mr) = 0.5427$$

$$ker'(mr) = -0.5923$$

$$kei'(mr) = 0.3451$$

To calculate φ using equation (3.45), we need to calculate the derivatives as a function of mq.

$$ber'(mq) = -0.0041$$

$$bei'(mq) = 0.2015$$

$$ker'(mq) = -2.1341$$

$$kei'(mq) = 0.2787$$

φ calculation:

$$\varphi = -\frac{[ber'(mq) + jbei'(mq)]}{ker'(mq) + jkei'(mq)} = -0.0140 + j0.0926$$

Calculating the internal impedance using equation (3.44), we have:

$$Z_{int} = R_{int} + jwL_{int} = 0.1483 + j0.0142\ \Omega/km$$

Carson's correction calculation:

$$\Delta R' = 8\pi f 10^{-4}\left\{\frac{\pi}{8} - b_1 k_i cos\theta_{mn'} + b_2\left[(c_2 - lnk_i)k_i^2 cos2\theta_{mn'} + \theta_{mn'}k_i^2 sen2\theta_{mn'}\right]\right.$$
$$+ b_3 k_i^3 cos3\theta_{mn'} - d_4 k_i^4 cos4\theta_{mn'} - b_5 k_i^5 cos5\theta_{mn'} + b_6\left[(c_6 - lnk_i)k_i^6 cos6\theta_{mn'}\right.$$
$$\left.+ \theta_{mn'}k_i^6 sen6\theta_{mn'} + b_7 k_i^7 cos7\theta_{mn'} - b_8 k_i^8 cos8\theta_{mn'} - \cdots\right\}$$

$$\Delta R'_A = 0.0577$$

$$\Delta R'_B = \Delta R'_C = 0.0580$$

$$\Delta X' = 8\pi f 10^{-4}\left\{\frac{1}{2}(0.6159315 - lnk_i) + b_1 k_i cos\theta_{mn'} - d_2 k_i^2 cos2\theta_{mn'} + b_3 k_i^3 cos3\theta_{mn'}\right.$$

$$- b_4\left[(c_4 - lnk_i)k_i^4 cos4\theta_{mn'} + \theta_{mn'}k_i^4 sen4\theta_{mn'}\right] + b_5 k_i^5 cos5\theta_{mn'}$$

$$\left. - d_6 k_i^6 cos6\theta_{mn'} + b_7 k_i^7 cos7\theta_{mn'} - b_8\left[(c_8 - lnk_i)k_i^8 cos8\theta_{mn'} + \theta_{mn'}k_i^8 sen8\theta_{mn'}\right] - \cdots\right\}$$

$$\Delta X'_A = 0.2816$$

$$\Delta X'_B = \Delta X'_C = 0.2975$$

Self-impedances:

$$Z_{AA} = (0.1483 + 0.0577) + j\left(4\pi 10^{-4} 60 ln\left(\frac{2x10.3667}{0.00883}\right) + 0.2816\right)$$

$$= 0.2060 + j0.8668\ \Omega/km$$

$$Z_{BB} = Z_{CC} = (0.1483 + 0.0580) + j\left(4\pi 10^{-4} 60 ln\left(\frac{2x8.3667}{0.00883}\right) + 0.2975\right)$$

$$= 0.2063 + j0.8665\ \Omega/km$$

Mutual impedances:

$$Z_{mn} = \Delta R' + j(4\pi 10^{-4} f ln\left(\frac{S_{mn'}}{d_{mn}}\right) + \Delta X')\ \Omega/km$$

$$\Delta R'_{AB} = 0.0578$$

$$\Delta R'_{AC} = 0.0579$$

$$\Delta R'_{BC} = 0.0580$$

$$\Delta X'_{AB} = 0.2891$$

$$\Delta X'_{AC} = 0.2879$$

$$\Delta X'_{BC} = 0.2960$$

$$Z_{AB} = 0.0578 + j\left(4\pi 10^{-4} 60 ln\left(\frac{18.7334}{2.0}\right) + 0.2891\right) = 0.0578 + j0.4578\ \Omega/km$$

$$Z_{AC} = 0.0579 + j\left(4\pi 10^{-4} 60 ln\left(\frac{19.0394}{3.9446}\right) + 0.2879\right) = 0.0579 + j0.4066\ \Omega/km$$

$$Z_{BC} = 0.0580 + j\left(4\pi 10^{-4} 60 ln\left(\frac{17.0753}{3.4}\right) + 0.2960\right) = 0.0580 + j0.4176\ \Omega/km$$

Impedance matrix with elements got with the mean.

Using equations (3.104) and (3.105), we have:

$$Z_D = \frac{0.2060 + j0.8668 + 0.2063 + j0.8665 + 0.2063 + j0.8665}{3}$$

$$= 0.2062 + j0.8666\ \Omega/km$$

$$Z_M = \frac{0.0578 + j0.4578 + 0.0579 + j0.4066 + 0.0580 + j0.4176}{3}$$

$$= 0.0579 + j0.4273\ \Omega/km$$

$$\left[Z_{imp}\right] = \begin{bmatrix} 0.2062 + j0.8666 & 0.0579 + j0.4273 & 0.0579 + j\,0.4273 \\ 0.0579 + j0.4273 & 0.2062 + j0.8666 & 0.0579 + j0.4273 \\ 0.0579 + j\,0.4273 & 0.0579 + j0.4273 & 0.2062 + j0.8666 \end{bmatrix}$$

Using equation (3.143), we have:

$$\left[Z_{012}\right] = \frac{1}{3}\begin{bmatrix} 1 & 1 & 1 \\ 1 & a & a^2 \\ 1 & a^2 & a \end{bmatrix} x$$

$$\begin{bmatrix} 0.2062 + j0.8666 & 0.0579 + j0.4273 & 0.0579 + j\,0.4273 \\ 0.0579 + j0.4273 & 0.2062 + j0.8666 & 0.0579 + j0.4273 \\ 0.0579 + j\,0.4273 & 0.0579 + j0.4273 & 0.2062 + j0.8666 \end{bmatrix} x \begin{bmatrix} 1 & 1 & 1 \\ 1 & a^2 & a \\ 1 & a & a^2 \end{bmatrix}$$

$$\left[Z_{012}\right] = \begin{bmatrix} 0.3220 + j1.7213 & 0 & 0 \\ 0 & 0.1483 + j0.4393 & 0 \\ 0 & 0 & 0.1483 + j0.4393 \end{bmatrix} \Omega/km$$

Considering the ideal ground $\rho = 100\ \Omega - m$, and TL with ground wire. In relation to the previous item, just calculate the terms that are related to the ground wire.

Internal resistance:

$$R_{int} = 4.909$$

Resistance correction:

$$\Delta R'_R = 0.0576$$

$$\Delta R'_{AR} = 0.0577$$

$$\Delta R'_{BR} = 0.0578$$

$$\Delta R'_{CR} = 0.0578$$

Reactance correction:

$$\Delta X'_R = 0.2773$$

$$\Delta X'_{AR} = 0.2792$$

$$\Delta X'_{BR} = 0.2865$$

$$\Delta X'_{CR} = 0.2865$$

Self-impedance:

$$Z_{RR} = (4.909 + 0.0576) + j\left(4\pi 10^{-4} 60 ln\left(\frac{2x11}{0.00355911}\right) + 0.2773\right)$$

$$= 4.9666 + j0.9354 \ \Omega/km$$

Mutual impedances:

$$Z_{AR} = 0.0577 + j\left(4\pi 10^{-4} 60 ln\left(\frac{21.4342}{1.8141}\right) + 0.2792\right) = 0.0577 + j0.4654 \ \Omega/km$$

$$Z_{BR} = Z_{CR} = 0.0578 + j\left(4\pi 10^{-4} 60 ln\left(\frac{19.4412}{3.1344}\right) + 0.2865\right) = 0.0578 + j0.4241 \ \Omega/km$$

$$\left[Z_{imp}\right] =$$

$$\begin{bmatrix} 0.2060 + j0.8668 & 0.0578 + j0.4578 & 0.0579 + j0.4066 & 0.0577 + j0.4654 \\ 0.0578 + j0.4578 & 0.2063 + j0.8665 & 0.0580 + j0.4176 & 0.0578 + j0.4241 \\ 0.0579 + j0.4066 & 0.0580 + j0.4176 & 0.2063 + j0.8665 & 0.0578 + j0.4241 \\ 0.0577 + j0.4654 & 0.0578 + j0.4241 & 0.0578 + j0.4241 & 4.9666 + j0.9354 \end{bmatrix} \Omega/km$$

$$\left[Z_{impred}\right] = \begin{bmatrix} 0.2455 + j0.8486 & 0.0937 + j0.4407 & 0.0937 + j0.3895 \\ 0.0937 + j0.4407 & 0.2388 + j0.8505 & 0.0905 + j0.4017 \\ 0.0937 + j0.3895 & 0.0905 + j0.4017 & 0.2388 + j0.8505 \end{bmatrix} \Omega/km$$

Impedance matrix with elements got with the mean:
Using equations (3.104) and (3.105), we have:

$$Z_D = \frac{0.2455 + j0.8486 + 0.2388 + j0.8505 + 0.2388 + j0.8505}{3}$$

$$= 0.2410 + j0.8499 \ \Omega/km$$

$$Z_M = \frac{0.0937 + j0.4407 + 0.0937 + j0.3895 + 0.0905 + j0.4017}{3}$$

$$= 0.0926 + j0.4106 \ \Omega/km$$

$$\left[Z_{imp}\right]=\begin{bmatrix} 0.2410+j0.8499 & 0.0926+j0.4106 & 0.0926+j0.4106 \\ 0.0926+j0.4106 & 0.2410+j0.8499 & 0.0926+j0.4106 \\ 0.0926+j0.4106 & 0.0926+j0.4106 & 0.2410+j0.8499 \end{bmatrix}$$

$$\left[Z_{012}\right]=\frac{1}{3}\begin{bmatrix} 1 & 1 & 1 \\ 1 & a & a^2 \\ 1 & a^2 & a \end{bmatrix}x$$

$$\begin{bmatrix} 0.2410+j0.8499 & 0.0926+j0.4106 & 0.0926+j0.4106 \\ 0.0926+j0.4106 & 0.2410+j0.8499 & 0.0926+j0.4106 \\ 0.0926+j0.4106 & 0.0926+j0.4106 & 0.2410+j0.8499 \end{bmatrix}x\begin{bmatrix} 1 & 1 & 1 \\ 1 & a^2 & a \\ 1 & a & a^2 \end{bmatrix}$$

$$\left[Z_{012}\right]=\begin{bmatrix} 0.4263+j1.6711 & 0 & 0 \\ 0 & 0.1484+j0.4393 & 0 \\ 0 & 0 & 0.1484+j0.4393 \end{bmatrix}\Omega/km$$

Using the ATPdraw LCC routine:

The data entered in ATPdraw are shown in Figures 3.29 and 3.30 and taken from the conductor tables.

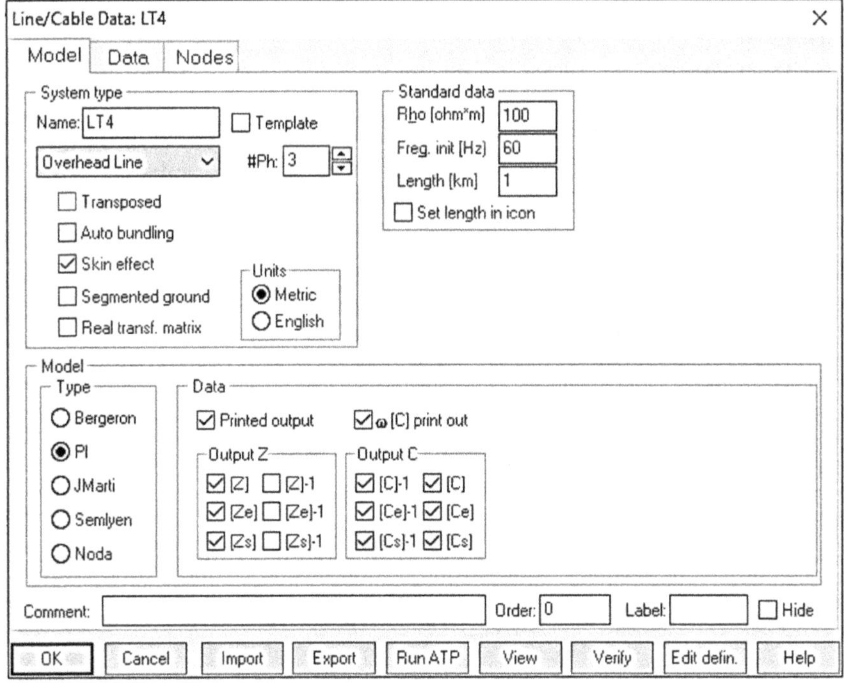

FIGURE 3.29
Model data for example 3.2.

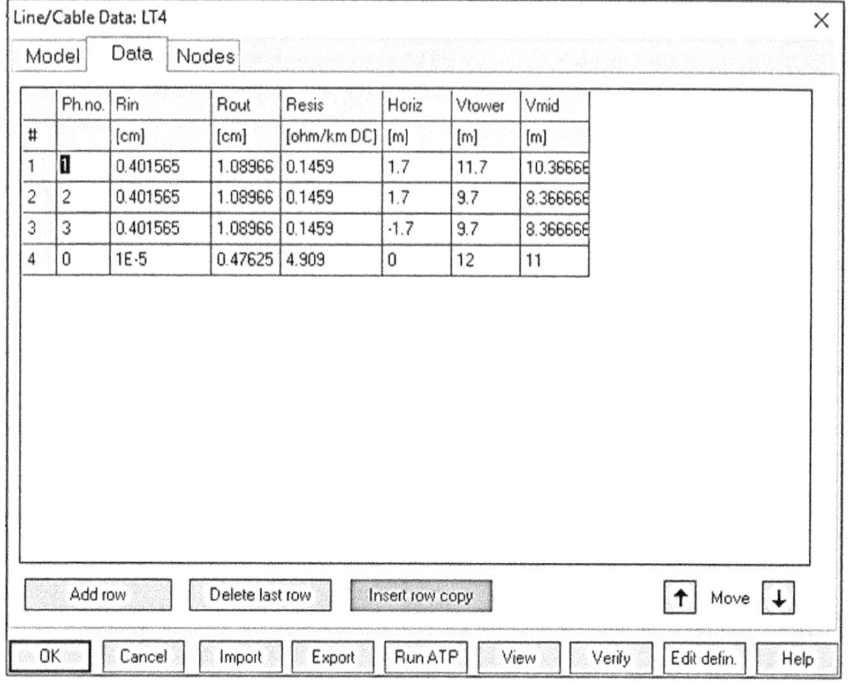

FIGURE 3.30
Data for example 3.2.

In the .lis file generated by ATPdraw the following results appear:

```
Sequence          Surge impedance          Attenuation   velocity     Wavelength   Resistance    Reactance
          magnitude(ohm)  angle(degr.)       db/km        km/sec          km          ohm/km        ohm/km
   Zero :   8.87486E+02 -7.14196E+00    2.09689E-03  1.95669E+05  3.26115E+03  4.25178E-01  1.67000E+00
Positive:   3.49714E+02 -9.23991E+00    1.84242E-03  2.89127E+05  4.81879E+03  1.46435E-01  4.38162E-01
Request for flushing of punch buffer.                |$PUNCH
```

Table 3.4 shows a comparison between the results obtained for zero, positive, and negative sequence impedances.

The methodology presented for $\rho = 100$ with GW has slight differences from ATPdraw results: $erro_{resizero} = 0.2587\%$; $erro_{reatzero} = 0.066\%$;

TABLE 3.4 Sequence Impedances

Impedances Ω/km	ρ = 0 without GW	ρ = 0 with GW	ρ = 100 without GW	ρ = 100 with GW	ATPdraw LCC
Zero	0.1459	0.1601	0.3220 +	0.4263	0.4252
	+j0.8472	+j0.8452	j1.7213	+j1.6711	+j1.6700
Positive	0.1459	0.1460	0.1483	0.1484	0.1464
	+j0.4380	+j0.4381	+j0.4393	+j0.4393	+j0.4382
Negative	0.1459	0.1460	0.1483	0.1484	0.1464
	+j0.4380	+j0.4381	+j0.4393	+j0.4393	+j0.4382

$erro_{resipos} = 1.37\%$; $erro_{reatpos} = 0.2510\%$. Therefore, errors occur in the third decimal place and are negligible.

Example 3.4

Repeat the previous example for the ground wire TL and $\rho = 100\ \Omega - m$, using Carson's approximate method.

SOLUTION:

The internal resistance for the phase conductor and ground wire calculated in example 3.3 are respectively:

$$R_{int} = 0.1483\ \Omega/km$$

$$R_{int} = 4.909\ \Omega/km$$

Using Table 3.2, it comes $D_e = 850\ m$.

The geometric mean radii of conductor and lightning conductor are respectively:

$$RMG_1 = 0.00883\ m$$

$$RMG_r = 0.00355911\ m$$

Distances between conductors:

$$d_{AB} = 2.0\ m$$

$$d_{BC} = 3.4\ m$$

$$d_{AC} = 3.944616584663204\ m$$

$$d_{AR} = 1.814141975394674\ m$$

$$d_{BR} = d_{CR} = 3.134396979785567\ m$$

Using equations (3.89) and (3.90), we have:

$$Z_{mm} = R_{int} + 0.05928 + j0.0753984 ln\left(\frac{D_e}{RMG_m}\right)\ \Omega/km$$

$$Z_{mn} = 0.05928 + j0.0753984 ln\left(\frac{D_e}{d_{mn}}\right)\ \Omega/km$$

Self-impedances:

$$Z_{aa} = Z_{bb} = Z_{cc} = 0.1483 + 0.05928 + j0.0753984 ln\left(\frac{850}{0.00883}\right)$$

$$= 0.2076 + j0.8652\ \Omega/km$$

$$Z_{rr} = 4.909 + 0.05928 + j0.0753984ln\left(\frac{850}{0.00355911}\right) = 4.9683 + j0.9337\ \Omega/km$$

Mutual impedances:

$$Z_{ab} = 0.05928 + j0.0753984ln\left(\frac{850}{2.0}\right) = 0.0593 + j0.4563\ \Omega/km$$

$$Z_{bc} = 0.05928 + j0.0753984ln\left(\frac{850}{3.4}\right) = 0.0593 + j0.4163\ \Omega/km$$

$$Z_{ac} = 0.05928 + j0.0753984ln\left(\frac{850}{3.944616584663204}\right) = 0.0593 + j0.4051\ \Omega/km$$

$$Z_{ar} = 0.05928 + j0.0753984ln\left(\frac{850}{3.134396979785567}\right) = 0.0593 + j0.4224\ \Omega/km$$

$$Z_{br} = Z_{cr} = 0.05928 + j0.0753984ln\left(\frac{850}{1.814141975394674}\right)$$

$$= 0.0593 + j0.4637\ \Omega/km$$

Impedance matrix:

$$\left[Z_{imp}\right] =$$

$$\begin{bmatrix} 0.2076 + j0.8652 & 0.0593 + j0.4563 & 0.0593 + j0.4050 & 0.0593 + j0.4224 \\ 0.0593 + j0.4563 & 0.2076 + j0.8652 & 0.0593 + j0.4163 & 0.0593 + j0.4637 \\ 0.0593 + j0.4050 & 0.0593 + j0.4163 & 0.2076 + j0.8652 & 0.0593 + j0.4637 \\ 0.0593 + j0.4224 & 0.0593 + j0.4637 & 0.0593 + j0.4637 & 4.9683 + j0.9337 \end{bmatrix} \Omega/km$$

Reduced impedance matrix:

$$\left[Z_{imp}\right] = \begin{bmatrix} 0.2398 + j0.8491 & 0.0948 + j0.4391 & 0.0948 + j0.3878 \\ 0.0948 + j0.4391 & 0.2467 + j0.8468 & 0.0984 + j0.3979 \\ 0.0948 + j0.3878 & 0.0984 + j0.3979 & 0.2467 + j0.8468 \end{bmatrix}$$

Balanced matrix:

$$\left[Z_{imp}\right] = \begin{bmatrix} 0.2444 + j0.8475 & 0.0960 + j0.4082 & 0.0960 + j0.4082 \\ 0.0960 + j0.4082 & 0.2444 + j0.8475 & 0.0960 + j0.4082 \\ 0.0960 + j0.4082 & 0.0960 + j0.4082 & 0.2444 + j0.8475 \end{bmatrix}$$

Sequence impedance matrix:

$$[Z_{012}] = \frac{1}{3}\begin{bmatrix} 1 & 1 & 1 \\ 1 & a & a^2 \\ 1 & a^2 & a \end{bmatrix} x$$

$$\begin{bmatrix} 0.2444 + j0.8475 & 0.0960 + j0.4082 & 0.0960 + j0.4082 \\ 0.0960 + j0.4082 & 0.2444 + j0.8475 & 0.0960 + j0.4082 \\ 0.0960 + j0.4082 & 0.0960 + j0.4082 & 0.2444 + j0.8475 \end{bmatrix} x \begin{bmatrix} 1 & 1 & 1 \\ 1 & a^2 & a \\ 1 & a & a^2 \end{bmatrix}$$

$$[Z_{012}] = \begin{bmatrix} 0.4364 + j1.6640 & 0 & 0 \\ 0 & 0.1484 + j0.4393 & 0 \\ 0 & 0 & 0.1484 + j0.4393 \end{bmatrix} \Omega/km$$

Comparing with ATPdraw results: $erro_{resizero} = 2.6341\%$; $erro_{reatzero} = 0.3593\%$; $erro_{resipos} = 1.37\%$; $erro_{reatpos} = 0.2510\%$. Therefore, errors in zero sequence parameters were slightly larger than in the case of complete Carson.

3.9 Capacitance

Consider a long straight cylindrical conductor in a uniform medium, such as air, and isolated from other loads, as shown in Figure 3.31.

The electric flow density is given by:

$$D = \frac{q}{2\pi x} \; C/m^2 \tag{3.154}$$

where:

q = the load of the insulated conductor in C

x = the distance from the center of the conductor to the point at which the density of the electric flux is calculated in m

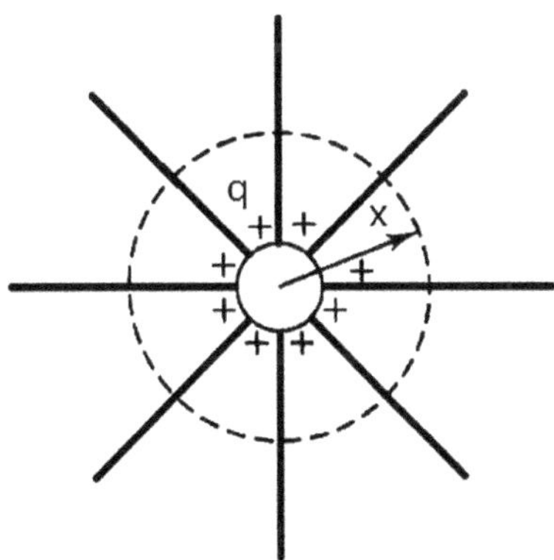

FIGURE 3.31
Isolated cylindrical conductor.

The intensity of the electric field is given by:

$$E = \frac{D}{\varepsilon} = \frac{q}{2\pi\varepsilon x} \; V/m \tag{3.155}$$

where:

D = the electric flow density

ε = the permissiveness of the medium and is $\varepsilon = \varepsilon_0 \varepsilon_r$ with $\varepsilon_0 = 8.859.10^{-12} \; F/m$ and for the air $\varepsilon_r = 1$

We can determine the voltage drop between two points by calculating the tension between equipotential surfaces passing through P_1 and P_2 by integrating the field strength over a radial path between equipotential surfaces, as shown in Figure 3.32.

$$V_{12} = \int_{D_1}^{D_2} E dx = \int_{D_1}^{D_2} \frac{q}{2\pi\varepsilon x} dx = \frac{q}{2\pi\varepsilon} \int_{D_1}^{D_2} \frac{1}{x} dx = \frac{q}{2\pi\varepsilon} ln \frac{D_2}{D_1} \; (V) \tag{3.156}$$

It should be noted in equation (3.92) that point P_2 is the farthest point from the center of the conductor.

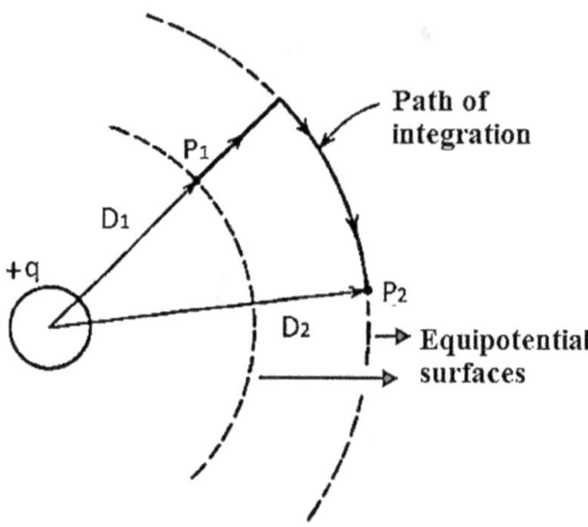

FIGURE 3.32
Integration path for calculating potential difference between two points.

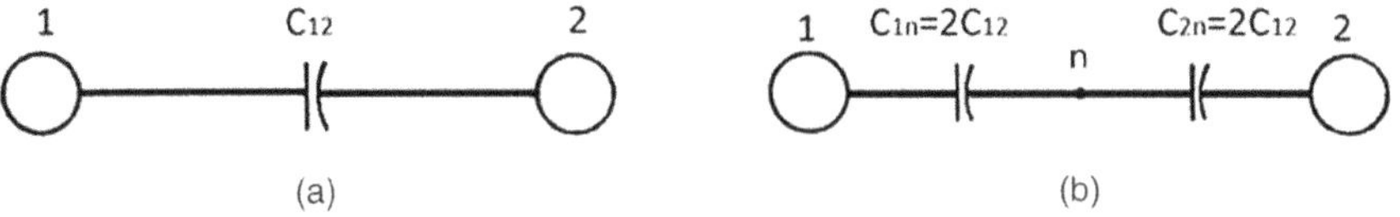

FIGURE 3.33
(a) Conductor capacitance. (b) Capacitance between the conductor and the neutral.

The capacitance of a two-conductor line 1 and 2, shown in Figure 3.33 (a), is defined as the conductor load per unit of potential difference between them.

$$C_{12} = \frac{q}{V_{12}} \text{ F/m} \tag{3.157}$$

3.9.1 Capacitance for Neutral

If the TL is energized from a grounded center tape transformer, the potential difference between each conductor and ground is half the potential difference between two conductors (equation (3.158)) and the ground or neutral capacitance is given by equation (3.159) and shown in Figure 3.33 (b).

$$V_{1n} = \frac{V_{12}}{2} \tag{3.158}$$

$$C_{1n} = C_{2n} = 2C_{12} \tag{3.159}$$

3.9.2 Capacitance Calculation with Earth Effect

The earth alters the capacitance of the three-phase TLs because its presence alters the electric field of the lines. To take into account the effect of the earth on the capacitance calculation, the earth plane is supposed to be a perfect conductor in the form of a horizontal plane with infinite extension. Thus, to calculate capacitance, the earth plane is replaced by a dummy conductor below the earth's surface at a distance equal to that of the air conductor above the surface. This conductor has an equal charge and opposite signal to the original conductor and is called the image conductor.

Before calculating the grounding capacitance, let us extend the calculation of the potential difference for N solid and cylindrical conductors, as shown in Figure 3.34, disregarding the distortion of the electric field in the other's vicinity conductors, caused because the other conductors themselves are constant potential surfaces.

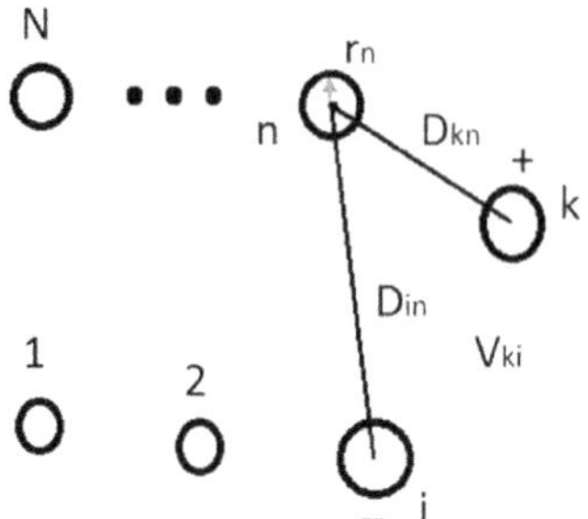

FIGURE 3.34
Solid cylinder conductor group.

where:

$$D_{nn} = r_n \text{ when } k = n \text{ or } i = n.$$

Suppose each conductor has a load q_n (C/m) evenly distributed across the conductor. The potential difference between two conductors k and i, because of the presence of the single load q_n, is given by: $V_{kin} = \frac{q_n}{2\pi\varepsilon} ln \frac{D_{in}}{D_{kn}}$.

Using superposition, the potential difference between two conductors k and i, because of the presence of all loads, is given by:

$$V_{ki} = \frac{1}{2\pi\varepsilon} \sum_{n=1}^{N} q_n ln\left(\frac{D_{in}}{D_{kn}}\right) (V) \tag{3.160}$$

Figure 3.35 shows a three-phase line with n bundled conductors per phase and N neutral conductors. Each conductor has an image conductor. The conductors a, b, c, $n1$, $n2$, …., N have loads $q_a, q_b, q_c, q_{n1}, q_{n2}, \dots q_{nN}$ and the image conductors have loads $-q_a, -q_b, -q_c, -q_{n1}, -q_{n2}, \dots -q_{nN}$. First, let's consider that each three-phase TL conductor a, b, and c and their images a', b', and c' have no bundled conductors. Applying equation (3.160) to determine the potential difference $V_{kk'}$ between a conductor k and its image k', we have:

$$V_{kk'} = \frac{1}{2\pi\varepsilon}\left[\sum_{m=a}^{nN} q_m ln\left(\frac{H_{km}}{D_{km}}\right) - \sum_{m=a}^{nN} q_m ln\left(\frac{D_{km}}{H_{km}}\right)\right] \tag{3.161}$$

Performing the operation on the logarithm, we have:

$$V_{kk'} = \frac{2}{2\pi\varepsilon} \sum_{m=a}^{nN} q_m ln\left(\frac{H_{km}}{D_{km}}\right) \tag{3.162}$$

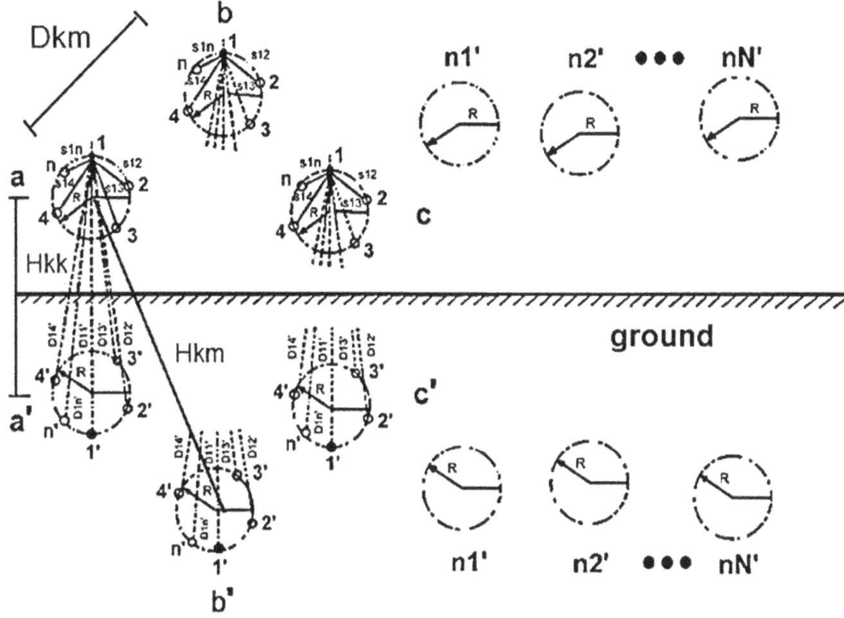

FIGURE 3.35
Three-Phase TL with neutral conductors and ground plane replaced with image conductors.

where:

$D_{kk} = r_k$

D_{km} = the distance between conductors k and m

H_{km} = the distance between the conductor k and the image of the conductor m

According to equation (3.158), the potential difference between conductor k and ground is half of $V_{kk'}$. Therefore:

$$V_{kn} = \frac{V_{kk'}}{2} = \frac{1}{2\pi\varepsilon}\sum_{m=a}^{nN} q_m ln\left(\frac{H_{km}}{D_{km}}\right) \tag{3.163}$$

where: $k = a, b, c, n1, n2, \ldots, nN$ and $m = a, b, c, n1, n2, \ldots, nN$

According to equation (3.160), we can write equation (3.164) in the following matrix form:

$$\begin{bmatrix} V_{an} \\ V_{bn} \\ V_{cn} \\ \vdots \\ V_{nN} \end{bmatrix} = \begin{bmatrix} A_{aa} & A_{ab} & \cdots & A_{anN} \\ A_{ba} & A_{bb} & \ddots & A_{bnN} \\ \vdots & & \cdots & \vdots \\ A_{nNa} & A_{nNb} & \ldots & A_{nNnN} \end{bmatrix} \begin{bmatrix} q_a \\ q_b \\ q_c \\ \vdots \\ q_{nN} \end{bmatrix} \tag{3.164}$$

The diagonal terms:

$$A_{kk} = \frac{1}{2\pi\varepsilon} ln\left(\frac{H_{kk}}{D_{kk}}\right) \tag{3.165}$$

are called self-field coefficients.

The terms out of diagonal:

$$A_{km} = \frac{1}{2\pi\varepsilon} ln\left(\frac{H_{km}}{D_{km}}\right) \tag{3.166}$$

are called mutual field coefficients.

If we replace the value of $\varepsilon = \varepsilon_0 \varepsilon_r$ with $\varepsilon_0 = 8.8541878176.10^{-12}\ F/m$ and for the air $\varepsilon_r = 1$, the self-field coefficients and mutual field coefficients take the form of equations (3.167) and (3.168).

$$A_{kk} = 17.975103.10^6 ln\left(\frac{H_{kk}}{D_{kk}}\right) km/F \tag{3.167}$$

$$A_{km} = 17.975103.10^6 ln\left(\frac{H_{km}}{D_{km}}\right) km/F \tag{3.168}$$

Let us then consider that each phase conductor of Figure 3.35 is composed of n bundled conductors uniformly arranged over a circle of radius R and whose center is at a height $H_{kk}/2$ above the ground and that $\frac{H_{kk}}{2} \gg R$. Therefore, the distance from each bundled conductor to its image is equal to H_{kk}. We assume that the electric charges are equal in each bundled conductor and evenly distributed over their surfaces, such that $q = q_1 = q_2 = q_3 = \cdots q_n = \frac{q_a}{n} = \frac{q_b}{n} = \cdots = \frac{q_n}{n}$. Each bundled conductor has the same radius r.

The self and mutual field coefficients for each bundled conductor are:

$$A_{11} = A_{22} = \ldots = A_{nn} = 17.975103.10^6 ln\left(\frac{H_{kk}}{r}\right) \tag{3.169}$$

$$\begin{aligned} A_{12} &= 17.975103.10^6 ln\left(\frac{H_{kk}}{s_{12}}\right);\ A_{13} = 17.975103.10^6 ln\left(\frac{H_{kk}}{s_{13}}\right);\ldots;A_{1n} \\ &= 17.975103.10^6 ln\left(\frac{H_{kk}}{s_{1n}}\right) \end{aligned} \tag{3.170}$$

Therefore:

$$V_{an} = (A_{11} + A_{12} + \ldots + A_{1n})\frac{q}{n} = (17.975103.10^6 \, ln\left(\frac{H_{kk}}{r}\right)$$

$$+ 17.975103.10^6 \, ln\left(\frac{H_{kk}}{s_{12}}\right) + \ldots + 17.975109.10^6 \, ln\left(\frac{H_{kk}}{s_{1n}}\right)$$

$$= 17.975103.10^6 \frac{q}{n}(nlnH_{kk} + ln\left(\frac{1}{rs_{12}\ldots s_{1n}}\right) = 17.975103.10^6 \, qln\left(\frac{H_{kk}}{\sqrt[n]{rs_{12}\ldots s_{1n}}}\right) \tag{3.171}$$

If we compare this equation with equation (3.167) we conclude that its denominator represents the radius of the bundled cylindrical conductor.

We can then consider the conductor sags, change the notation like that in Figure 3.35, and write the equations of the self and mutual field coefficients as:

$$A_{kk} = 17.975103.10^6 \, ln\left(\frac{H_{kk}}{\sqrt[n]{rs_{12}\ldots s_{1n}}}\right) = 17.975103.10^6 \, ln\left(\frac{H_{kk}}{D_{SS}}\right)$$
$$= 17.975103.10^6 \, ln\left(\frac{2h_m}{D_{SS}}\right) \; km/F \tag{3.172}$$

$$A_{km} = 17.975103.10^6 \, ln\left(\frac{H_{km}}{D_{km}}\right) = 17.975103.10^6 \, ln\left(\frac{S_{mn'}}{d_{mn}}\right) \; km/F \tag{3.173}$$

where:

$$D_{SS} = \sqrt[n]{rs_{12}\ldots s_{1n}}$$

For one TL with:

- Two bundled conductors with radius r: 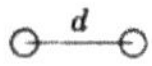

$$D_{SS2} = \sqrt[2x2]{r(d)r(d)} = \sqrt[4]{[(r)(d)]^2} = \sqrt{(r)(d)} \tag{3.174}$$

- Three bundled conductors: 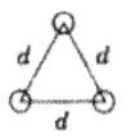

$$D_{SS3} = \sqrt[3x3]{r(d)(d)r(d)(d)r(d)(d)} = \sqrt[9]{[(r)(d)(d)]^3} = \sqrt[3]{r(d)^2} \tag{3.175}$$

- Four bundled conductors:

$$D_{SS4} = \sqrt[4x4]{r(d)(d)(d\sqrt{2})r(d)(d)r(d)(d)(d\sqrt{2})r(d)(d)(d\sqrt{2})} = \sqrt[16]{[(r)(d)(d)(d\sqrt{2})]^4}$$
$$= 1.0905\sqrt[4]{r(d)^3} \tag{3.176}$$

3.9.3 Operationalization of the Image Method

As in the operationalization of impedances, we will consider four cases: TL without ground wire, TL with one ground wire, TL with two ground wires, and TL with two parallel circuits and two ground wires.

Case 1: TL without ground wire

The matrix of potential coefficients has the following form:

$$\left[A_p\right] = \begin{bmatrix} A_{aa} & A_{ab} & A_{ac} \\ A_{ba} & A_{bb} & A_{bc} \\ A_{ca} & A_{cb} & A_{cc} \end{bmatrix} \tag{3.177}$$

Where: The elements of the potential coefficient matrix are calculated by expressions (3.165) and (3.166).

Case 2: TL with one ground wire

The matrix of potential coefficients has the following form:

$$\left[A_p\right] = \begin{bmatrix} A_{aa} & A_{ab} & A_{ac} & | & A_{ar} \\ A_{ba} & A_{bb} & A_{bc} & | & A_{br} \\ A_{ca} & A_{cb} & A_{cc} & | & A_{cr} \\ -- & -- & -- & & -- \\ A_{ra} & A_{rb} & A_{rc} & | & A_{rr} \end{bmatrix} \tag{3.178}$$

In order for the matrix to be 3 × 3 in size, the matrix of potential coefficients must be reduced using the Kron reduction and, the line and column for the ground wire are eliminated and equation (3.178) is in the form of equation (3.177).

Case 3: TL with two ground wires

The matrix of potential coefficients has the following form:

$$[A_p]=\begin{bmatrix} A_{aa} & A_{ab} & A_{ac} & | & A_{ar1} & A_{ar2} \\ A_{ba} & A_{bb} & A_{bc} & | & A_{br1} & A_{br2} \\ A_{ca} & A_{cb} & A_{cc} & | & A_{cr1} & A_{cr2} \\ -- & -- & -- & & ---- & -- \\ A_{r1a} & A_{r1b} & A_{r1c} & | & A_{r1r1} & A_{r1r2} \\ A_{r2a} & A_{r2b} & A_{r2c} & | & A_{r2r1} & A_{r2r2} \end{bmatrix} \tag{3.179}$$

For the matrix of potential coefficients with two lightning conductors to be 3×3 in size, it must also be reduced using the Kron reduction.

In cases 2 and 3 the Kron reduction is applied and in each of the three cases, the capacitance matrix is obtained by the inverse of the potential coefficient matrix:

$$[C]=[A_p]^{-1}=\begin{bmatrix} C_{aa} & C_{ab} & C_{ac} \\ C_{ba} & C_{bb} & C_{bc} \\ C_{ca} & C_{cb} & C_{cc} \end{bmatrix} F/km \tag{3.180}$$

If TL is completely transposed:

$$C_D=\frac{C_{aa}+C_{bb}+C_{cc}}{3}\ F/km \tag{3.181}$$

$$C_M=\frac{C_{ab}+C_{bc}+C_{ca}}{3}\ F/km \tag{3.182}$$

The new impedance matrix with the transposed conductors is equal to:

$$[C_T]=\begin{bmatrix} C_D & C_M & C_M \\ C_M & C_D & C_M \\ C_M & C_M & C_D \end{bmatrix} F/km \tag{3.183}$$

The shunt admittance matrix is given by:

$$[Y]=jw[C_T]\ S/km \tag{3.184}$$

Case 4: TL with two parallel circuits and two ground wires

The matrix of potential coefficients has the following form:

$$
\left[A_p\right]=\left[\begin{array}{cccccc|cc}
A_{aa} & A_{ab} & A_{ac} & A_{aa'} & A_{ab'} & A_{ac'} & A_{ar1} & A_{ar2} \\
A_{ba} & A_{bb} & A_{bc} & A_{ba'} & A_{bb'} & A_{bc'} & A_{br1} & A_{br2} \\
A_{ca} & A_{cb} & A_{cc} & A_{ca'} & A_{cb'} & A_{cc'} & A_{cr1} & A_{cr2} \\
A_{a'a} & A_{a'b} & A_{a'c} & A_{a'a'} & A_{a'b'} & A_{a'c'} & A_{a'r1} & A_{a'r2} \\
A_{b'a} & A_{b'b} & A_{b'c} & A_{b'a'} & A_{b'b'} & A_{b'c'} & A_{b'r1} & A_{b'r2} \\
A_{c'a} & A_{c'b} & A_{c'c} & A_{c'a'} & A_{c'b'} & A_{c'c'} & A_{c'r1} & A_{c'r2} \\
- & - & - & - & - & - & - & - \\
A_{r1a} & A_{r1b} & A_{r1c} & A_{r1a'} & A_{r1b'} & A_{r1c'} & A_{r1r1} & A_{r1r2} \\
Z_{r2a} & Z_{r2b} & Z_{r2c} & Z_{r2a'} & Z_{r2b'} & Z_{r2c'} & Z_{r2r1} & Z_{r2r2}
\end{array}\right] \tag{3.185}
$$

The matrix of reduced potential coefficients is 6×6 in size.

$$
\left[A_p\right]=\begin{bmatrix}
A_{11} & A_{12} & A_{13} & A_{14} & A_{15} & A_{16} \\
A_{21} & A_{22} & A_{23} & A_{24} & A_{25} & A_{26} \\
A_{31} & A_{32} & A_{33} & A_{34} & A_{35} & A_{36} \\
A_{41} & A_{42} & A_{43} & A_{44} & A_{45} & A_{46} \\
A_{51} & A_{52} & A_{53} & A_{54} & A_{55} & A_{56} \\
A_{61} & A_{62} & A_{63} & A_{64} & A_{65} & A_{66}
\end{bmatrix} \tag{3.186}
$$

The capacitance matrix is given by the inverse of the potential coefficient matrix.

$$
[C]=\left[A_p\right]^{-1}=\begin{bmatrix}
C_{11} & C_{12} & C_{13} & C_{14} & C_{15} & C_{16} \\
C_{21} & C_{22} & C_{23} & C_{24} & C_{25} & C_{26} \\
C_{31} & C_{32} & C_{33} & C_{34} & C_{35} & C_{36} \\
C_{41} & C_{42} & C_{43} & C_{44} & C_{45} & C_{46} \\
C_{51} & C_{52} & C_{53} & C_{54} & C_{55} & C_{56} \\
C_{61} & C_{62} & C_{63} & C_{64} & C_{65} & C_{66}
\end{bmatrix} \tag{3.187}
$$

3.10 Sequence Admittances

For a simple and transposed TL, the admittance matrix in symmetrical components is calculated as:

$$[Y_{012}] = \frac{jw}{3}\begin{bmatrix} 1 & 1 & 1 \\ 1 & a & a^2 \\ 1 & a^2 & a \end{bmatrix}\begin{bmatrix} C_D & C_M & C_M \\ C_M & C_D & C_M \\ C_M & C_M & C_D \end{bmatrix}\begin{bmatrix} 1 & 1 & 1 \\ 1 & a^2 & a \\ 1 & a & a^2 \end{bmatrix} \tag{3.188}$$

$$[Y_{012}] = \begin{bmatrix} Y_0 & 0 & 0 \\ 0 & Y_1 & 0 \\ 0 & 0 & Y_2 \end{bmatrix} = \begin{bmatrix} Y_D + 2Y_M & 0 & 0 \\ 0 & Y_D - Y_M & 0 \\ 0 & 0 & Y_D - Y_M \end{bmatrix} \tag{3.189}$$

In the case of double circuit TL, the sequence admittances are obtained through equation (3.190).

$$[Y_{012}] = \begin{bmatrix} [A]^{-1} & [0] \\ [0] & [A]^{-1} \end{bmatrix}\begin{bmatrix} [Y_{11}] & [Y_{12}] \\ [Y_{21}] & [Y_{22}] \end{bmatrix}\begin{bmatrix} [A] & [0] \\ [0] & [A] \end{bmatrix} \tag{3.190}$$

where:

$$[C_T] = \begin{bmatrix} \begin{bmatrix} C_{D1} & C_{M1} & C_{M1} \\ C_{M1} & C_{D1} & C_{M1} \\ C_{M1} & C_{M1} & C_{D1} \end{bmatrix} & \begin{bmatrix} C_{D12} & C_{M12} & C_{M12} \\ C_{M12} & C_{D12} & C_{M12} \\ C_{M12} & C_{M12} & C_{D12} \end{bmatrix} \\ \begin{bmatrix} C_{D12} & C_{M12} & C_{M12} \\ C_{M12} & C_{D12} & C_{M12} \\ C_{M12} & C_{M12} & C_{D12} \end{bmatrix} & \begin{bmatrix} C_{D2} & C_{M2} & C_{M2} \\ C_{M2} & C_{D2} & C_{M2} \\ C_{M2} & C_{M2} & C_{D2} \end{bmatrix} \end{bmatrix} \tag{3.191}$$

$$C_{D1} = \frac{C_{11} + C_{22} + C_{33}}{3} \tag{3.192}$$

$$C_{D2} = \frac{C_{44} + C_{55} + C_{66}}{3} \tag{3.193}$$

$$C_{M1} = \frac{C_{12} + C_{13} + C_{23}}{3} \tag{3.194}$$

$$C_{M2} = \frac{C_{45} + C_{46} + C_{56}}{3} \tag{3.195}$$

$$C_{D12} = \frac{C_{14} + C_{25} + C_{36}}{3} \tag{3.196}$$

$$C_{M12} = \frac{C_{15} + C_{16} + C_{26}}{3} \tag{3.197}$$

And the submatrices $[Y_{11}]$, $[Y_{12}]$, $[Y_{21}]$, and $[Y_{22}]$ are given by:

$$[Y_{11}] = jw\begin{bmatrix} C_{D1} & C_{M1} & C_{M1} \\ C_{M1} & C_{D1} & C_{M1} \\ C_{M1} & C_{M1} & C_{D1} \end{bmatrix} \tag{3.198}$$

$$[Y_{12}] = jw\begin{bmatrix} C_{D12} & C_{M12} & C_{M12} \\ C_{M12} & C_{D12} & C_{M12} \\ C_{M12} & C_{M12} & C_{D12} \end{bmatrix} \tag{3.199}$$

$$[Y_{21}] = jw\begin{bmatrix} C_{D12} & C_{M12} & C_{M12} \\ C_{M12} & C_{D12} & C_{M12} \\ C_{M12} & C_{M12} & C_{D12} \end{bmatrix} \tag{3.200}$$

$$[Y_{22}] = jw\begin{bmatrix} C_{D2} & C_{M2} & C_{M2} \\ C_{M2} & C_{D2} & C_{M2} \\ C_{M2} & C_{M2} & C_{D2} \end{bmatrix} \tag{3.201}$$

Example 3.5

Calculate the positive, negative, and zero sequence susceptances for the line of example 3.4 without a ground wire and with a ground wire and compare the results got with the ATPdraw program.

SOLUTION:

From example 3.4:

$$h_A = 10.3667\ m$$

$$h_B = h_C = 8.3667\ m$$

$$d_{AB} = 2.0\ m$$

$$d_{BC} = 3.4\ m$$

$$d_{AC} = 3.9446\ m$$

$$S_{AC'} = 19.0394\ m$$

$$S_{BC'} = 17.0753\ m$$

$$S_{AB'} = 18.7334\ m$$

$$h_r = 11.0\ m$$

$$d_{AR} = 1.8141\ m$$

$$d_{BR} = d_{CR} = 3.1344\ m$$

$$S_{AR'} = 21.4342\ m$$

$$S_{BR'} = S_{CR'} = 19.4412\ m$$

From cable diameter type 477 CAA - 26/7 MCM, we have:

$$r = \frac{D}{2} = \frac{0.0217932}{2} = 0.0108966\ m$$

$$D_{SSA} = D_{SSB} = D_{SSC} = r = 0.0108966\ m$$

From the diameter of 7-wire galvanized steel cable, we get:

$$D_{SSR} = r_{para-raios} = 0.0047625\ m$$

Using equations (3.172) and (3.173), we have:

$$A_{kk} = 17.975103.10^6 ln\left(\frac{2h_m}{D_{SS}}\right)\ km/F$$

$$A_{km} = 17.975103.10^6 ln\left(\frac{S_{mn'}}{d_{mn}}\right)\ km/F$$

Self-potential coefficients:

Numbers with more than four decimal places were calculated using MATLAB.

$$A_{aa} = 17.975103.10^6 ln\left(\frac{2h_a}{D_{SSA}}\right) = 17.975103.10^6 ln\left(\frac{2x10.36666667}{0.0108966}\right)$$

$$= 1.357308507538349x10^8\ km/F$$

$$A_{bb} = A_{cc} = 17.975103.10^6 ln\left(\frac{2h_b}{D_{SSB}}\right) = 17.975109.10^6 ln\left(\frac{2x8.36666667}{0.0108966}\right)$$

$$= 1.318780676626714x10^8\ km/F$$

$$A_{rr} = 17.975109.10^6 ln\left(\frac{2h_r}{D_{SSR}}\right) = 17.975103.10^6 ln\left(\frac{2x11.0}{0.0047625}\right)$$

$$= 1.516743683457544x10^8 \ km/F$$

Mutual potential coefficients:

$$A_{ab} = 17.975103.10^6 ln\left(\frac{S_{AB'}}{d_{AB}}\right) = 17.975103.10^6 ln\left(\frac{18.7334}{2}\right)$$

$$= 0.4021319664132173x10^8 \ km/F$$

$$A_{ac} = 17.975103.10^6 ln\left(\frac{S_{AC'}}{d_{AC}}\right) = 17.975103.10^6 ln\left(\frac{19.0394}{3.9446}\right)$$

$$= 0.2829574136769197x10^8 \ km/F$$

$$A_{bc} = 17.975103.10^6 ln\left(\frac{S_{BC'}}{d_{BC}}\right) = 17.975103.10^6 ln\left(\frac{17.0753}{3.4}\right)$$

$$= 0.2900925556682357x10^8 \ km/F$$

$$A_{ar} = 17.975103.10^6 ln\left(\frac{S_{AR'}}{d_{AR}}\right) = 17.975103.10^6 ln\left(\frac{21.4342}{1.8141}\right)$$

$$= 0.4438768879504780x10^8 \ km/F$$

$$A_{br} = A_{cr} = 17.975103.10^6 ln\left(\frac{S_{BR'}}{d_{BR}}\right) = 17.975103.10^6 ln\left(\frac{19.4412}{3.1344}\right)$$

$$= 0.3280378568326689x10^8 \ km/F$$

The potential matrices without ground wires and with ground wire are as follows:

$$\left[A_{without\ ground\ wire}\right] =$$

$$10^8 \begin{bmatrix} 1.357308507538349 & 0.4021319664132173 & 0.2829574136769197 \\ 0.4021319664132173 & 1.318780676626714 & 0.2900925556682357 \\ 0.2829574136769197 & 0.2900925556682357 & 1.318780676626714 \end{bmatrix} km/F$$

$$\left[A_{with\ ground\ wire}\right] =$$

$$\left[\begin{array}{ccc|c} 1.357308507538349 & 0.4021319664132173 & 0.2829574136769197 & 0.4438768879504780 \\ 0.4021319664132173 & 1.318780676626714 & 0.2900925556682357 & 0.3280378568326689 \\ 0.2829574136769197 & 0.2900925556682357 & 1.318780676626714 & 0.3280378568326689 \\ -- & -- & -- & -- \\ 0.4438768879504780 & 0.3280378568326689 & 0.3280378568326689 & 1.516743683457544 \end{array}\right] 10^8 \ km/F$$

$$\left[A_{with\ ground\ wire\ reduced}\right] = \begin{bmatrix} 1.2274 & 0.3061 & 0.1870 \\ 0.3061 & 1.2478 & 0.2191 \\ 0.1870 & 0.2191 & 1.2478 \end{bmatrix} 10^8\ km/F$$

The capacitance matrices without ground wire and with ground wire are:

$$\left[C_{without\ ground\ wire}\right] = 10^{-8} \begin{bmatrix} 0.8302 & -0.2248 & -0.1287 \\ -0.2248 & 0.8577 & -0.1404 \\ -0.1287 & -0.1404 & 0.8168 \end{bmatrix} F/km$$

$$\left[C_{with\ ground\ wire}\right] = 10^{-8} \begin{bmatrix} 0.8790 & -0.1986 & -0.0968 \\ -0.1986 & 0.8718 & -0.1233 \\ -0.0968 & -0.1233 & 0.8376 \end{bmatrix} F/km$$

Impedance matrix with elements obtained with the mean:
Using equations (3.181) and (3.182), we have:
Capacitances without ground wire:

$$C_D = \frac{(0.8302 + 0.8577 + 0.8168)10^{-8}}{3} = 0.8349x10^{-8}\ F/km$$

$$C_M = \frac{(-0.2248 - 0.1287 - 0.1404)10^{-8}}{3} = -0.1646x10^{-8}\ F/km$$

Capacitances with ground wire:

$$C_D = \frac{(0.8790 + 0.8718 + 0.8376)10^{-8}}{3} = 0.8628x10^{-8}\ F/km$$

$$C_M = \frac{(-0.1986 - 0.0968 - 0.1233)10^{-8}}{3} = -0.1396x10^{-8}\ F/km$$

Sequence admittances without ground wire:

$$\left[Y_{012}\right] = j2\pi 60 \frac{1}{3} \begin{bmatrix} 1 & 1 & 1 \\ 1 & a & a^2 \\ 1 & a^2 & a \end{bmatrix} x$$

$$10^{-8} \begin{bmatrix} 0.8349 & -0.1646 & -0.1646 \\ -0.1646 & 0.8349 & -0.1646 \\ -0.1646 & -0.1646 & 0.8349 \end{bmatrix} x \begin{bmatrix} 1 & 1 & 1 \\ 1 & a^2 & a \\ 1 & a & a^2 \end{bmatrix}$$

$$\left[Y_{012}\right] = \begin{bmatrix} j1.906 & 0 & 0 \\ 0 & j3.768 & 0 \\ o & 0 & j3.768 \end{bmatrix} 10^{-6}\ S/km$$

TABLE 3.5

Sequence Admittances

Admittances S/km	Without GW	With GW CPR	ATPdraw LCC
Zero	1.906×10^{-6}	2.200×10^{-6}	2.18792×10^{-6}
Positive	3.768×10^{-6}	3.779×10^{-6}	3.77747×10^{-6}
Negative	3.768×10^{-6}	3.779×10^{-6}	3.77747×10^{-6}

Sequence admittances with ground wire:

$$[Y_{012}] = j2\pi60\frac{1}{3}\begin{bmatrix} 1 & 1 & 1 \\ 1 & a & a^2 \\ 1 & a^2 & a \end{bmatrix} x$$

$$10^{-8}\begin{bmatrix} 0.8628 & -0.1396 & -0.1396 \\ -0.1396 & 0.8628 & -0.1396 \\ -0.1396 & -0.1396 & 0.8628 \end{bmatrix} x \begin{bmatrix} 1 & 1 & 1 \\ 1 & a^2 & a \\ 1 & a & a^2 \end{bmatrix}$$

$$[Y_{012}] = \begin{bmatrix} j2.200 & 0 & 0 \\ 0 & j3.779 & 0 \\ o & 0 & j3.779 \end{bmatrix} 10^{-6}\ S/km$$

ATPdraw results:

```
Sequence        Surge impedance          Attenuation   velocity    Wavelength   Resistance    Reactance    Susceptance
          magnitude(ohm) angle(degr.)      db/km        km/sec        km          ohm/km        ohm/km        mho/km
   Zero :  8.87486E+02 -7.14196E+00  2.09689E-03  1.95669E+05  3.26115E+03  4.25178E-01  1.67000E+00  2.18792E-06
Positive:  3.49714E+02 -9.23991E+00  1.84242E-03  2.89127E+05  4.81879E+03  1.46435E-01  4.38162E-01  3.77747E-06
```

Table 3.5 shows a comparison between the results obtained for zero, positive, and negative sequence impedances.

The methodology presented with GW has slight differences from ATPdraw results: $erro_{admitzero} = 0.55\%$; $erro_{admitpos} = 0.041\%$. Therefore, errors are negligible.

3.11 Problems

3.11.1 A 60-kHz 230-kV three-phase TL has a double circuit with one conductor per phase, as shown in Figure 3.36, where the distances between the conductors and the corresponding heights in meters are indicated. The single conductor cables are type 636 CAA - 26/7 MCM, and the ground wire is made of 7-wire HS galvanized steel with a 3/8″ nominal diameter. The conductor arrow is 2.0 m

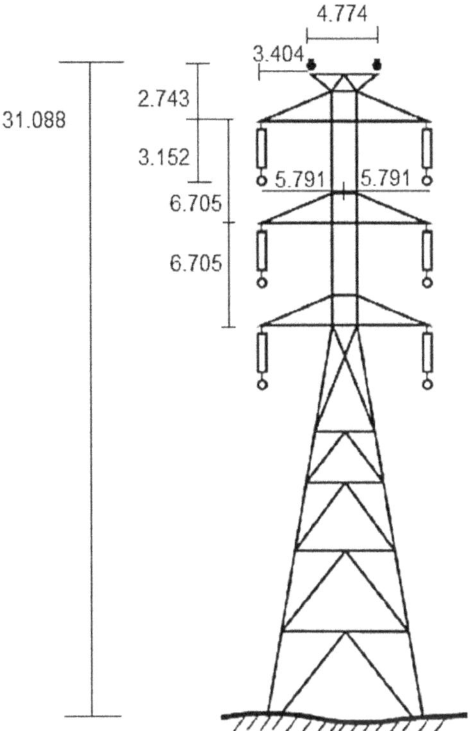

FIGURE 3.36
Double-circuit tower.

and the arrester cable arrow is 1.5 m. The soil resistivity is $\rho = 439$ Ω-m, and the temperature is 50°C.

Calculate positive, negative, and null sequence impedances using Carson's full method and ATPdraw's LCC routine.

Answers:

$Z_0 = 0.275106 + j1.72000\ \Omega/km$; $Z_+ = 0.101171 + j0.505599\ \Omega/km$; $Z_- = Z_+$.

3.11.2 Calculate the zero, positive, and negative sequence susceptances for the line in example 3.11.1.

Answers: $Y_0 = 2.22696\ \mu mho/km$; $Y_+ = 3.28860\ \mu mho/km$; $Y_- = Y_+$.

3.11.3 A 60-kHz 500-kV three-phase TL has a single horizontal circuit with four 46 cm-spaced conductors per phase, as shown in Figure 3.37, where the distances between the conductors and the corresponding heights in meters are indicated. The single conductor cables are of type 636 CAA - 26/7 MCM and the ground wire is made of 7-wire EHS galvanized steel with a nominal diameter 3/8″. The conductor arrow is 2.0 m and the arrester cable arrow is 1.5 m. The soil resistivity is $\rho = 1500\ \Omega$-m, and the temperature is 50°C.

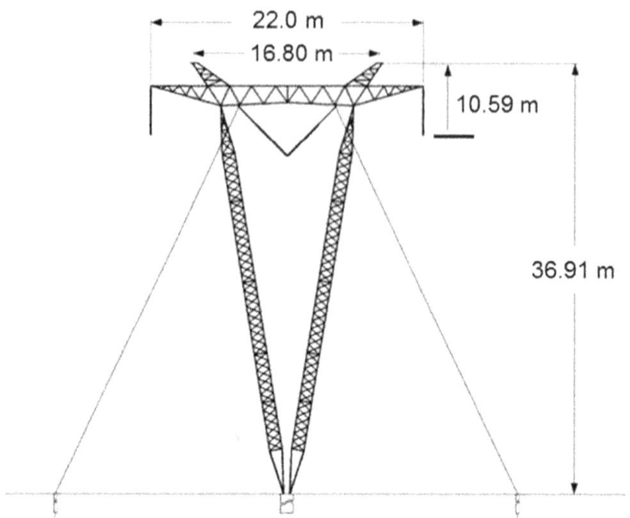

FIGURE 3.37
Tower with cable-stayed structure.

Calculate positive, negative, and null sequence impedances using Carson's full method and ATPdraw LCC routine.
Answers:
$Z_0 = 0.200020 + j1.56228\ \Omega/km$; $Z_+ = 0.0253086 + j0.32173\ \Omega/km$; $Z_- = Z_+$.

3.11.4 Calculate the zero, positive, and negative sequence susceptances for the line in example 3.11.3.

Answers: $Y_0 = 2.95202\ \mu mho/km$; $Y_+ = 5.10769\ \mu mho/km$; $Y_- = Y_+$.

3.11.5 In power transmission lines with high voltages of the order of 500 kV, the approximate series reactance is:

a. Same as series resistance

b. 10 to 20 times lower than the series resistance

c. Double the series resistance

d. 20 to 30 times greater than the series resistance

e. Half of the resistance series

Answer: d.

3.11.6 The geometric mean distance is a quantity used in studies of:

a. Synchronous machines

b. Transmission lines

c. Asynchronous machines

d. DC machines

e. Transformers

Answer: b.

3.11.7 The skin effect occurs on transmission lines. The same:

a. Occurs when the conductor is traversed by a direct current

b. Occurs in poorly grounded systems

c. Occurs on lines with impedance mismatch

d. Occurs when the conductor is crossed by an AC

Answer: d.

3.11.8 The phase-to-ground capacitance of a transposed three-phase transmission line is equal to:

a. Positive sequence capacitance of the circuit

b. Circuit zero sequence capacitance

c. Sum of positive and zero sequence capacitance

d. Difference between positive sequence and zero capacitance

e. Mean between positive and zero sequence capacitance

Answer: b.

3.11.9 For a given transmission line, the impedance matrix per unit length for a portion of the transmission line is given by:

$$j\begin{bmatrix} 0.5 & 0.25 & 0.2 \\ 0.25 & 0.6 & 0.15 \\ 0.2 & 0.15 & 0.55 \end{bmatrix} \Omega/km$$

Positive and zero sequence impedances (Ω/km) are given respectively by:

a. j0,95 e j0,35

b. j0,35 e j0,95

c. j0,55 e j0,35

d. j0,55 e j0,2

e. j0,35 e j0,2

Answer: d.

3.11.10 A 120 km transmission line has its self-serial impedance equal to 0.02 + j0.05 [Ω/km] and mutual impedance between phases of j0.02 [Ω/km]. The direct sequence impedance for this line, in ohms, is:

a. 2.4 + j3.6

b. 2.4 + 6.0

c. 1.2 + j3.6

d. 1.2 + 6.0

e. 1.2 + j8.4

Answer: a.

4

Operation of Transmission Lines at Steady-State

4.1 Introduction

For the calculation of tensions and currents, at the beginning and the end of transmission lines (TLs), in steady-state, it is necessary to develop models that represent TLs. These models are obtained for short-length, medium-length, and long-length TLs.

4.2 General Equations for TLs

The general equations of a TL are developed using a distributed parameter circuit.

4.2.1 Circuit of Distributed Parameters

A circuit of distributed parameters is a finite-length circuit in which any constituent element, however small, causes a voltage variation in the longitudinal direction and a current branch in the transverse direction. Figure 4.1 shows this circuit.

In practice, all real circuits are distributed parameters. But it is possible to neglect the transverse current derivation, due to the value of the current through the longitudinal elements, or the variation of the longitudinal voltage, because of the potential difference between the conductors of the circuit. In the first case, the current that crosses the longitudinal elements of the circuit is the same, being possible to replace these elements by an equivalent element, as shown in Figure 4.2.

In the second case, the simplification allows to consider that all constituent elements of the circuit are in parallel, subjected to the same voltage, and can be replaced by a single equivalent element, as shown in Figure 4.3.

In both cases, the circuits are called concentrated parameter circuits.

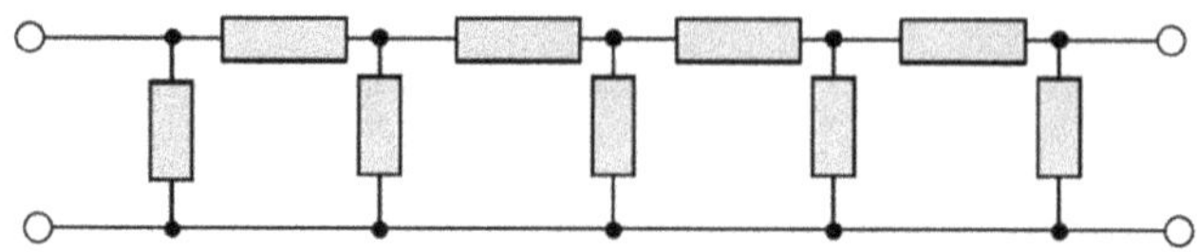

FIGURE 4.1
Circuit with distributed parameters.

FIGURE 4.2
Reduced longitudinal elements.

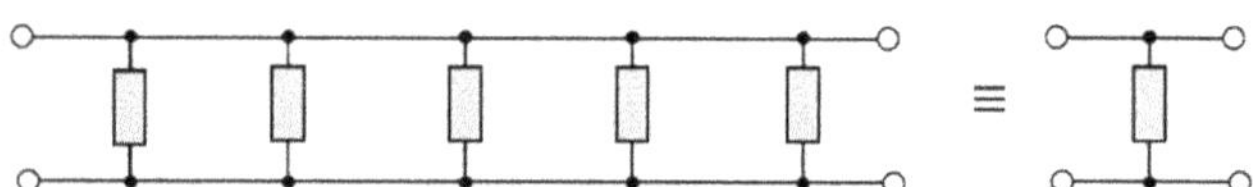

FIGURE 4.3
Reduced transverse elements.

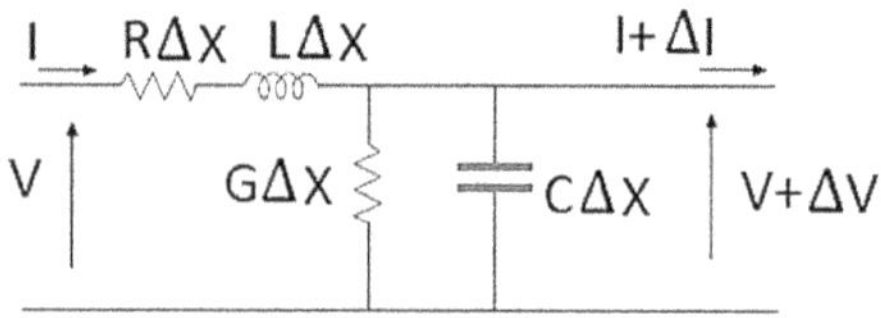

FIGURE 4.4
TL infinitesimal circuit.

4.2.2 General Propagation Equations of a TL

The distributed parameters of a TL are the resistance R (Ω / km), the inductance L (H/km), the conductance (S/km), and the capacitance (F/km).

We consider an infinitesimal element of a TL of length Δx, as shown in Figure 4.4.

The instantaneous magnitudes of voltage and current are functions of distance and time, that is:

$$i = i(t, x) \tag{4.1}$$

$$v = v(t, x) \tag{4.2}$$

Applying the law of Kirchhoff tensions to the circuit of Figure 4.4 and using partial derivative for the terms of the voltage as a function of two variables, we get:

$$v = (R\Delta x)i + (L\Delta x)\frac{\partial i}{\partial t} + v + \Delta v \tag{4.3}$$

Isolating Δv, it comes:

$$\Delta v = -(R\Delta x)i - (L\Delta x)\frac{\partial i}{\partial t} \tag{4.4}$$

Applying the law of Kirchhoff currents to the circuit of Figure 4.4, we have:

$$i = (G\Delta x)(v + \Delta v) + (C\Delta x)\frac{\partial(v + \Delta v)}{\partial t} + i + \Delta i \tag{4.5}$$

Isolating Δi, it comes:

$$\Delta i = -(G\Delta x)(v + \Delta v) - (C\Delta x)\frac{\partial(v + \Delta v)}{\partial t} \tag{4.6}$$

Substituting the value of Δv given in equation (4.4), it follows:

$$\Delta i = -(G\Delta x)v + (G\Delta x^2)\left(Ri + L\frac{\partial i}{\partial t}\right) - (C\Delta x)\frac{\partial v}{\partial t} + (C\Delta x^2)\left(Ri\frac{\partial i}{\partial t} + L\frac{\partial^2 i}{\partial t^2}\right) \tag{4.7}$$

Equations (4.4) and (4.7) are divided by Δx and then the limit is set when $\Delta x \to 0$. Therefore, from equation (4.4), we have:

$$\lim_{\Delta x \to 0}\left(\frac{\Delta v}{\Delta x}\right) = -Ri - L\frac{\partial i}{\partial t} \tag{4.8}$$

$$\frac{\partial v}{\partial x} = -Ri - L\frac{\partial i}{\partial t} \tag{4.9}$$

And of equation (4.7), it comes:

$$\lim_{\Delta x \to 0}\left(\frac{\Delta i}{\Delta x}\right) = \lim_{\Delta x \to 0} -(Gv) + (G\Delta x)\left(Ri + L\frac{\partial i}{\partial t}\right) - C\frac{\partial v}{\partial t} + (C\Delta x)\left(Ri\frac{\partial i}{\partial t} + L\frac{\partial^2 i}{\partial t^2}\right) \tag{4.10}$$

$$\frac{\partial i}{\partial x} = -(Gv) - C\frac{\partial v}{\partial t} \tag{4.11}$$

The fundamental propagation equations are then equations (4.9) and (4.11).

The equations of the telegraphists are obtained by deriving once again the equations (4.9) and (4.11).

$$\frac{\partial^2 v}{\partial x^2} = -R\frac{\partial i}{\partial x} - L\frac{\partial}{\partial t}\left(\frac{\partial i}{\partial x}\right) \tag{4.12}$$

$$\frac{\partial^2 i}{\partial x^2} = -G\frac{\partial v}{\partial x} - C\frac{\partial}{\partial t}\left(\frac{\partial v}{\partial x}\right) \tag{4.13}$$

Substituting (4.11) into (4.12), and (4.9) into (4.13), we have:

$$\frac{\partial^2 v}{\partial x^2} = -R\left[-(Gv) - C\frac{\partial v}{\partial t}\right] - L\frac{\partial}{\partial t}\left[-(Gv) - C\frac{\partial v}{\partial t}\right]$$

$$= RGv + (RC + LG)\frac{\partial v}{\partial t} + LC\frac{\partial^2 v}{\partial t^2} \tag{4.14}$$

$$\frac{\partial^2 i}{\partial x^2} = -G\left[-Ri - L\frac{\partial i}{\partial t}\right] - C\frac{\partial}{\partial t}\left[-Ri - L\frac{\partial i}{\partial t}\right]$$

$$= RGi + (RC + LG)\frac{\partial i}{\partial t} + LC\frac{\partial^2 i}{\partial t^2} \tag{4.15}$$

The system of differential equations (4.14) and (4.15) is general for a transmission line of distributed parameters.

4.3 Equations of the TL at Steady-State

The solution of equations (4.14) and (4.15) in the frequency domain (time dependence is implicit) takes the form of ordinary differential equations. Figure 4.5 shows the schematic diagram of the TL.

The direction in which x grows is not important, since the variable x is eliminated during the solution of the equations, when V_S and I_S are expressed in terms of V_R and I_R.

We consider the phasors:

$$V = Ve^{(j\omega t)} \tag{4.16}$$

$$I = Ie^{(j\omega t+\theta)} \tag{4.17}$$

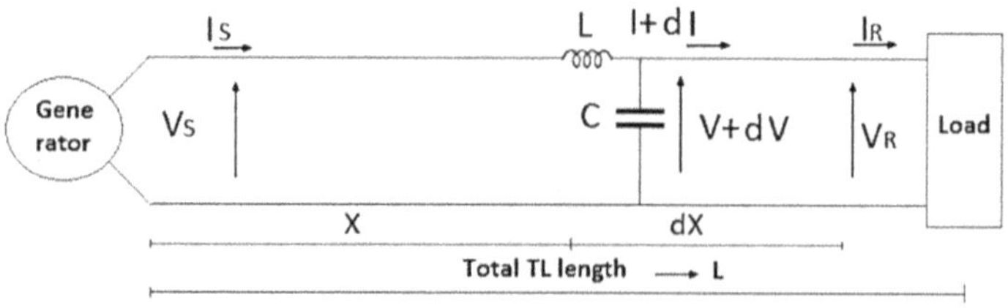

FIGURE 4.5
Schematic diagram of the TL.

derivatives of which are given by:

$$\frac{d\mathbf{V}}{dt} = V\left(e^{(jwt)}\right) jw = \mathbf{V} jw \tag{4.18}$$

$$\frac{d\mathbf{I}}{dt} = I\left(e^{(jwt+\theta)}\right) jw = \mathbf{I} jw \tag{4.19}$$

$$\frac{d^2\mathbf{V}}{dt^2} = V\left(e^{(jwt)}\right)(jw)^2 = \mathbf{V}(jw)^2 \tag{4.20}$$

$$\frac{d^2\mathbf{I}}{dt^2} = I\left(e^{(jwt+\theta)}\right)(jw)^2 = \mathbf{I}(jw)^2 \tag{4.21}$$

The differential equations are given by:

$$\frac{d^2\mathbf{V}}{dx^2} = RG\mathbf{V} + (RC + LG)\frac{d\mathbf{V}}{dt} + LC\frac{d^2\mathbf{V}}{dt^2} \tag{4.22}$$

$$\frac{d^2\mathbf{I}}{dx^2} = RG\mathbf{I} + (RC + LG)\frac{d\mathbf{I}}{dt} + LC\frac{d^2\mathbf{I}}{dt^2} \tag{4.23}$$

Substituting equations (4.18) and (4.20) into equation (4.22), and equations (4.19) and (4.21) into equation (4.23), we get:

$$\frac{d^2\mathbf{V}}{dx^2} = RG\mathbf{V} + (RC + LG)\mathbf{V}jw + LC\mathbf{V}(jw)^2 = (R + jwL)(G + jwC)\mathbf{V} = zy\mathbf{V} \tag{4.24}$$

$$\frac{d^2\mathbf{I}}{dx^2} = RG\mathbf{I} + (RC + LG)\mathbf{I}jw + LC\mathbf{I}(jw)^2 = (R + jwL)(G + jwC)\mathbf{I} = zy\mathbf{I} \tag{4.25}$$

The solutions of the differential equations must be expressions that differentiate twice regarding x resulting in the original expression multiplied by zy. This suggests an exponential form of solution. Suppose the solution of equation (4.24) is:

$$V(x) = A_1 e^{x\sqrt{zy}} + A_2 e^{-x\sqrt{zy}} \tag{4.26}$$

When calculating the second derivative regarding x, we obtain:

$$\frac{d^2V(x)}{dx^2} = zy\left(A_1 e^{x\sqrt{zy}} + A_2 e^{-x\sqrt{zy}}\right) \tag{4.27}$$

Therefore, equation (4.26) is the solution of equation (4.24).

Equation (4.9) in the frequency domain can be written as:

$$\frac{dV(x)}{dx} = -RI - L\frac{dI}{dt} \tag{4.28}$$

But from equation (4.19), with x growing from the load to the generator, it comes:

$$\frac{dV(x)}{dx} = RI + LI(x)jw = I(x)\left(R + jwL\right) = zI(x) \tag{4.29}$$

Deriving equation (4.26), we have:

$$\frac{dV(x)}{dx} = \sqrt{ZY}A_1e^{x\sqrt{zy}} - \sqrt{ZY}A_2e^{-x\sqrt{zy}} \tag{4.30}$$

Equating the equations (4.29) and (4.30), it comes:

$$zI(x) = \sqrt{ZY}A_1e^{x\sqrt{zy}} - \sqrt{ZY}A_2e^{-x\sqrt{zy}} \tag{4.31}$$

Therefore:

$$I(x) = \frac{1}{\sqrt{\frac{Z}{Y}}}A_1e^{x\sqrt{zy}} - \frac{1}{\sqrt{\frac{Z}{Y}}}A_2e^{-x\sqrt{zy}} \tag{4.32}$$

To calculate A_1 and A_2, we do $x = 0 \rightarrow V(x) = V_R$ and $I(x) = I_R$ in the equations (4.26) and (4.32). Soon,

$$V_R = A_1 + A_2 \tag{4.33}$$

$$I_R = \frac{A_1 - A_2}{\sqrt{\frac{z}{y}}} \tag{4.34}$$

Solving the system of equations (4.33) and (4.34), we have:

$$A_1 = \frac{V_R + I_R Z_C}{2} \tag{4.35}$$

$$A_2 = \frac{V_R - I_R Z_C}{2} \tag{4.36}$$

Where Z_C is called the characteristic line impedance (CLI).

$$Z_C = \sqrt{\frac{z}{y}}(\Omega) \tag{4.37}$$

Substituting equations (4.35) and (4.36) into (4.26) and (4.32), we get:

$$\gamma = \sqrt{zy} = \alpha + j\beta \ (1/m) \tag{4.38}$$

where γ is called the propagation constant (PC): the real part of the propagation constant α is called the attenuation constant and the imaginary part β is called the phase constant.

We have:

$$V(x) = \frac{V_R + I_R Z_C}{2} e^{x\gamma} + \frac{V_R - I_R Z_C}{2} e^{-x\gamma} \tag{4.39}$$

$$I(x) = \frac{\frac{V_R}{Z_C} + I_R}{2} e^{x\gamma} - \frac{\frac{V_R}{Z_C} - I_R}{2} e^{-x\gamma} \tag{4.40}$$

Thus, with the line parameters, voltage and current at the line receiver being known, equations (4.39) and (4.40) calculate the voltage and current effective values in modulus and angle, at any specific point in the line, at a distance from the receiver to the specific point.

The first term of equation (4.39) $\frac{V_R + I_R Z_C}{2} e^{x\alpha} e^{xj\beta}$ grows in magnitude and advances in phase as the distance increases from the receiving end. When we consider the distance increasing from the emitter end, the same term decreases in modulo and delays in phase. This characteristic is of a traveling wave. The first term is called incident voltage. The second term of equation (4.39) $\frac{V_R - I_R Z_C}{2} e^{-x\alpha} e^{-xj\beta}$ decreases in modulo and delays in phase as the distance increases from the receiving end and is called the reflected voltage. Similarly, the current equation has the terms of incident current $\frac{\frac{V_R}{Z_C} + I_R}{2} e^{x\alpha} e^{xj\beta}$ and reflected current $-\frac{\frac{V_R}{Z_C} - I_R}{2} e^{x\alpha} e^{xj\beta}$.

The voltage and current equations at the beginning of the TL are:

$$V_S = \frac{V_R + I_R Z_C}{2} e^{l\gamma} + \frac{V_R - I_R Z_C}{2} e^{-l\gamma} \tag{4.41}$$

$$I_S = \frac{\frac{V_R}{Z_C} + I_R}{2} e^{l\gamma} - \frac{\frac{V_R}{Z_C} - I_R}{2} e^{-l\gamma} \tag{4.42}$$

4.3.1 Analysis of CLI and PC

The characteristic impedance is independent of the TL length, since impedance and admittance are values per unit length.

But

$$Z = R + jwL = |Z| \angle \varphi_z = \sqrt{R^2 + (wL)^2} \angle \text{arctg} \frac{wL}{R} \tag{4.43}$$

and

$$Y = G + jwC = |Y|\angle\varphi_y = \sqrt{G^2 + (wC)^2}\angle\text{arctg}\frac{wC}{G} \quad (4.44)$$

give:

$$Z_C = |Z_C|\angle\varphi_C = \sqrt[4]{\frac{R^2 + (wL)^2}{G^2 + (wC)^2}}\angle(\varphi_z - \varphi_y)/2 \quad (4.45)$$

The characteristic impedance module and phase are characterized by typical values. In balanced operation, three-phase overhead lines have characteristic impedance modulus in the range of 250 Ω to 450 Ω.

The value of the angle φ_C is always small and negative, since φ_y is very close to $\pi/2$, since G almost always is negligible with respect to φ_C. Thus, we can approximate the angle φ_C as:

$$\varphi_C = -\frac{\left(\frac{\pi}{2} - \varphi_z\right)}{2} \quad (4.46)$$

For underground cables, the proximity of the conductors causes L to decrease and C to increase and the existence of a solid dielectric in the cable increases the parameter C even more. Typical characteristic impedance values in this case range from 30 Ω to 60 Ω.

The PC, in terms of the line parameters, is given by:

$$\gamma = \sqrt{zy} = \alpha + j\beta = \sqrt{(R + jwL)(G + jwC)} = |\gamma|\angle\theta_\gamma \quad (4.47)$$

Therefore:

$$\gamma = \sqrt[4]{(R^2 + (wL)^2)(G^2 + (wC)^2)}\angle(\varphi_z + \varphi_y)/2 \quad (4.48)$$

We can write α *and* β as:

$$\alpha = \gamma cos\theta_\gamma \quad (4.49)$$

$$\beta = \gamma sen\theta_\gamma \quad (4.50)$$

For the calculation of the typical values, we must calculate α and β as a function of the line parameters.

From equation (4.47), we have:

$$\begin{aligned}\gamma^2 &= (\alpha + j\beta)^2 = \alpha^2 + j2\alpha\beta + (j\beta)^2 = (R + jwL)(G + jwC) \\ &= RG + jwRC = JwLG - w^2LC\end{aligned} \quad (4.51)$$

Equating the real and imaginary parts, it comes:

$$\alpha^2 - \beta^2 = RG - w^2LC \tag{4.52}$$

From equations (4.49) and (4.50):

$$\alpha^2 + \beta^2 = \gamma^2 \tag{4.53}$$

Solving the system of equations (4.52) and (4.53), we find the values of α and β.

$$\alpha = \sqrt{\frac{\gamma^2}{2} + \frac{RG - w^2LC}{2}} \tag{4.54}$$

$$\beta = \sqrt{\frac{\gamma^2}{2} - \frac{RG - w^2LC}{2}} \tag{4.55}$$

Considering the value of $|\gamma| = \sqrt[4]{(R^2 + (wL)^2)(G^2 + (wC)^2)}$ of equation (4.48) and considering the negligible conductance, we have:

$$\gamma^2 = wc\sqrt{R^2 + (wL)^2} = w^2LC\sqrt{1 + \frac{R^2}{(wL)^2}} \tag{4.56}$$

Recalling that a binomial series is of the form:

$$(1+x)^k = 1 + kx + \frac{k(k-1)}{2!}x^2 + \cdots + \frac{k(k-1)\ldots(k-n+1)}{n!}x^n + \cdots \tag{4.57}$$

Where $|x| < 1$, for every real number k.

Therefore, as in TLs $\frac{R}{wl} < 1$, the term of the square root can be developed in a binomial series.

$$\left(1 + \frac{R^2}{(wL)^2}\right)^{1/2} = 1 + \frac{1}{2}\frac{R^2}{(wL)^2} + \cdots \tag{4.58}$$

Neglecting the terms of the series larger than the second, equation (4.56) can be written as:

$$\gamma^2 = w^2LC\left(1 + \frac{R^2}{2(wL)^2}\right) = w^2LC + \frac{CR^2}{2L} \tag{4.59}$$

Substituting the value of γ^2 into equations (4.54) and (4.55) and neglecting G results in equation (4.60).

$$\alpha \cong \sqrt{\frac{R^2C}{4L}+\frac{w^2LC}{2}-\frac{w^2LC}{2}}=\frac{R}{2}\sqrt{\frac{C}{L}}\cong\frac{R}{2Z_C}\left(\frac{nepers}{m}\right) \tag{4.60}$$

$$\beta \cong \sqrt{\frac{w^2LC}{2}+\frac{w^2LC}{2}}=w\sqrt{LC}\ (rad/m) \tag{4.61}$$

From equation (4.61), we can determine the propagation velocity of the waves and the wavelength as:

$$V_0=\lambda f=\frac{1}{\sqrt{LC}}=\frac{w}{\beta}\left(\frac{m}{s}\right) \tag{4.62}$$

where $\lambda = \frac{2\pi}{\beta}$ (m) is the wavelength, which is the distance required to change the voltage or current phase by 2π radians.

The theoretical propagation velocity of an electromagnetic wave in a transmission line is 300.000 km / s.

In overhead transmission lines, as $250 < Z_C < 450$:

$$\frac{R}{2\times 450}<\alpha<\frac{R}{2\times 250} \tag{4.63}$$

and

$$\beta=\frac{2\pi 60}{300000}=1.26x10^{-3}\left(\frac{1}{km}\right) \tag{4.64}$$

For underground cables, we have:

$$\frac{R}{2\ x60}<\alpha<\frac{R}{2\ x30} \tag{4.65}$$

and

$$V_0=\frac{300000}{\sqrt{\varepsilon}}=\frac{300000}{\sqrt{3.6}}\cong 157000\left(\frac{km}{s}\right) \tag{4.66}$$

Where: ε is the absolute permittivity of the dielectric used in the cable (for impregnated paper $\varepsilon \cong 3.6$).

Therefore:

$$\beta=\frac{2\pi 60}{157000}\cong 2.4x10^{-3}\left(\frac{1}{km}\right) \tag{4.67}$$

Example 4.1

An energy transmission line of 500 kV, 60 Hz, with 294.84 km has the following parameters:

$$R = 0.025\frac{\Omega}{\text{km}}$$

$$L = 0.854 \text{ mH/km}$$

$$C = 13.65 \text{ nF/km}$$

$$G = 0.0375 \ \mu \text{ S/km}$$

The TL fulfills a load of 550 MW with a power factor of 0.85 inductive. Calculate the characteristic impedance, propagation constant, attenuation constant, phase constant, propagation velocity, wavelength, the sending-end voltage, and current.

SOLUTION:

Calculation of the TL impedance:

$$Z = R + jwl = \left(0.025 + j377x0.854x10^{-3}\right) = 0.3229\angle 85.5598°\frac{\Omega}{km} = 7.371 + j94.9261\ \Omega = 95.2118\angle 85.5598°\ \Omega$$

Calculation of the TL admittance:

$$Y = G + jwC = \left(0.0375\ x\ 10^{-6} + j377\ x\ 0.01365\ x\ 10^{-6}\right) = 0.0000 +\ j5.1553x10^{-6}\ S/km = 0.00152\angle 90°\ S$$

Calculation of the characteristic impedance:

$$Z_C = \sqrt{\frac{Z}{Y}} = \sqrt{\frac{95.2118\angle 85.5598°}{0.00152\angle 90°}} = 250.0906 - j9.6954 = 250.2785\angle -2.2201°\ \Omega$$

Calculation of propagation, attenuation, and phase constants:

$$\gamma = \alpha + j\beta = \sqrt{zy} = \sqrt{(0.3229\angle 85.5598°)\ (j5.1553x10^{-6})} = 1.2902x10^{-3}\angle 87.7799° = 0.04998x10^{-3} + j1.2892x10^{-3}\left(\frac{1}{km}\right)$$

$$\gamma l = 0.3804\angle 87.7799° = 0.0147 + j0.3801\ pu$$

Calculation of speed and wavelength:

$$\lambda = \frac{2\pi}{\beta} = \frac{2\pi}{1.2892x10^{-3}} = 4873.7087\ km$$

$$V_0 = \lambda f = 4873.7087x60 = 292422.522\ km/s$$

Calculation of the sending-end voltage and current:

$$V_R = \frac{500}{\sqrt{3}} \angle 0° = 288675.1346\ V$$

$$I_R = \frac{550x10^6}{500x10^3\sqrt{3}x0.85} \angle 0° - arccos 0.85 = 635.0855 - j393.5902$$
$$= 747.1592 \angle -31.7883°\ A$$

$$V_S = V(l) = \frac{V_R + I_R Z_C}{2} e^{\gamma l} + \frac{V_R - I_R Z_C}{2} e^{-\gamma l}$$
$$= \frac{288675.1346 + (250.0906 - j9.6954)(635.0855 - j393.5902)}{2} e^{(0.0147 + j0.3801)}$$
$$+ \frac{288675.1346 - (250.0906 - j9.6954)(635.0855 - j393.5902)}{2} e^{-(0.0147 + j0.3801)}$$
$$= 309025.480643 + j57664.7278 = 314359.6166 \angle 10.5699°\ V$$

Line voltage is calculated as:

$$V_{SLL} = 314359.6166\sqrt{3} \angle 10.5699° = 544486.8278 \angle 10.5699°\ V$$

$$I_S = I(l) = \frac{\frac{V_R}{Z_C} + I_R}{2} e^{\gamma l} - \frac{\frac{V_R}{Z_C} - I_R}{2} e^{-\gamma l}$$
$$= \frac{\left(\left(\frac{288675.1346}{(250.0906 - j9.6954)}\right) + (635.0855 - j393.5902)\right)}{2} e^{(0.0147 + j0.3801)}$$
$$- \frac{\left(\left(\frac{288675.1346}{(250.0906 - j9.6954)}\right) - (635.0855 - j393.5902)\right)}{2} e^{-(0.0147 + j0.3801)}$$
$$= 591.1229 + j66.1932 = 594.8175 \angle 6.3893°\ A$$

4.3.2 Variation of Propagation Parameters with Frequency

The characteristic propagation parameters, which are characteristic impedance, propagation constant, attenuation constant, phase constant, and wave propagation speed, are functions of the frequency. The analysis of these parameters with frequency is relevant for the study of transient lines.

According to equation (4.45) and neglecting the conductance G, the characteristic impedance modulus as a function of frequency is given by equation (4.68).

$$|Z_C| = \sqrt[4]{\left(\frac{\frac{R^2}{(2\pi f)^2} + L^2}{C^2}\right)} \tag{4.68}$$

Therefore, we can conclude that the characteristic impedance decreases slowly with the frequency.

From equation (4.60) $\alpha \cong \frac{R}{2Z_C}$, we can conclude that the attenuation constant increases slowly with frequency.

From equation (4.61) $\beta \cong w\sqrt{LC}$, we can conclude that the phase constant varies almost linearly with frequency.

The propagation speed according to the line and frequency parameters is developed next.

As: $V_o = \frac{w}{\beta}$ and beta is given by equation (4.55), $\beta = \sqrt{\frac{\gamma^2}{2} - \frac{RG - w^2LC}{2}}$, then:

$$V_o = \frac{w}{\sqrt{\frac{\gamma^2}{2} - \frac{RG - w^2LC}{2}}} \tag{4.69}$$

Using equation (4.59) and neglecting terms with G, we have:

$$V_o = \frac{1}{\sqrt{\left(\frac{1}{2}\right)\left(\sqrt{\frac{R^2}{(2\pi f)^2} + L^2}\right)C + LC}} \tag{4.70}$$

We can conclude that the speed of propagation varies slightly with frequency and tends to the value $1/\sqrt{LC}$, when $f \to \infty$.

4.4 Hyperbolic Form of TL Equations

A more convenient way of expressing equations (4.39) and (4.40) for the calculation of voltages and currents in a TL is expressed using the hyperbolic sine and cosine functions. These functions, in exponential form, are given by:

$$senh(\gamma l) = \frac{e^{\gamma x} - e^{-\gamma x}}{2} \tag{4.71}$$

$$cosh(\gamma l) = \frac{e^{\gamma x} + e^{-\gamma x}}{2} \tag{4.72}$$

with:

$$senh(\alpha x + j\beta x) = senh(\alpha x)\cos(\beta x) + jcosh(\alpha x)\,sen(\beta x) \tag{4.73}$$

$$\cosh(\alpha x + j\beta x) = cosh(\alpha x)\cos(\beta x) + jsenh(\alpha x)\,sen(\beta x) \tag{4.74}$$

From equation (4.39), we have:

$$V(x) = \frac{V_R(e^{x\gamma} + e^{-x\gamma})}{2} + \frac{I_R Z_C(e^{x\gamma} - e^{-x\gamma})}{2} = V_R \cosh(\gamma x) + I_R Z_C senh(\gamma x) \quad (4.75)$$

From equation (4.40), we have:

$$I(x) = \frac{\frac{V_R}{Z_C} + I_R}{2} e^{x\gamma} - \frac{\frac{V_R}{Z_C} - I_R}{2} e^{-x\gamma} = \frac{I_R\left(e^{x\gamma} + e^{-x\gamma}\right)}{2} + \frac{V_R}{Z_C}\frac{\left(e^{x\gamma} - e^{-x\gamma}\right)}{2}$$

$$= I_R \cosh(\gamma x) + \frac{V_R}{Z_C} senh(\gamma x) \quad (4.76)$$

When $x = l$, the sending-end voltage and the current are given by:

$$V_S = V_R \cosh(\gamma l) + I_R Z_C senh(\gamma l) \quad (4.77)$$

$$I_S = I_R \cosh(\gamma l) + \frac{V_R}{Z_C} senh(\gamma l) \quad (4.78)$$

4.5 Passive Quadrupole

A circuit or a circuit element having "n" free terminals is called a multipole. A particular case is when the terminals are grouped in pairs, in each of which Kirchhoff's first law is separately verified. Then each pair of terminals is called a port, and the multipole is called multiport.

A passive quadrupole has no dependent or independent sources.

Applying the superposition theorem in the circuit of Figure 4.6, we have:

$$\mathbf{V_1} = Z_{11}\mathbf{I_1} + Z_{12}\mathbf{I_2} \quad (4.79)$$

$$\mathbf{V_2} = Z_{21}\mathbf{I_1} + Z_{22}\mathbf{I_2} \quad (4.80)$$

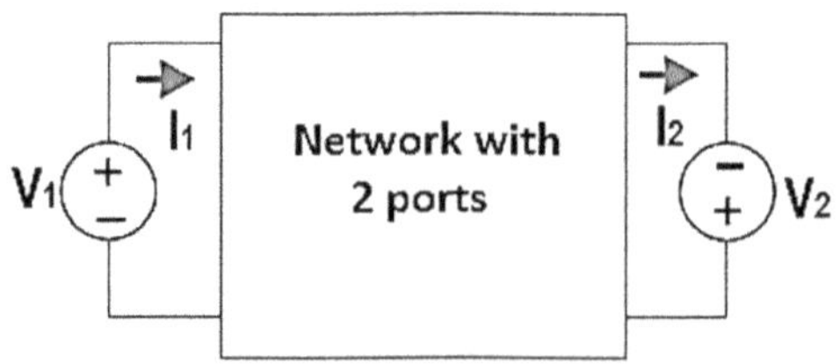

FIGURE 4.6
Passive quadrupole.

Or:

$$V_1 = AI_1 + BI_2 \tag{4.81}$$

$$V_2 = CI_1 + DI_2 \tag{4.82}$$

Putting equations (4.81) and (4.82) in matrix form, we have:

$$\begin{bmatrix} V_1 \\ V_2 \end{bmatrix} = \begin{bmatrix} A & B \\ C & D \end{bmatrix} \begin{bmatrix} I_1 \\ I_2 \end{bmatrix} \tag{4.83}$$

where:

$$A = \frac{V_1}{I_1}\{para\ I_2 = 0\} \tag{4.84}$$

A is the input impedance with the open circuit output.

$$B = \frac{V_2}{I_1}\{para\ I_2 = 0\} \tag{4.85}$$

B is the open circuit direct transfer impedance.

$$C = \frac{V_1}{I_2}\{para\ I_1 = 0\} \tag{4.86}$$

C is the open circuit reverse transfer impedance.

$$D = \frac{V_2}{I_2}\{para\ I_1 = 0\} \tag{4.87}$$

D is the open circuit output impedance.

4.5.1 Association of Quadrupoles

In practice, it is common for a circuit to divide into several interconnected blocks. In this way, it is easy to study each of the constituent parts of the circuit and the set. In Table 4.1, four-wire circuits are shown and the connection forms of two or more quadrilles with the corresponding equations.

Equations (4.77) and (4.78) can be written in terms of constants *A*, *B*, *C*, and *D*, as follows:

$$\begin{bmatrix} V_S \\ I_S \end{bmatrix} = \begin{bmatrix} A & B \\ C & D \end{bmatrix} \begin{bmatrix} V_R \\ I_R \end{bmatrix} \tag{4.88}$$

TABLE 4.1

Association of Quadrupoles

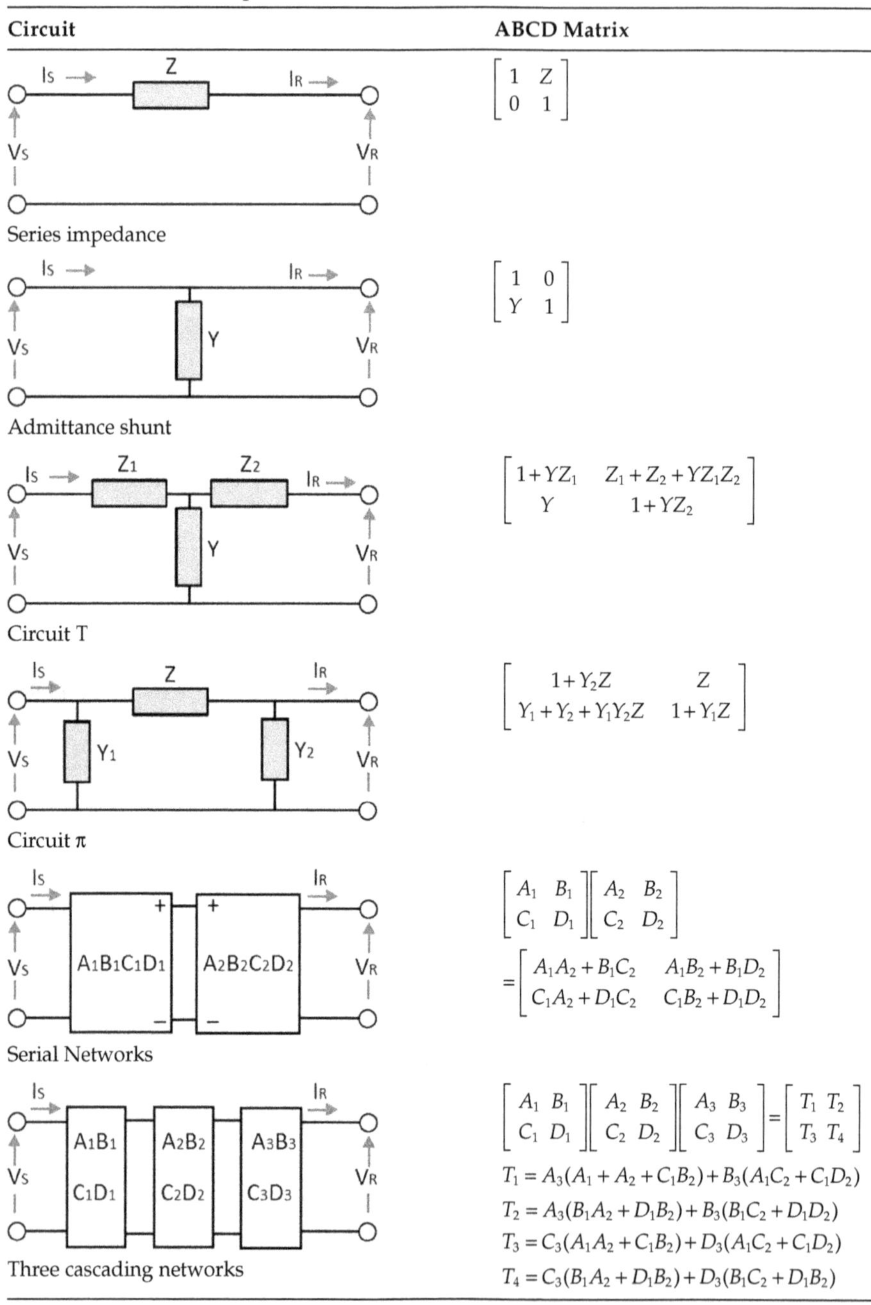

Circuit	ABCD Matrix
Series impedance	$\begin{bmatrix} 1 & Z \\ 0 & 1 \end{bmatrix}$
Admittance shunt	$\begin{bmatrix} 1 & 0 \\ Y & 1 \end{bmatrix}$
Circuit T	$\begin{bmatrix} 1+YZ_1 & Z_1+Z_2+YZ_1Z_2 \\ Y & 1+YZ_2 \end{bmatrix}$
Circuit π	$\begin{bmatrix} 1+Y_2Z & Z \\ Y_1+Y_2+Y_1Y_2Z & 1+Y_1Z \end{bmatrix}$
Serial Networks	$\begin{bmatrix} A_1 & B_1 \\ C_1 & D_1 \end{bmatrix}\begin{bmatrix} A_2 & B_2 \\ C_2 & D_2 \end{bmatrix} = \begin{bmatrix} A_1A_2+B_1C_2 & A_1B_2+B_1D_2 \\ C_1A_2+D_1C_2 & C_1B_2+D_1D_2 \end{bmatrix}$
Three cascading networks	$\begin{bmatrix} A_1 & B_1 \\ C_1 & D_1 \end{bmatrix}\begin{bmatrix} A_2 & B_2 \\ C_2 & D_2 \end{bmatrix}\begin{bmatrix} A_3 & B_3 \\ C_3 & D_3 \end{bmatrix} = \begin{bmatrix} T_1 & T_2 \\ T_3 & T_4 \end{bmatrix}$ $T_1 = A_3(A_1+A_2+C_1B_2)+B_3(A_1C_2+C_1D_2)$ $T_2 = A_3(B_1A_2+D_1B_2)+B_3(B_1C_2+D_1D_2)$ $T_3 = C_3(A_1A_2+C_1B_2)+D_3(A_1C_2+C_1D_2)$ $T_4 = C_3(B_1A_2+D_1B_2)+D_3(B_1C_2+D_1B_2)$

where:

$$A = \cosh(\gamma l) \tag{4.89}$$

$$B = Z_C senh(\gamma l) \tag{4.90}$$

$$C = \frac{1}{Z_C} senh(\gamma l) \tag{4.91}$$

$$D = \cosh(\gamma l) \tag{4.92}$$

We note that $A = D$ and $AD - BC = 1$.

Hyperbolic functions can be represented in Maclaurin series. Making $\gamma = \sqrt{zy}$, we have:

$$senh(\gamma l) = l\sqrt{zy} + \frac{(l\sqrt{zy})^3}{3!} + \frac{(l\sqrt{zy})^5}{5!} + \cdots + \frac{\left(l\sqrt{zy}\right)^{2n+1}}{(2n+1)!} + \cdots \tag{4.93}$$

$$cosh(\gamma l) = 1 + \frac{(l\sqrt{zy})^2}{2!} + \frac{(l\sqrt{zy})^4}{4!} + \cdots + \frac{\left(l\sqrt{zy}\right)^{2n}}{(2n)!} + \cdots \tag{4.94}$$

4.6 Short-Length Lines

Considering only the first term of each series in equations (4.93) and (4.94), we have:

$$senh(\gamma l) = l\sqrt{zy} = \sqrt{ZY} \tag{4.95}$$

where:

$$Z = zl \text{ and } Y = yl$$

and

$$cosh(\gamma l) = 1 \tag{4.96}$$

Substituting equations (4.95) and (4.96) into equations (4.77) and (4.78), respectively, we get:

Voltage equation:

$$V_S = V_R x 1 + I_R Z_C \sqrt{ZY} = V_R + I_R \sqrt{\frac{Z}{Y}}\left(\sqrt{ZY}\right) = V_R + I_R Z \tag{4.97}$$

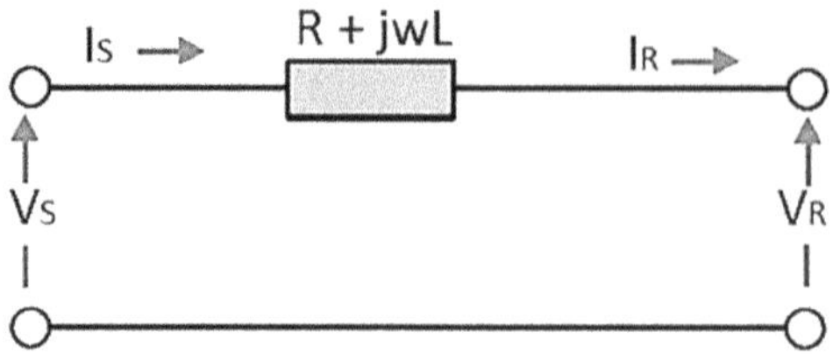

FIGURE 4.7
Short-length line equivalent circuit.

Current equation:

$$I_S = I_R x 1 + \frac{V_R \sqrt{ZY}}{Z_C} = I_R + YV_R \tag{4.98}$$

Neglecting conductance and capacitance, we have:

$$I_S = I_R \tag{4.99}$$

Equations (4.97) and (4.98) represent a mathematical model of a series circuit, as shown in Figure 4.7.

The voltage and current equations of the short-length line in terms of parameters *A*, *B*, *C*, and *D* are:

$$V_S = AV_R + BI_R \tag{4.100}$$

$$I_S = CV_R + DI_R \tag{4.101}$$

with $A = 1,\ B = Z,\ C = 0,\ D = 1.$

4.7 Medium-Length Lines

Considering only the first two terms of the series in equations (4.93) and (4.94), we have:

$$senh(\gamma l) = \sqrt{ZY} + \frac{\left(\sqrt{ZY}\right)^3}{3!} = \sqrt{ZY} + \frac{\left(\sqrt{ZY}\right)^3}{6} \tag{4.102}$$

$$cosh(\gamma l) = 1 + \frac{\left(\sqrt{ZY}\right)^2}{2!} = 1 + \frac{ZY}{2} \tag{4.103}$$

Substituting equations (4.102) and (4.103) into equations (4.75) and (4.76), respectively, we get:

Voltage equation:

$$V_S = V_R\left(1+\frac{ZY}{2}\right)+I_R Z_C\left(\sqrt{ZY}+\frac{(\sqrt{ZY})^3}{6}\right)=V_R\left(1+\frac{ZY}{2}\right)+I_R Z\left(1+\frac{ZY}{6}\right) \tag{4.104}$$

Current equation:

$$I_S = I_R\left(1+\frac{ZY}{2}\right)+\frac{V_R}{Z_C}\left(\sqrt{ZY}+\frac{(\sqrt{ZY})^3}{6}\right)=I_R\left(1+\frac{ZY}{2}\right)+V_R Y\left(1+\frac{ZY}{6}\right) \tag{4.105}$$

We look for an equivalent circuit that represents equations (4.104) and (4.105). For this, we will develop the equations of a circuit π, as shown in Figure 4.8.

The shunt conductance per unit length representing the leakage current of the insulators and the corona effect is negligible, thus $G = 0$.

The Kirchhoff current law applied to the circuit of Figure 4.8 results in:

$$I_L = I_R + \frac{Y}{2}V_R \tag{4.106}$$

From Kirchhoff's law of voltages, the voltage at the sending-end of the line is equal to:

$$V_S = V_R + ZI_L = V_R + Z\left(I_R + \frac{Y}{2}V_R\right) = V_R\left(1+\frac{ZY}{2}\right)+ZI_R \tag{4.107}$$

By the Kirchhoff current law, the current at the beginning of the line is equal to:

$$I_S = I_L + \frac{Y}{2}V_S \tag{4.108}$$

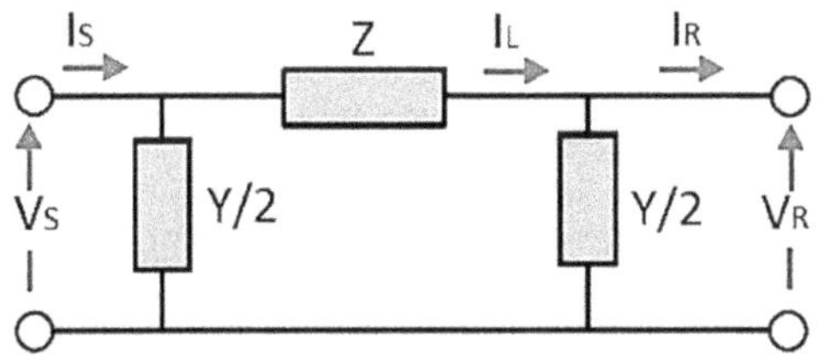

FIGURE 4.8
Equivalent circuit π.

Substituting equations (4.106) and (4.107) into (4.108), we have:

$$I_S = I_R + \frac{Y}{2}V_R + \frac{Y}{2}\left[V_R\left(1+\frac{ZY}{2}\right)+ZI_R\right] = YV_R\left(1+\frac{ZY}{4}\right)+\left(1+\frac{ZY}{2}\right)I_R \quad (4.109)$$

Observing equation (4.103) with (4.107) and equation (4.104) with (4.109), we note that they are approximate.

The voltage and current equations of the medium-length line in terms of ABCD parameters are:

$$V_S = A1V_R + B1I_R \quad (4.110)$$

$$I_S = C1V_R + D1I_R \quad (4.111)$$

With: $A1=\left(1+\frac{ZY}{2}\right)$, $B1=Z$, $C1=Y\left(1+\frac{ZY}{4}\right)$, $D1=\left(1+\frac{ZY}{2}\right)$.

We can develop the equations of a circuit T. However, in studies of power systems, such as load flow and stability studies, the matrix of admittances is used, and the number of elements of this matrix is directly proportional to the bus numbers of the power system. In the case of the circuit T, for each circuit we will have three buses, while for each circuit π we will have only two buses. Thus, the circuit π is preferred for the representation of the TLs.

Let us now consider the calculation of the sending-end voltages and currents for three TLs with the following data:

a. Line of 69 kV powering a load of $P = 52.77628093$ MW, power factor 0.92 inductive, 60 Hz, one conductor per phase, with the following parameters: $R = 0.219\frac{\Omega}{km}$, $L = 1.233\frac{mH}{km}$ and $C = 9.38\frac{nF}{km}$.
b. Line of 230 kV powering a load of $P = 285.87152170$ MW, power factor 0.92 inductive, 60 Hz, four conductors per phase, with the following parameters: $R = 0.031\frac{\Omega}{km}$, $L = 0.8435\frac{mH}{km}$ and $C = 13.91\frac{nF}{km}$.
c. Line of 500 kV powering a load of $P = 621.45982980$ MW, power factor 0.92 inductive, 60 Hz, four conductors per phase, with the following parameters: $R = 0.025\frac{\Omega}{km}$, $L = 0.8541\frac{mH}{km}$ and $C = 13.65\frac{nF}{km}$.

The length of the lines is varied from 40 km to 320 km and the sending-end voltages and currents are calculated using the exact equations, the equations of the π, model, the short-length line equations, the medium-length line equations, considering the $senh(\gamma l)$ e $cosh(\gamma l)$, and the short-length-line equations, considering only one term of the series $senh(\gamma l)$ e $cosh(\gamma l)$. The results of voltage and current errors calculated in relation to the exact equations are shown in Table 4.2.

TABLE 4.2 Errors of the Receiving-end Voltages and Currents in Relation to the Exact Voltages and Currents

L	Voltage (kV)	Error Model π (%)		Error Model Medium-Length Line with Power Series (%)		Error Model Short-Length Line (%)		Error Model Short-Length Line with Power Series (%)	
(km)	V_R	$\|V_S\|$	$\|I_S\|$	$\|V_S\|$	$\|I_S\|$	$\|V_S\|$	$\|I_S\|$	$\|V_S\|$	$\|I_S\|$
40	69	0.0100	9.2253e −005	−0.8416	−0.7764	−0.0919	−0.1316	−0.0919	−0.5885
	230	0.0020	−0.0011	−0.3853	−0.3545	−0.1256	−0.1353	−0.1256	−1.5003
	500	8.7417e −004	−0.0021	−0.3121	−0.3994	−0.1293	−0.1362	−0.1293	−2.9496
80	69	0.0685	0.0011	−1.8551	−1.9332	−0.2539	−0.5272	−0.2539	−1.4367
	230	0.0169	−0.0073	−1.2186	−1.1828	−0.4698	−0.5484	−0.4698	−3.2084
	500	0.0074	−0.0120	−1.0269	−1.4332	−0.5045	−0.5527	−0.5045	−5.7695
120	69	0.2003	0.0048	−2.7986	−3.4803	−0.3884	−1.1899	−0.3884	−2.5508
	230	0.0591	−0.0201	−2.3958	−2.5086	−0.9826	−1.2493	−0.9826	−5.1318
	500	0.0265	−0.0209	−2.1131	−3.1220	−1.1071	−1.2430	−1.1071	−8.3515
160	69	0.4158	0.0144	−3.5977	−5.4313	−0.4531	−2.1249	−0.4531	−3.9389
	230	0.1435	−0.0365	−3.8212	−4.3536	−1.6148	−2.2455	−1.6148	−7.2744
	500	0.0661	0.0022	−3.5411	−5.4303	−1.9186	−2.1694	−1.9186	−10.5582
200	69	0.7190	0.0343	−4.2366	−7.8030	−0.4302	−3.3392	−0.4302	−5.6112
	230	0.2848	−0.0472	−5.4094	−6.7353	−2.3200	−3.5409	−2.3200	−9.6346
	500	0.1354	0.1137	−5.2821	−8.2532	−2.9206	−3.2569	−2.9206	−12.2277
240	69	1.1105	0.0711	−4.7184	−10.6155	−0.3129	−4.8413	−0.3129	−7.5790
	230	0.4970	−0.0358	−7.0860	−9.6642	−3.0555	−5.1333	−3.0555	−12.2025
	500	0.2450	0.3933	−7.3077	−11.4127	−4.0940	−4.3924	−4.0940	−13.1865
280	69	1.5896	0.1332	−5.0517	−13.8916	−0.0991	−6.6405	−0.0991	−9.8546
	230	0.7926	0.0235	−8.7871	−13.1394	−3.7826	−7.0111	−3.7826	−14.9561
	500	0.4063	0.936	−9.5891	−14.6687	−5.4189	−5.4324	−5.4189	−13.2715
320	69	2.1545	0.2318	−5.2461	−17.6557	0.2105	−8.7462	0.2105	−12.4502
	230	1.1826	0.1680	−10.4581	−17.1426	−4.4670	−9.1488	−4.4670	−17.8568
	500	0.6320	1.8413	−12.0966	−17.7466	−6.8734	−6.2206	−6.8734	−12.3594

The negative sign in the values in Table 4.2 indicates that the calculated value is greater than the exact value in the sending-end bus.

Observing the results of Table 4.2, we can conclude that:

a. The worst results are obtained with the equations expressed in the series of powers, equations (4.95), (4.101), and (4.102).

Considering the short and medium line models:

b. For $0 < l \leq 80$ km, the voltage and current errors obtained with the short-length-line model are less than 0.6%.
c. For $80 < l \leq 240$ km, the largest voltage error was 1.11% and the largest current error was 0.3933% obtained with the model π and 4.0940% for the voltage and 5.1333% for the current, considering the short-length line model.
d. For $l > 240$ km, the largest voltage error was 2.1545% and the largest current error was 1.8413% obtained with the model π and 6.8734% for the voltage and 9.1488% for the current, considering the short-length line model.

The general conclusion is that we can use the series circuit for short-length lines where $0 < l \leq 80$ km, the circuit π for medium-length lines where $80 < l \leq 240$ km, and the complete equations for lines where $l > 240$ km.

4.8 Long-Length Line Equivalent Circuit

We can find an equivalent π model for the long-length line.

For this, it is enough to compare equation (4.75) with the voltage equation of the nominal π model, with the modified impedance and admittance:

$$V_S = V_R \cosh(\gamma l) + I_R Z_C senh(\gamma l) \tag{4.112}$$

$$V_S = V_R\left(1 + \frac{Z'Y'}{2}\right) + Z'I_R \tag{4.113}$$

Then, we can conclude that:

$$\cosh(\gamma l) = \left(1 + \frac{Z'Y'}{2}\right) \tag{4.114}$$

$$Z' = Z_C senh(\gamma l) = \sqrt{\frac{z}{y}} senh(\gamma l) = zl \frac{senh(\gamma l)}{\sqrt{zy}l} = Z \frac{senh(\gamma l)}{\gamma l} \tag{4.115}$$

Replacing the impedance Z' in equation (4.114), we have:

$$\cosh(\gamma l) = \left(1 + \frac{Z_C senh(\gamma l) Y'}{2}\right) \tag{4.116}$$

$$\frac{Y'}{2} = \frac{\cosh(\gamma l) - 1}{Z_C senh(\gamma l)} \tag{4.117}$$

Using a trigonometric identity:

$$tanh\left(\frac{\gamma l}{2}\right) = \frac{\cosh(\gamma l) - 1}{senh(\gamma l)} \tag{4.118}$$

we have:

$$\frac{Y'}{2} = \frac{1}{Z_C} tanh\left(\frac{\gamma l}{2}\right) = \frac{Y}{2} \frac{\left(tanh\left(\frac{\gamma l}{2}\right)\right)}{(\gamma l/2)} \tag{4.119}$$

Therefore, the circuit π equivalent for a long-length TL is shown in Figure 4.9.

The equations of the long-length line in the matrix form are:

$$\begin{bmatrix} V_S \\ I_S \end{bmatrix} = \begin{bmatrix} A & B \\ C & D \end{bmatrix} \begin{bmatrix} V_R \\ I_R \end{bmatrix} \tag{4.120}$$

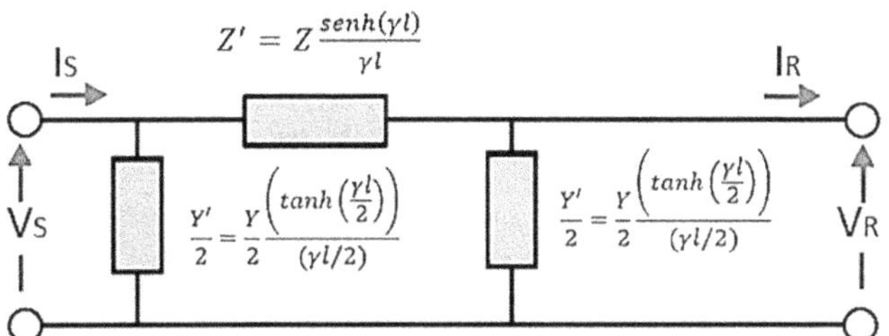

FIGURE 4.9
Circuit π equivalent for a long-length TL.

4.9 Voltage Regulation

The voltage regulation (VR) of a TL is the receiving-end voltage change when the load varies from empty to full load, with a specified power factor, while the receiving-end voltage is kept constant. The setting is expressed as a percentage of the full load voltage.

$$\text{VR}(\%) = \frac{V_{receiving-end\ no\ load} - V_{receiving-end\ full\ load}}{V_{receiving-end\ full\ load}} \times 100 \quad (4.121)$$

As:

$$V_{receiving-end\ \ no\ \ load} = \frac{|V_S|}{|A|} \text{ and } V_{receiving-end\ \ full\ \ load} = |V_R|, \text{ then:}$$

$$VR\ (\%) = \frac{\frac{|V_S|}{|A|} - |V_R|}{|V_R|} x100 \quad (4.122)$$

Example 4.2

A three-phase 60 Hz, 40 km, and 69 kV transmission line has a phase resistance of 0.219 Ω/km and a phase inductance of 1,233 mH/km. The capacitance is negligible. Use the short-length-line model to calculate:

a. The sending-end voltage and power.
b. The VR and efficiency.

When the line is supplying a three-phase load of:

1. 52 MVA with pf = 0.8 inductive at 69 kV.
2. 52 MVA with pf = 0.8 capacitive at 68 kV.

SOLUTION:

Series impedance per phase is:

$$Z = (R + jwL)l = (0.219 + j2\pi 60x1.233x10^{-3})40 = 8.7600 + j18.5932\ \Omega$$

The receiving-end voltage per phase is given by:

$$V_R = \frac{69}{\sqrt{3}} \angle\ 0° = 39.8372 \angle\ 0°\ kV$$

Three-phase complex power is calculated as:

$$S_{3\varphi} = 52 \angle\ arccos0.8° = 52 \angle\ 36.8699° = 41.6 + j31.2\ MVA$$

Calculation of current per phase:

$$I_R = \frac{S_{3\varphi}^*}{3V_R^*} = \frac{52x10^3 \angle -36.8699°}{3x39.8372 \angle 0°} = 435.1042 \angle -36.8699° \ A$$

From equation (4.100), with $A = 1$ e $B = Z$, we have the receiving-end voltage, in kV:

a. $$V_S = V_R + ZI_R = 39.8372 \angle 0° + (8.7600 + j18.5932)435.1042x10^{-3}$$
$$\angle -36.8699° = 47.7404 + j4.1851 = 47.9235 \angle 5.0077°$$

The sending-end line voltage:

$$V_{SL} = \sqrt{3}V_S = \sqrt{3}x47.9235 \angle 5.0077° = 83 \angle 5.0077° \ kV$$

The receiving-end complex power in MVA:

$$S_{S3\varphi} = 3V_S I_S^* = 3x(47.9235 \angle 5.0077°)x10^3 x435.1042x10^{-3} \angle 36.8699°$$
$$= 62.5551 \angle 41.8776° = 46.5768 + j41.7581 \text{ MVA}$$

From equation (4.122), we have the VR:

$$RT\ (\%) = \frac{47.9235 - 39.8372}{39.8372} \times 100 = 20.2984\%$$

Transmission efficiency:

$$\eta = \frac{P_{R3\varphi}}{P_{S3\varphi}} \times 100 = \frac{41.6}{46.5768} \times 100 = 89.3149\%$$

Calculations for the power factor 0.8 capacitive:
Current per phase:

$$I_R = \frac{S_{3\varphi}^*}{3V_R^*} = \frac{52x10^3 \angle 36.8699°}{3x39.8372 \angle 0°} = 435.1042 \angle 36.8699° \ A$$

The sending-end voltage:

$$V_S = V_R + ZI_R = 39.8372 \angle 0° + (8.7600 + j18.5932)435.1042x10^{-3} \angle 36.8699°$$
$$= 38.0324 + j8.7589 = 39.0280 \angle 12.9718° \ kV$$

The sending-end line voltage:

$$V_{SL} = \sqrt{3}V_S = \sqrt{3}x39.0280 \angle 12.9718° = 67.5985 \angle 12.9718° \ kV$$

The sending-end complex power in MVA:

$$S_{S3\varphi} = 3\boldsymbol{V}_S\boldsymbol{I}_S^* = 3\times(39.0280\ \angle\ 12.9718°)\times 10^3 \times 435.1042\times 10^{-3}$$
$$\angle -36.8699° = 50.944\ \angle -23.8981° = 46.5764 - j20.6379 \text{ MVA}$$

Voltage regulation:

$$RT\ (\%) = \frac{39.0280 - 39.8372}{39.8372}\times 100 = -2.0313\%$$

Transmission efficiency:

$$\eta = \frac{P_{R3\varphi}}{P_{S3\varphi}}\times 100 = \frac{41.6}{46.5764}\times 100 = 89.3149\%$$

Example 4.3

A three-phase 150 kW, 230 kV, 60 Hz TL has a phase resistance of 0.031 Ω/km and the phase inductance is 0.8435 mH/km. The shunt capacitance is 13.91 nF/km. The load at the end of the line is 285 MVA with pf = 0.8 inductive at 230 kV. Use the mid-line model and calculate the voltage, the complex power at the end and at the beginning of the line, voltage regulation, and line efficiency.

SOLUTION:

Series impedance per phase is given by:

$$Z = (R + jwL)l = (0.031 + j2\pi 60\times 0.8435x10^{-3})150$$
$$= 4.6500 + j47.6988\ \Omega$$

Parallel admittance per phase is calculated as:

$$Y = (G + jwC)l = (0 + j2\pi 60x13.91\times 10^{-9})150$$
$$= j0.00078659\ S$$

The receiving-end voltage bus per phase is given by:

$$\boldsymbol{V}_R = \frac{230}{\sqrt{3}}\angle\ 0° = 132.7906\ \angle\ 0°\ kV$$

Three-phase complex power is calculated as:

$$S_{3\varphi} = 285\ \angle\ arccos0.8° = 285\ \angle\ 36.8699°$$
$$= 228 + j171\ MVA$$

Calculation of current per phase:

$$\boldsymbol{I_R} = \frac{S_{3\varphi}^*}{3\boldsymbol{V_R^*}} = \frac{285\times10^3\ \angle -36.8699^\circ}{3\times132.7906\ \angle\ 0^\circ}$$
$$= 715.4140\ \angle -36.8699^\circ\ A$$

The sending-end voltage, in kV, using equation (4.107), is given by:

$$V_S = V_R\left(1+\frac{ZY}{2}\right)+ZI_R$$
$$= 132.7906\left(1+\frac{(4.6500+j47.6988)(j0.00078659)}{2}\right)$$
$$+(4.6500+j47.6988)715.4140(0.8-j0.6) = 153.44$$
$$+j25.546\ kV = 155.5520\ \angle\ 9.4538^\circ\ kV$$

The sending-end line voltage:

$$\boldsymbol{V_{SL}} = \sqrt{3}\boldsymbol{V_S} = \sqrt{3}x155.5520\ \angle\ 9.4538^\circ = 269.4239\ \angle\ 9.4538^\circ\ kV$$

The sending-end current, using equation (4.109), is given by:

$$I_S = YV_R\left(1+\frac{ZY}{4}\right)+\left(1+\frac{ZY}{2}\right)I_R$$
$$= (j0.00078659)132.7906\ x10^3\left(1+\frac{(4.6500+j47.6988)(j0.00078659)}{4}\right)$$
$$+\left(1+\frac{(4.6500+j47.6988)(j0.00078659)}{2}\right)715.4140(0.8-j0.6)$$
$$= 562.28-j316.68 = 645.3255\ \angle -29.3870^\circ\ A$$

The receiving-end complex power in MVA is:

$$S_{S3\varphi} = 3\times132.7906\times715.4140\ \angle\ 36.8699^\circ = 285\ \angle\ 36.8699^\circ$$
$$= 228+j171\ \text{MVA}$$

which verifies the receiving-end voltage and current calculations.

The sending-end complex power in MVA:

$$S_{S3\varphi} = 3\boldsymbol{V_S}\boldsymbol{I_S}^* = 3\times(155.5520\ \angle\ 9.4538^\circ)\times10^3\times645.3255\ \angle\ 29.3870^\circ$$
$$= 301.1450\ \angle\ 19.9332^\circ = 283.1036+j102.6677\ MVA$$

Voltage regulation:

$$A = \left(1+\frac{ZY}{2}\right) = \left(1+\frac{(4.6500+j47.6988)(j0.00078659)}{2}\right)$$
$$= 0.9812\ \angle\ 0.1068^\circ$$

$$RT\ (\%) = \frac{\frac{|V_S|}{|A|} - |V_R|}{|V_R|} \times 100 = \frac{\frac{155.5520}{0.9812} - 132.7906}{132.7906} \times 100 = 19.39\%$$

Transmission efficiency:

$$\eta = \frac{P_{R3\varphi}}{P_{S3\varphi}} \times 100 = \frac{228}{283.1036} \times 100 = 80.5359\%$$

Example 4.4

A three-phase transmission line with 320 km, at 60 Hz, supplies a load of 621 MVA, pf = 0.85 inductive at 500 kV. The line constants are $R = 0.025$ Ω/km, $L = 0.8541$ mH/km, and $C = 13.95$ nF/km.

Calculate:

a. The attenuation and phase constants;
b. Constants A, B, C, and D;
c. Voltage, current, power, and power factor at the beginning of the line;
d. Line losses;
e. Transmission efficiency:
f. Voltage regulation;
g. The charging current at the beginning of the line without load;
h. Value of the voltage increase without load if the voltage at the start of line is kept constant;
i. The equivalent circuit of the TL.

SOLUTION:

Calculation of series impedance per phase of the TL:

$$Z = R + jwl = \left(0.025 + j377 \times 0.8541x10^{-3}\right) = 0.3229 \angle 85.5598° \frac{\Omega}{km}$$
$$= 8 + j103.0386\ \Omega = 103.3487 \angle 85.5604°\ \Omega$$

Calculation of the parallel admittance per phase of the TL:

$$Y = G + jwC = \left(0.0375 \times 10^{-6} + j377 \times 0.01365 \times 10^{-6}\right)$$
$$= 0.0000 + j5.1553 \times 10^{-6}\ S/km = 0.00165 \angle 90°\ S$$

Calculation of the characteristic impedance:

$$Z_C = \sqrt{\frac{Z}{Y}} = \sqrt{\frac{103.3487 \angle 85.5604°}{0.00165 \angle 90°}} = 250.0832 - j9.6938$$
$$= 250.2710 \angle -2.2198°\ \Omega$$

Calculation of propagation, attenuation, and phase constants:

$$\gamma = \alpha + j\beta = \sqrt{zy} = \sqrt{(0.3229 \angle 85.5598°)\,(j5.1553 \times 10^{-6})}$$
$$= 1.2902 \times 10^{-3} \angle 87.7799°$$
$$= 0.04998 \times 10^{-3} + j1.2892 \times 10^{-3} \left(\frac{1}{km}\right)$$

$$\gamma l = 0.4128 \angle 87.7926° = 0.0159 + j0.4125\ pu$$

The receiving-end voltage per phase is given by:

$$V_R = \frac{500}{\sqrt{3}} \angle 0° = 288675.1346\ V$$

$$I_R = \frac{621x10^6}{500x10^3\sqrt{3}} \angle 0° - arccos 0.85 = 609.5087 - j377.7395$$
$$= 717.0690 \angle -31.7883°\ A$$

Calculation of constants A, B, C, and D.

$$A = \cosh(\gamma l) = \cosh(0.0159 + j0.4125) = 0.9162 + j0.0064 = 0.9163 \angle 0.3986°$$

$$B = Z_C senh(\gamma l) = (250.0832 - j9.6938) senh(0.0159 + j0.4125)$$
$$= 7.5297 + j100.1300 = 100.4128 \angle 85.6995°\ \Omega$$

$$C = \frac{1}{Z_C} senh(\gamma l) = \frac{1}{(250.0832 - j9.6938)} senh(0.0159 + j0.4125)$$
$$= -0.000003892294490 + j0.001603121792343 = 0.0016 \angle 90.1391°\ S$$

$$D = A$$

Calculation of the tensions and currents at the beginning of the line:
According to equation (4.120), we have:

$$\begin{bmatrix} V_S \\ I_S \end{bmatrix} = \begin{bmatrix} A & B \\ C & D \end{bmatrix} \begin{bmatrix} V_R \\ I_R \end{bmatrix}$$
$$= \begin{bmatrix} 0.9162 + j0.0064 & 7.5297 + j100.1300 \\ -0.000003892294490 + j0.001603121792343 & 0.9162 + j0.0064 \end{bmatrix} \times$$
$$\begin{bmatrix} 288675.1346 \\ 609.5087 - j377.7395 \end{bmatrix} = \begin{bmatrix} 306896.6321 + j60033.3619 \\ 559.7258 + j120.5973 \end{bmatrix}$$
$$= \begin{bmatrix} 312713.2030 \angle 11.0681°\ V \\ 572.5702 \angle 12.1589°\ A \end{bmatrix}$$

The line voltage at the beginning of the TL is:

$$V_{SL} = (312713.2030 \angle 11.0681°)\sqrt{3}\ V = 541.6352 \angle 11.0681°\ kV$$

Calculation of the power at the beginning of TL:

$$\begin{aligned} S = 3V_S I_S^* &= 3(312713.2030 \angle 11.0681°)\ (572.5702 \angle -12.1589°) \\ &= 537053472.8236 - j10226051.0826\ VA = 537.1508 \angle -1.0908°\ MVA \end{aligned}$$

Calculation of power factor:

$$\cos(11.0681° - 12.1589°) = 0.9998\ capacitivo$$

Losses in TL:

$$P_L = P_S - P_R = 537.0535 - 621 \times 0.85 = 9.2035\ MW$$

Transmission efficiency of the TL:

$$\eta = \frac{621x0.85}{537.0535} \times 100 = 98.29\%$$

Voltage regulation:

$$RT\ (\%) = \frac{\frac{|V_S|}{|A|} - |V_R|}{|V_R|} \times 100 = \frac{\frac{312713.2030}{0.9163} - 288675.1346}{288675.1346} \times 100 = 18.22\%$$

Charging current at start of the TL:

$$\begin{aligned} I_C &= \frac{YV_S}{2} = \frac{(0.00165 \angle 90°)(312713.2030 \angle 11.0681°)}{2} \\ &= 257.9884 \angle 101.0681°\ A \end{aligned}$$

Value of the voltage increase without load:

$$\begin{aligned} V_R = V_S - ZI_C &= (306896.6321 + j60033.3619) - (8 + j103.0386)257.9884 \\ &\angle 101.0681° = 339301.0798 \angle 10.7196°\ V \end{aligned}$$

Increased line voltage:

$$V_{RL} = \sqrt{3}(339301.0798 \angle 10.7196°)\ V = 587686.7093 \angle 10.7196°\ V$$

TL equivalent circuit:

$$Z' = Z\frac{senh(\gamma l)}{\gamma l} = 100.4128 \angle 85.6995° \ \Omega$$

$$\frac{Y'}{2} = \frac{Y}{2}\frac{\left(tanh\left(\frac{\gamma l}{2}\right)\right)}{(\gamma l/2)} = 7.5182\times10^{-7} + j8.3659\times10^{-4} \ S$$

4.10 Lossless Lines

The surge impedance is defined as follows:

For a lossless TL, R = 0 and G = 0:

$$z = jwL \ \Omega/m \tag{4.123}$$

$$y = jwC \ \Omega/m \tag{4.124}$$

$$Z_C = \sqrt{\frac{z}{y}} = \sqrt{\frac{jwL}{jwC}} = \sqrt{\frac{L}{C}} \ (\Omega) \tag{4.125}$$

and

$$\gamma = \sqrt{zy} = \sqrt{(jwL)(jwC)} = jw\sqrt{LC} = j\beta \ 1/m \tag{4.126}$$

where:

$$\beta = w\sqrt{LC}$$

The characteristic impedance, commonly called surge impedance for a lossless line, is purely real and the PC is purely imaginary.

The power supplied by a line to a pure resistive load equal to its surge impedance is known as line loading by the surge impedance (SIL) (surge impedance loading), as shown in Figure 4.10. SIL can be used to compare transmission line loading capacities. A line charged below the SIL generates reactive power, while a line charged above the SIL consumes reactive power.

The surge impedance is calculated as:

$$S = SIL = \sqrt{3}(V_L)(I_L) = \sqrt{3}(V_L)\frac{\frac{V_L}{\sqrt{3}}}{\sqrt{\frac{L}{C}}} = \frac{|V_L|^2}{\sqrt{\frac{L}{C}}} = \frac{|V_L|^2}{Z_C} \ (W) \tag{4.127}$$

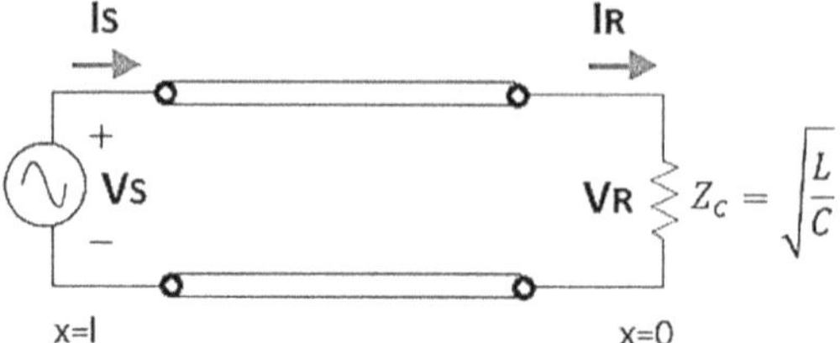

FIGURE 4.10
TL loaded by surge impedance.

The characteristic values for the SIL of TLs are shown in Table 4.3.

4.11 Voltage Profile

In practice, TLs do not end with surge impedance. Instead, charging may vary from a small fraction of the SIL under light load to multiples of SIL under heavy load, depending on the compensation and the length of the TL.

In the graph of Figure 4.11, the voltage profile of a 500 kV and 294 km TL is shown without reactive compensation, under no-load, under heavy load, short-length circuit, and charged by SIL.

We observe that the unladen TL shows a voltage increase in the receiving-end bus in relation to the sending-end. This effect is known as the Ferranti effect, in homage to the physicist who discovered it.

The consequences of this phenomenon are:

1. As there is an increase in voltage in the receiver bar, there is a need to increase the insulation voltage level of the lines and terminal equipment.

TABLE 4.3
Characteristic Values of SIL for Three-Phase TLs, 60 Hz

TL Voltage (kV)	SIL (MW)	Examples Conductor SIL
69	12 – 13	
138	47 – 52	1×556 (26/7) 52.0 1×795 (26/7) 53.5
230	134 – 145	1×795 (26/7) 132.9
345	325 – 425	2×795 (26/7) 412.7 2×954 (45/7) 414.6
500	850 – 1075	2×954 (45/7) 915.2
765	2200 – 2300	

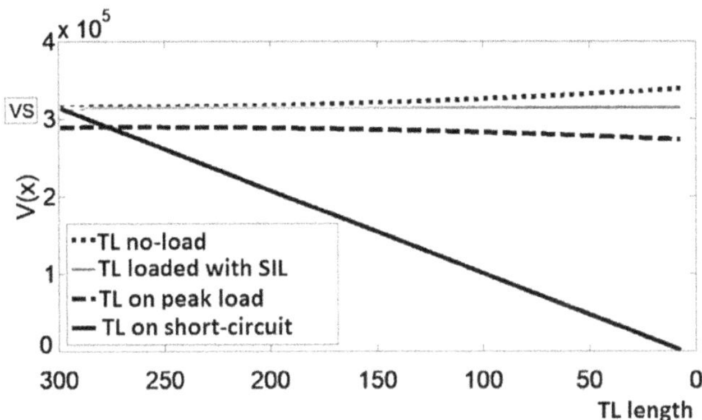

FIGURE 4.11
Voltage profile of a TL.

2. The dispersion losses (mainly corona effect) increase with the square of the voltage and reduce the overvoltage. However, radio interference and audible noise accompanying the corona effect also increase with increasing voltage. But to keep them within reasonable limits, you need to increase the gauge of the conductors with increased line cost.
3. The capacitive current in the light-load lines can cause self-excitation (the self-excitation in synchronous machines is an electrical instability that is associated to the increase of the flux links, being characterized mainly by the rapid increase of the terminal voltage of the machine, because of the capacitive load connected to their terminals) on synchronous machines, where such machines do not absorb such currents.

4.12 Stability Limit of Steady-State (PR)

The equivalent lossless TL circuit of Figure 4.9 can be used to obtain an active power equation. Assume that V_S and V_R are constants and that the phase angle between the voltages is δ.

$$I_R = \frac{V_S - V_R}{Z'} - \frac{V_R Y'}{2} \tag{4.128}$$

The complex power delivered to the receiving-end bus is calculated as:

$$S_R = V_R I_R^* = V_R \left(\frac{V_S - V_R}{Z'} - \frac{V_R Y'}{2} \right)^* \tag{4.129}$$

Putting the voltages in the polar form, we have:

$$\begin{aligned} S_R = P_R + jQ_R &= \frac{V_R V_S \angle -\delta^\circ - V_R^2}{-jX'} + \frac{jwC'lV_R^2}{2} \\ &= \frac{V_R V_S\left(cos\delta - jsen\delta\right) - V_R^2}{-jX'} + \frac{jwC'lV_R^2}{2} \\ &= \frac{jV_R V_S cos\delta + V_R V_S sen\delta - jV_R^2}{X'} + \frac{jwC'lV_R^2}{2} \\ &= \frac{V_R V_S sen\delta}{X'} + j\left(\frac{V_R V_S cos\delta - V_R^2}{X'} + \frac{wC'lV_R^2}{2}\right) \end{aligned} \tag{4.130}$$

Therefore, the receiving-end active power is equal to:

$$P_R = \frac{V_R V_S sen\delta}{X'} \tag{4.131}$$

And the maximum power is calculated when $sen\delta = 1$.

$$P_{Rmax} = \frac{V_R V_S}{X'} \tag{4.132}$$

Equation (4.132) represents the theoretical limit of steady-state stability for a lossless TL.

Figure 4.12 shows the theoretical stability limit.

The theoretical stability limit can be set in terms of the SIL.

Using equation (4.132), equation (4.115), and equation (4.125), we can write:

$$P_{Rmax} = \frac{V_R V_S}{\sqrt{\frac{L}{C}}\, sen(\beta l)} \tag{4.133}$$

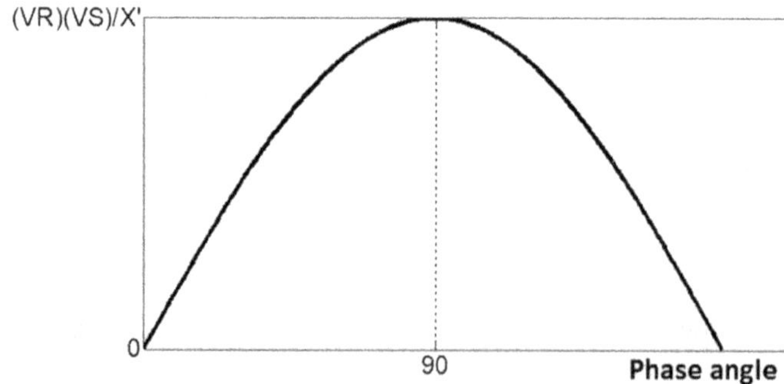

FIGURE 4.12
Theoretical steady-state stability limit.

Substituting the value of $\beta = \frac{2\pi}{\lambda}$ and calculating V_R e V_S in p.u, with the line voltage basis, we have:

$$P_{Rmax} = \frac{V_R V_S}{\sqrt{\frac{L}{C}}\, sen\left(\frac{2\pi}{\lambda} l\right)} = \left(\frac{V_R}{V_{Lnominal}}\right)\left(\frac{V_S}{V_{Lnominal}}\right)\left(\frac{V_{Lnominal}^2}{\sqrt{\frac{L}{C}}}\right)\frac{1}{sen\left(\frac{2\pi}{\lambda} l\right)} \quad (4.134)$$

Substituting the value of SIL, according to equation (4.127), we get:

$$P_{Rmax} = \frac{V_{Rpu}\, V_{Spu} SIL}{sen\left(\frac{2\pi}{\lambda} l\right)} \ (W) \quad (4.135)$$

In equation (4.132), we see that two factors affect the theoretical stability limit:

a. It increases with the square of the voltage.
b. It decreases with the length of the line.

At the practical stability limit, to maintain stability during transient disturbances, we consider:

$$V_S = 1.0\ pu,\ V_R = 0.95\ pu,\ \delta \cong 30^\circ$$

Therefore:

$$P_{R1} = \frac{0.95 V_{Rpu}\, V_{Spu} SIL}{sen\left(\frac{2\pi}{\lambda} l\right)} sen(30^\circ) = \frac{0.475 V_{Rpu}\, V_{Spu} SIL}{sen\left(\frac{2\pi}{\lambda} l\right)} \ (W) \quad (4.136)$$

Figure 4.13 shows the limits of theoretical and practical stability of a 500 kV TL, with L = 0.8541 mH/km and C = 13.95 nF/km, from 50 km.

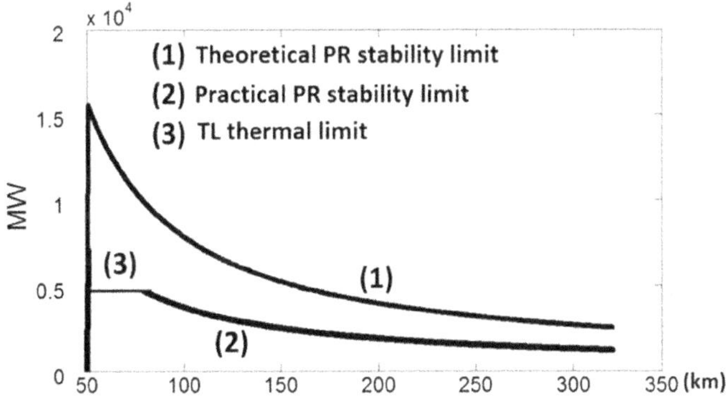

FIGURE 4.13
Theoretical and practical stability limit for 500 kV LT.

The thermal limit is determined by the type of conductor. For example, the thermal limit for three conductors ACSR 1113 kcmil is $3 \times 1.11 = 3.33$ kA / phase.

4.13 Maximum Power Flow for Line with Loss

The active and reactive power flow equations for a TL with losses, in terms of the parameters *A*, *B*, *C*, and *D*, are developed below. Polar notation is used, that is,

$$A = \cosh(\gamma l) = |A| \angle \theta_A{}^\circ$$

$$B = Z' = |Z'| \angle \theta_Z{}^\circ$$

$$V_S = |V_S| \angle \delta^\circ$$

$$V_R = |V_R| \angle 0^\circ$$

From equation (4.85), we have:

$$I_R = \frac{V_S - AV_R}{B} = \frac{|V_S| \angle \delta^\circ - |A| \angle \theta_A{}^\circ (V_R \angle 0^\circ)}{|Z'| \angle \theta_Z{}^\circ} \quad (4.137)$$

The receiving-end complex power is calculated as:

$$S_R = P_R + jQ_R = V_R I_R^* = V_R \left(\frac{|V_S| \angle \delta^\circ - |A| \angle \theta_A{}^\circ (V_R \angle 0^\circ)}{|Z'| \angle \theta_Z{}^\circ} \right)^* \quad (4.138)$$

Using Euler's identity $1 \angle \alpha^\circ = cos\alpha + jsen\alpha$ and separating the real and imaginary parts, we have:

$$P_R = \left(\frac{|V_S||V_R|}{|Z'|} \right) \cos(\theta_Z{}^\circ - \delta^\circ) - \left(\frac{|V_S| V_R^2}{|Z'|} \right) \cos(\theta_Z{}^\circ - \theta_A{}^\circ) \quad (4.139)$$

$$Q_R = \left(\frac{|V_S||V_R|}{|Z'|} \right) \mathrm{sen}(\theta_Z{}^\circ - \delta^\circ) - \left(\frac{|V_S| V_R^2}{|Z'|} \right) \mathrm{sen}(\theta_Z{}^\circ - \theta_A{}^\circ) \quad (4.140)$$

4.14 Reactive Compensation

At the outset of commercial power systems, power systems were poorly interconnected, generators were close to loads and were used to regulate local voltage through their excitation systems. Therefore, there was no need for special reactive compensation equipment. As power systems began to expand with the construction of new power plants and an increasing distance from the loads, there was an increase in the transmission level, the amount of transmitted power, and there was a considerable increase in reactive compensation equipment in number and importance.

Shunt capacitors have been used increasingly in industrial consumer, distribution and sub-transmission, and then with the development of isolated capacitors for high voltage levels; they are also used in high voltage transmission.

Shunt reactors have become decisive for the compensation of long high voltage TLs with intermediate substations.

Synchronous compensators have become a common component for continuous control and as a supplementary source of reactive power during emergencies, also contributing to stabilize the power system when disturbances occur. The synchronous compensators, normally installed near the large load centers, in the alternating current transmission, and also for the voltage control and increase of the level of the short circuit in direct current terminals, were gradually replaced, from the 1970s, by the static compensators.

The TLs generate, depending on their loading, different situations for the power system in terms of additional reactive compensation. Let us cite two very common problems: (1) if in the heavy's hour load an important line is withdrawn from operation, there will be a redistribution of flow by the others, increasing the loading and consequently the reactive consumption, not only for the greater consumption in the reactances of the other lines, but also because of the reduction of the reactive generated by the load of the lines, which will be smaller because of the voltage drop resulting from the increase of the load that the TLs in operation should meet. Thus, a reactive power support is required. This support usually comes via synchronous or static compensators; (2) Another type of problem is the tendency to withdraw or transform shunt reactors planned for the initial years of operation of transmission systems with long lines and sectioned at intermediate points into switchable ones.

Of course, over the years, these trunks will carry and the reactors, which originally served to adjust the tension resulting from the loading of the lines, begin to become unnecessary and inconvenient. If the rejection studies confirm that they are dispensable, they can be simply switched off or converted into a switch with the installation of a switch with its own circuit breaker if it is convenient to keep them available for voltage control at light load or for energizing.

There are two types of compensation equipment: static and rotary. Static equipment is built by separate capacitor banks and inductive reactors, while the rotary units are built by synchronous motors. Let's look at each of the traditional compensation equipment next.

4.14.1 Series Capacitors

The inductive reactance of the line changes the angle of power of the TL and, consequently, the degree of stability, besides interfering in the voltage drop of the TL. The series capacitor banks are applied in transmission systems to decrease the serial reactance of the TLs without changing the terminal voltages. This reduces the parallel compensation for the voltage control and the electrical distance between the end bars.

The typical degree of compensation of the series capacitors in TLs is of the order of 40% to 50%. The advantages of series compensation are as follows:

- Improve overall load distribution and TL losses;
- Improve TL voltage regulation;
- More economical solution to improve static and transient stability limits;
- Helps maintain the balance of reactive energy.

The disadvantages of series compensation are as follows:

- Cost is very high (the ideal operation requires installation in the middle of the line, but the cost requires the installation at its ends);
- High short-circuit currents requiring greater insulation;
- Ferro resonance (overvoltage);
- Difficulty in coordinating protection.

In Figure 4.14, the operating principle of a series capacitor bank can be explained.

Under normal operating conditions, they insert the series capacitor bank in the TL with the circuit-breaker open. Under short-circuit conditions in the power system, the current reaches high values and the voltage at the terminals of the series capacitor bank is limited by the varistor. When the dissipated energy in the varistor becomes excessive, the circuit breaker is automatically closed, removing the bank of the capacitor series of operation.

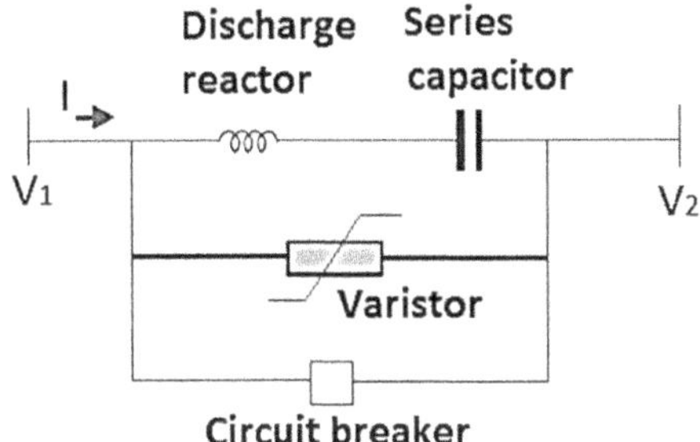

FIGURE 4.14
Series capacitor scheme in TL.

The reactor or damping circuit is used to reduce the transient discharge currents when the circuit-breaker is closed.

4.14.2 Capacitor Switchable to Thyristors

This is a set of capacitor bank modules in series with an antiparallel thyristor connection. Each module, individually, can be turned on or off, in order to allow a discontinuous control of the power generated by the assembly. In the manual control mode, the firing angle is fixed, therefore the degree of compensation of the line, too. In automatic control mode, the operator adjusts the desired flow of active power or current in the line, and the control system automatically varies the trigger angle to keep the current or current power flow constant and close to the set value.

The modules must be connected to the high voltage through a transformer of their own, being able to control the voltage on both the high and low side of the transformer.

A schematic with three equal power capacitor modules forming the thyristor switched capacitor (TSC) is shown in Figure 4.15.

Each stage takes an average half-cycle time to be turned on and off. Since there are practically no harmonics in the current wave, there is no filter placement.

The operation of the stages of this equipment is shown in Figure 4.16.

The main applications of TSC are:

- Damping of low frequency power oscillations between systems;
- Interconnection of power systems through long-length lines;
- Active power flow control in TLs;
- Control of TLs loading.

A practical example of the TSC application is Brazil's North-South interconnection line, which is one of the largest and most modern electricity supply systems in the world, extending 1,276 km from the Samambaia substation

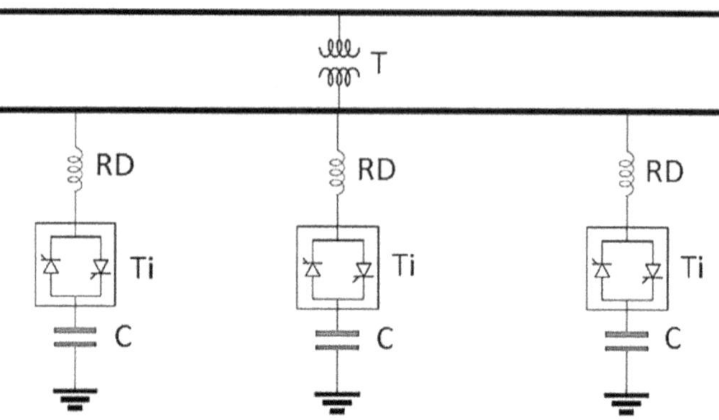

FIGURE 4.15
Capacitor switchable to thyristors.

T – Transformer
RD – Discharge reactor
Ti – Thyristors
C – Capacitors

in Brasília-Brazil, to Imperatriz-Brazil, where two units are operating at the 500 kV level, enabling exchanges of up to 1,000 MW.

4.14.3 Shunt Reactor

It is used in transmission systems to absorb reactive power, controlling voltage at levels set by standards. It can be installed in a transformer tertiary, in busbars, or directly in the LT. When this equipment is installed in the tertiary of the transformer, it is used to control the voltage at the light load. For this, it has a specific circuit breaker. The busbar reactor can be fixed or

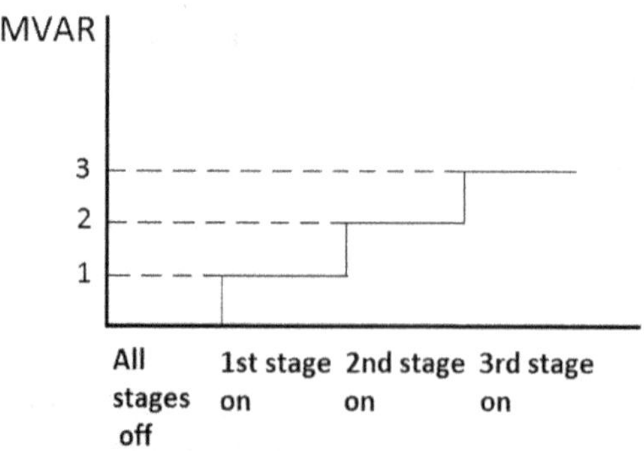

FIGURE 4.16
Operation of stages.

switchable, depending on the convenience of withdrawing it from operation under heavy load. The TL reactor is connected directly to the line, entering and leaving operation with the TL.

Shunt reactors are often used in long-length lines to compensate for the Ferranti effect of the open or lightweight terminal. Sometimes, especially in long-length TLs, they are indispensable to avoid high overvoltage in load rejections.

The shunt reactors have the characteristic that the reactive power consumed is proportional to the square of the voltage. Thus, when the voltage increases or decreases, the reactive power increases or decreases with the square of the voltage.

4.14.4 Thyristor-Controlled Reactor

The current in the reactor is controlled by thyristors (thyristor-controlled reactor-TCR) placed in series with the reactor, in an antiparallel connection, that allows the flow of current in two directions. The thyristors allow conduction only during the interval at which they are receiving the pulse. The conduction angle can be continuously varied and the actuation time of the trigger angle variation is half cycle. The current waveform, however, is completely distorted. The fundamental component decreases as the driving angle is reduced from 180° to 0°. Thus, the TCR requires filters and normally the 5th and 7th harmonics are filtered, but it is sometimes necessary to use filters for the 3rd harmonic and higher frequency harmonics.

Figure 4.17 shows the TCR scheme.

4.14.5 Saturated Reactor

The saturated reactor, unlike the thyristor-controlled reactor, has an iron core and a special saturation characteristic that gives it a voltage × current curve that is adequate to control the voltage of the rod to which it is connected.

As with the controlled reactor, the saturated reactor requires a transformer and is connected to the secondary reactor. The control characteristic of the reactor, however, because it is inherent in the equipment, allows the control of the voltage directly in the low voltage bar, and the control of the high voltage bar is done only indirectly.

The response of the equipment is very fast, since it depends only on the constant inherent to the electromagnetic phenomenon.

Because it is an equipment that has an iron core around which a winding develops, the assembly being inside a tank, the thermal characteristics of the saturated reactor are similar to those of a transformer, giving that equipment a high capacity of short-time overload, particularly in cases of load rejection.

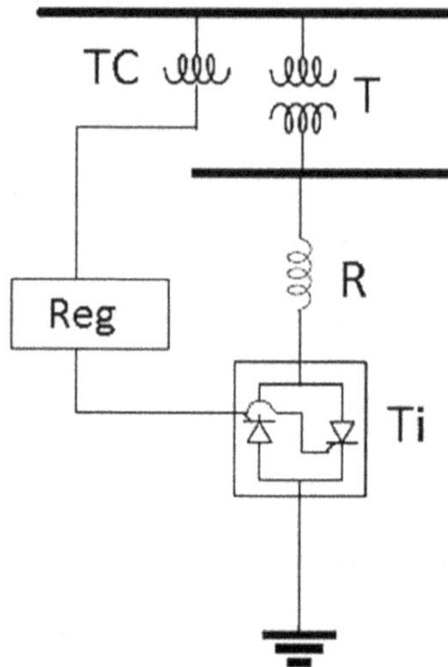

FIGURE 4.17
Thyristor-controlled reactor.

T – Transformer
R – Reactor
Ti –Thyristors
TC – Current Transformer
Reg – Regulator

4.14.6 Shunt Capacitor

Shunt capacitor were initially widely used in distribution and by industrial consumers for voltage support and power factor compensation. Even today, these are an effective and economical means of doing reactive capacitive compensation.

With the advent of isolated capacitors for higher voltage levels, they are also used in the transmission, often using the tertiary transformers, or even connecting them directly to the 138-kV level or above. However, their installation typically occurs at near consumer points of supply.

The capacitors are normally switchable, having a range with its own circuit breaker that allows them to be switched off during light load times.

The shunt capacitors, as well as the shunt reactors, have the inconvenience of generating a reactive proportional to the square of the voltage of the point at which they are connected. Thus, if there is a voltage decrease at a time when capacitive reactive power is needed, the capacitor bank will contribute less than expected. In this situation, it would be necessary to connect additional capacitor banks to improve the voltage profile of the power system. This situation reveals the need for the static compensator of the thyristor-type capacitors.

4.14.7 Synchronous Compensators

Unlike shunt capacitors, which only generate reactive power, and shunt reactors that only absorb reactive power, the synchronous compensators are rotary machines that generate and absorb reactive power. To do this, these devices have sensors and a logic of actuation performed by the automatic voltage regulator, which increases or decreases the excitation of the machine field to meet the needs of the system.

Due to its capacity to generate and absorb reactive power in a continuous range, which ranges from absorbing about 75% of its nominal power to generate 100% nominal power, the synchronous compensators allow fine voltage control and are useful to network emergency conditions.

Synchronous compensators require a specific transformer, since they are not insulated. Maintenance is frequent and costly since it is a rotary machine.

Synchronous compensators have short-time overload capability that contributes to the first few moments of a more severe emergency. They are also important when, besides the reactive medium-length, there is a need to raise the short-circuit power of the system. In this application, they are useful in the inverter terminals of the direct current links, making it possible to get an adequate short-circuit relationship.

Synchronous compensators, and generators, are subject to self-excitation, a phenomenon in which the machine loses control of its terminal voltage. This phenomenon can occur if a high capacitance remains connected to the machine terminals after a load rejection.

A synchronous compensator is shown in Figure 4.18.

FIGURE 4.18
Synchronous compensator.

The use of synchronous compensators has the following advantages over discrete control equipment:

- It provides continuous voltage control;
- It increases system short-circuit power;
- It responds automatically during disturbances in the power system;
- It has great transient overload capability.

Practical examples of synchronous compensators are those of the Camaçari II (Brazil)—230 kV substations, whose nominal power in MVAr is 2x (–105 to 150), Bom Jesus da Lapa (Brazil)—230 kV, with nominal power in MVAr of (–15 to 30), and the Irecê (Brazil)—230 kV substation, also with nominal power in MVAr of (–15 to 30).

4.14.8 Static Compensator

From the controlled reactor to thyristors, the saturated reactor and the capacitor switchable to thyristors, it is possible to get different sets we call static compensator.

They are used to mitigate power quality disturbances, such as short-term voltage variation and voltage fluctuations, power factor correction, and harmonic reduction.

The static compensators can be of the following types: SR/FC (fixed capacitor saturated reactor), TSC/FR (capacitor switched to thyristors with fixed reactor), TCR/FC (thyristor-controlled reactor with a fixed capacitor), and TCR/TSC (controlled to thyristors with capacitor switched to thyristors).

The advantages of the static compensator when compared to the synchronous compensator are:

- Reduced response time;
- Lower maintenance cost, since it has no moving parts;
- It takes up less space than a synchronous of the same power;
- Greater reliability of operation.

4.14.8.1 Static Compensator of the Type SR/FC

The association of a fixed capacitor in parallel with the saturated reactor gives the set the possibility to act continuously from a capacitive range, within the limit of the power of the installed capacitor, until an inductive range in the difference's limit between the maximum power of the reactor and the power of the fixed capacitor.

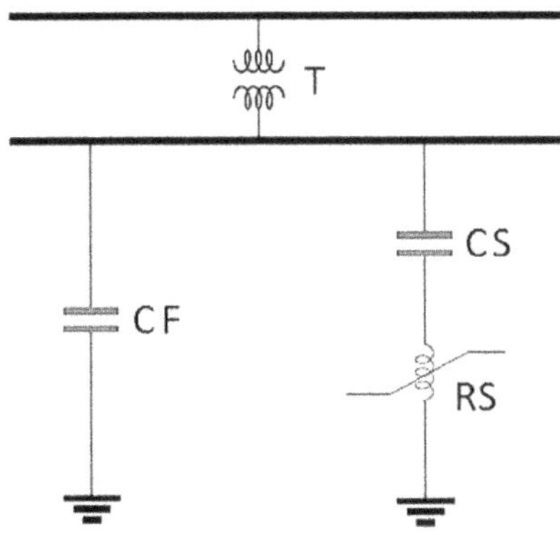

FIGURE 4.19
SR/FC type static compensator.

T – Transformer
CF – Fixed Capacitor
CS – Serie Capacitor
RS –Saturated reactor

Figure 4.19 shows the schematic of the SR/FC type static compensator.

In Figure 4.19, the fixed capacitor is installed on the low voltage side of the transformer, along with the saturated reactor. Voltage control on the high voltage side is done indirectly via the transformer tape.

4.14.8.2 Static Compensator Type TSC/FR

This static compensator is obtained by conjugating a set of capacitor modules switchable to thyristors with a fixed reactor. In this way, this equipment acts in the capacitive and inductive range with a step control. In the inductive range, the limit is the power of the fixed reactor in the situation where all the capacitor modules are off. In the capacitive range, the limit is the difference between the sum of the powers of the fixed capacitor banks and the fixed reactor.

Figure 4.20 shows the schematic of the set constituting a TSC/FR, with two switchable capacitor modules.

In the diagram shown, the fixed reactor is placed on the low voltage side, but it can also be installed on the high voltage side. The reactor may have its own circuit breaker, which reduces the number of capacitor banks needed, or even its power.

The maximum reactive power that can be generated is the difference between the sum of the capacitor banks and the power of the reactor in case the reactor is fixed. If it is switchable, it becomes the sum of the power capacitor banks, which reduces the need for capacitor banks to a desired power.

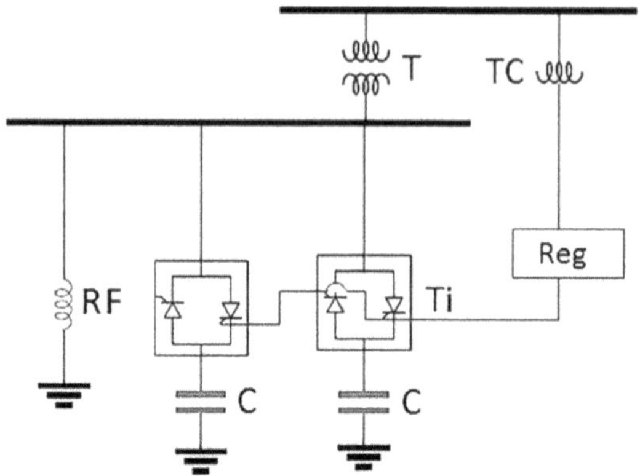

FIGURE 4.20
Static compensator type TSC/FR.

T – Transformer
C – Capacitor switchable to thyristor
TC – Current transformer
RF – Fixed reactor
Reg – Voltage and trip regulator

4.14.8.3 TCR/FC Compensator

A static compensator is obtained when a fixed capacitor is associated with the thyristor-controlled reactor; this allows the set to operate constantly in a capacitive range up to the maximum power value of the capacitor bank and in an inductive range up to the power difference limit (maximum conduction angle) subtracted from the power of the capacitor bank (Figure 4.21).

4.14.8.4 Compensator Type TCR/TSC

This static compensator is obtained by combining the two others, using controlled reactor and capacitor switchable. This compensator is mostly versatile because it uses all the advantages of combining the equipment with the thyristors; this allows continuous control over the entire reactive and capacitive range, taking advantage of the maximum power of each component.

The level of harmonics generated by the TCR-controlled reactor is greatly reduced by the compositional flexibility, which allows a marked reduction in the level of losses. Other important advantages are the minimization of the number of capacitors and reactors needed and the modularity and redundancy, which enable high reliability.

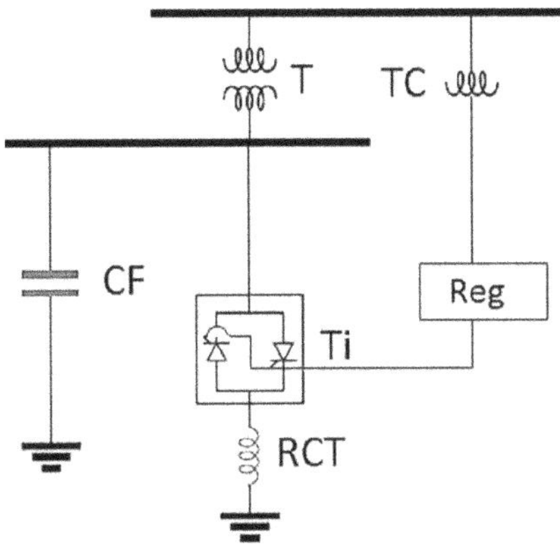

FIGURE 4.21
TCR/FC compensator.

T – Transformer
C – Fixed Capacitor
TC – Current Transformer
RTC –Thyristor-controlled reactor
Reg –Voltage and trip regulator

A detailed schematic of a TCR/TSC type static compensator is shown in Figure 4.22.

There are two ways to operate a static TCR/TSC compensator:

- Manual operation: In this situation, the operator adjusts the firing angle and consequently the reactive power of the static compensator by a predetermined value. Thus, the compensator operates as a fixed capacitor bank or a fixed reactor and does not regulate system voltage.
- Automatic operation: In this situation, the operator selects the desired voltage value in the 230-kV bus and the compensator will supply the chosen value. Here, the static compensator works as voltage control equipment.

Table 4.4 summarizes the operating ranges of a TCR/TSC type static compensator.

TABLE 4.4 TCR/TSC Compensator Operation

Faring Angle (degrees)	Reactive Power (MVAr)	Condition of Compensator
92.5	–140	Fully inserted reactors
120	0	Reactors balancing capacitors
170	200	Totally withdrawn reactors

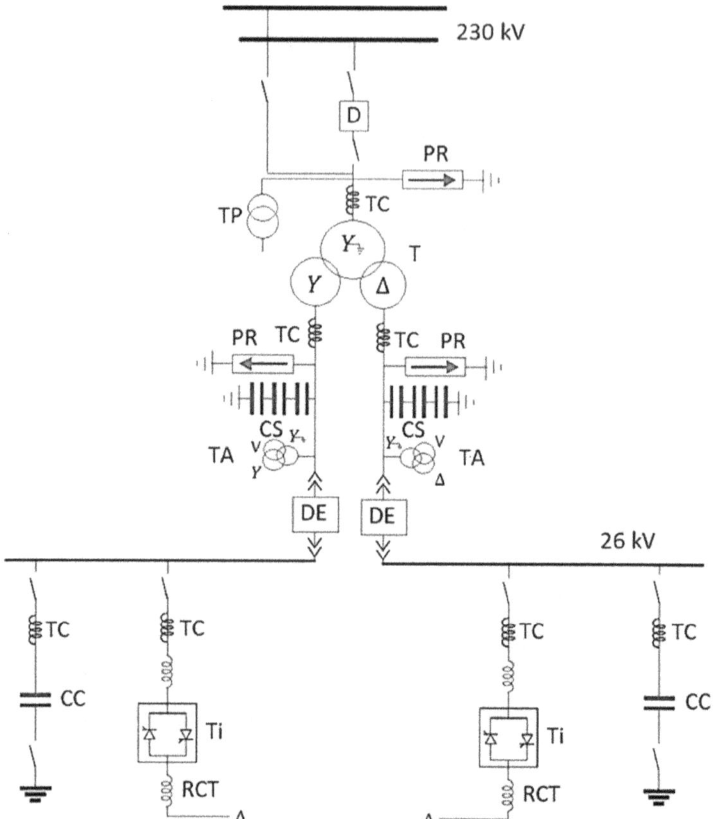

FIGURE 4.22
TCR / TSC compensator.

D – 230 kV circuit breaker
T – Three single-phase transformers 3×66.67 MVA = 200 MVA, 230/26/26 kV with two secondary, one delta connected and one star with floating neutral
PR – Lightning arrester
TP – Transformer of potential
TC – Transformer of current
CS – Surge capacitors, which attenuate the rate of growth of the incident voltage wave, from maneuvering surges or atmospheric discharges
TA – Ground transformer – transformers used in a delta system to provide a ground source in such a way that a ground current relay can detect or isolate ground fault lines in the system. In Figure 4.22, both transformers have their primary windings connected in a zigzag with grounded neutral. Two secondary open circuits connected to the 220 V supply the auxiliary services of the substation. The grounding transformer is responsible for the artificial 26 kV industry neutral.
DE – Removable circuit breaker
CC – Switchable capacitor – two banks of capacitors of 100 MVAr, connected in a star with floating neutral, fed in 26 kV
TI – Thyristors (12 pulses, sized to operate, also, 6 pulses)
RTC – Thyristor controlled reactor – two three-phase reactors of 170 MVAr each, connected in delta to the secondary of the transformer to cancel the harmonic currents of 3rd, 5th, 9th, and 15th order. The sectioning of the windings is intended to limit the short-circuit, which would be subjected to the thyristor valve.

From Table 4.4, we can conclude that the reduction of the firing angle causes an increase of inductive reactive power and, the increase of the firing angle decreases the inductive reactive power injection.

4.15 Compensation of TLs

An equivalent circuit for the compensation of a TL is shown in Figure 4.23. It is assumed that half the compensation is installed on each side of the line.

According to Table 4.1, the equivalent quadrupoles for the TL, the inductor compensation, and the capacitor compensation are respectively:

$$q_{LT} = \begin{bmatrix} A_1 & B_1 \\ C_1 & D_1 \end{bmatrix} \tag{4.141}$$

$$q_{shunt} = \begin{bmatrix} 1 & 0 \\ Y_{ind} & 1 \end{bmatrix} \tag{4.142}$$

$$q_{serie} = \begin{bmatrix} 1 & Z_{cap} \\ 0 & 1 \end{bmatrix} \tag{4.143}$$

where:

$$A_1 = 1 + ZY'/2$$

$$B_1 = Z$$

$$C_1 = Y'[1 + ZY'/4]$$

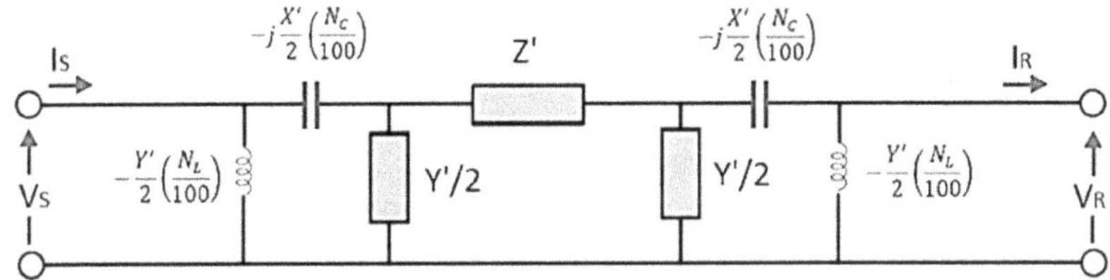

FIGURE 4.23
Equivalent circuit for TL compensation.

N_C is the amount of capacitive series compensation expressed as a percentage of the positive sequence impedance of the line;
N_L is the amount of reactive compensation shunt expressed as a percentage of the transmission line positive sequence admittance.

$$D_1 = 1 + ZY'/2$$

$$Y_{ind} = -\frac{Y'}{2}\left(\frac{N_L}{100}\right)$$

$$Z_{cap} = -j\frac{X'}{2}\left(\frac{N_C}{100}\right)$$

Therefore, the equivalent quadrupole is obtained by the product of the quadrupole of each component:

$$\begin{aligned} q_{equivalente} &= (q_{shunt})(q_{serie})(q_{LT})(q_{serie})(q_{shunt}) \\ &= \begin{bmatrix} 1 & 0 \\ Y_{ind} & 1 \end{bmatrix} \begin{bmatrix} 1 & Z_{cap} \\ 0 & 1 \end{bmatrix} \begin{bmatrix} A_1 & B_1 \\ C_1 & D_1 \end{bmatrix} \begin{bmatrix} 1 & Z_{cap} \\ 0 & 1 \end{bmatrix} \begin{bmatrix} 1 & 0 \\ Y_{ind} & 1 \end{bmatrix} \\ &= \begin{bmatrix} A & B \\ C & D \end{bmatrix} \end{aligned} \tag{4.144}$$

where:

$$A = A_1\left(1 + Z_{cap}Y_{ind}\right) + B_1Y_{ind} + Z_{cap}[C_1\left(1 + Z_{cap}Y_{ind}\right) + D_1Y_{ind}] \tag{4.145}$$

$$B = A_1Z_{cap} + B_1 + Z_{cap}(C_1Z_{cap} + D_1) \tag{4.146}$$

$$\begin{aligned} C = Y_{ind}\Big\{&A_1\left(1 + Z_{cap}Y_{ind}\right) + B_1Y_{ind} + Z_{cap}\left[C_1\left(1 + Z_{cap}Y_{ind}\right) + D_1Y_{ind}\right] \\ &+ C_1\left(1 + Z_{cap}Y_{ind}\right) + D_1Y_{ind}\Big\} \end{aligned} \tag{4.147}$$

$$D = Y_{ind}\left[A_1Z_{cap} + B_1 + Z_{cap}\left(C_1Z_{cap} + D_1\right)\right] + C_1Z_{cap} + D_1 \tag{4.148}$$

Equations (4.144) to (4.148) are general for the calculation of the shunt compensation and the compensation series.

The procedure for calculating the total reactive, which partially compensates for the no-load line, knowing the sending-end line voltages V_S and the receiving-end voltage V_R, is described next.

As $V_S = AV_R$, making $A = k = \frac{|V_S|}{|V_R|}$ e $Z_{cap} = 0$, in equation (4.145), we have:

$$k = A_1 + B_1Y_{ind} \tag{4.149}$$

Calculating Y_{ind}, we have:

$$Y_{ind} = -Im\left\{\frac{k - A_1}{B_1}\right\} S/fase \tag{4.150}$$

For the admittance shunt is worth $Y = y_1 - jY_{ind}$ and *Im* is the imaginary part of the complex expression.

Therefore, the three-phase reactive capacity of the reactors to be placed at each end of the TL is calculated by equation (4.151), where V_L is the line voltage of the TL in kV.

$$Q_{ind} = V_L^2 Y_{ind} \ MVAr \tag{4.151}$$

The procedure for calculating the total of reactive, which partially compensates for the series line, is described next.

From equation (4.146), writing the real and imaginary parts, with the imaginary part of B equal to $k1 = (vl)(imag(B1))$ with vl ranging from 0 to 1, we have:

$$\begin{aligned} b_r + jk1 = & -jX_C\left(a1_r + ja1_i + d1_r + jd1_i\right) \\ & +\left(c1_r + jc1_i\right)\left(-jX_C\right)^2 + b1_r + jb1_i \end{aligned} \tag{4.152}$$

As $A_1 = D_1$, we have:

$$b_r + jk1 = j2X_C\left(a1_r + ja1_i\right) + \left(c1_r + jc1_i\right)\left(jX_C\right)^2 + b1_r + jb1_i \tag{4.153}$$

$$b_r + jk1 = \left(-2X_C a1_i - X_C^2 c1_r + b1_r\right) + j(2X_C a1_r + X_C^2 c1_i + b1_i) \tag{4.154}$$

Equating the imaginary parts, we get:

$$k1 = (2X_C a1_r + X_C^2 c1_i + b1_i) \tag{4.155}$$

$$c1_i X_C^2 + 2a1_r X_C + b1_i - k1 = 0 \tag{4.156}$$

Solving the equation of the second degree:

$$X_C = \frac{+2a1_r \pm \sqrt{(2a1_r)^2 + 4c1_i\ (b1_i - k1)}}{2c1_i} \tag{4.157}$$

We must use the negative root because the calculation is of the capacitive reactance.

The three-phase reactive power of the capacitor bank shall be equal to:

$$Q_C = 3X_C I_L^2 \ MVAr \tag{4.158}$$

Where I_L is the line current of the TL in kA.

Example 4.5

A 500-kV TL is fed by a constant voltage bus equal to 500 kV and supplies passive loads. The TL has a length of 305 km and the following parameters:

$$R = 0.025\ \Omega/km$$

$$L = 0.322\ mH/km$$

$$C = 13.65\ nF/km$$

Calculate:

a. The value of the voltage at the end of the TL when it operates in no-load.
b. Assuming that in no-load operation the voltage at the receiver is at most equal to 515 kV, what power of the reactors should be placed in the TL?
c. What are the constants of the compensated TL?

SOLUTION:

TL impedance and admittance calculation:

$$Z' = (0.0025 + j0.322 \times 377 \times 0.001)305 = 7.6250 + j37.0252\ \Omega$$

$$Y' = j305(13.65 \times 377 \times 0.000000001 = j0.0016\ S$$

Calculation of the quadrupole parameters:

$$A1 = 1 + \frac{Z'Y'}{2} = 1 + \frac{(7.6250 + j37.0252)(0.0016j)}{2} = 0.9709 + j0.0060$$

$$B1 = Z = 7.6250 + j37.0252$$

$$C1 = Y\left(1 + \frac{ZY}{4}\right) = j0.0015$$

$$D1 = A1$$

a. $V_R = \frac{V_S}{|A1|} = \frac{500}{\sqrt{0.9709^2 + 0.0060^2}} = \frac{500}{0.9710} = 514.9532\ kV$

b. Using equation (4.150), we get $k = \frac{V_S}{V_R} = \frac{500}{515} = 0.9709$

$$Y_{ind} = -Im\left\{\frac{k - A_1}{B_1}\right\} = -Im\left\{\frac{0.9709 - (0.9709 + j0.0060)}{7.6250 + j37.0252}\right\} = 3.0119 \times 10^{-5}$$

Power of the reactors:
Using equation (4.151):

$$Q_{ind} = V_L^2 Y_{ind} = (500)^2\, 3.0119x10^{-5} = 7.5297\ MVAr$$

c. Constants of the compensated TL: using equation (4.144):

$$q_{equivalente} = \begin{bmatrix} 1 & 0 \\ Y_{ind} & 1 \end{bmatrix} \begin{bmatrix} A_1 & B_1 \\ C_1 & D_1 \end{bmatrix} \begin{bmatrix} 1 & 0 \\ Y_{ind} & 1 \end{bmatrix}$$

$$= \begin{bmatrix} 1 & 0 \\ j3.0119x10^{-5} & 1 \end{bmatrix} \begin{bmatrix} 0.9709 + j0.0060 & 7.6250 + j37.0252 \\ j0.0015 & 0.9709 + j0.0060 \end{bmatrix}$$

$$\times \begin{bmatrix} 1 & 0 \\ j3.0119x10^{-5} & 1 \end{bmatrix}$$

$$= \begin{bmatrix} 0.9698 + j0.0062 & 7.6250 + j37.0252 \\ -0.0000 + j0.0016 & 0.9698 + j0.0062 \end{bmatrix}$$

Example 4.6

Calculate the value of the capacitive reactance, the capacitive reactive power required to make a 30% compensation with a current of 1,500 A and the new compensated TL constants.

SOLUTION:

Using equation (4.157), with $k1 = 0.3Im\{B1\} = 11.1076$

$$X_C = \frac{2a1_r - \sqrt{(2a1_r)^2 + 4c1_i\,(b1_i - k1)}}{2c1_i}$$

$$= \frac{2(0.9709\,) - \sqrt{(2 \times 0.9709\,)^2 + 4 \times 0.0015(37.0252 - 11.1076)}}{2 \times 0.0015}$$

$$= -13.2077\ \Omega$$

$$Q_C = 3X_C I_L^2 = 3(13.2077)1500^2 = 89.1519\ MVAr$$

New constants of the TL: using equation (4.144), we have:

$$q_{equivalente} = \begin{bmatrix} 1 & Z_{cap} \\ 0 & 1 \end{bmatrix} \begin{bmatrix} A_1 & B_1 \\ C_1 & D_1 \end{bmatrix} \begin{bmatrix} 1 & Z_{cap} \\ 0 & 1 \end{bmatrix}$$

$$= \begin{bmatrix} 1 & -j13.2077 \\ 0 & 1 \end{bmatrix} \begin{bmatrix} 0.9709 + j0.0060 & 7.6250 + j37.0252 \\ j0.0015 & 0.9709 + j0.0060 \end{bmatrix}$$

$$\times \begin{bmatrix} 1 & -j13.2077 \\ 0 & 1 \end{bmatrix}$$

$$= \begin{bmatrix} 0.9914 + j0.0060 & 7.7839 + j11.1076 \\ -0.0000 + j0.0015 & 0.9914 + j0.0060 \end{bmatrix}$$

4.16 Introduction to the Electrical Design of a TL

The electrical design of a TL at the steady-state has several stages. Among them, we can cite:

A. Getting necessary data:
- A1- The receiving-end three-phase power;
- A2- The receiving-end three-phase voltage;
- A3- Frequency of the TL operation;
- A4- The receiving-end load power factor;
- A5- Number of TL circuits;
- A6- Number of sub-conductors per phase;
- A7- TL length;
- A8- Conductors' arrangement;
- A9- Wire gauge;
- A10- Operating temperature of the conductors.

B. Calculations:
- B1- The receiving-end phase voltage;
- B2- Charging current of the receiving bus;
- B3- Calculation of distances;
- B4- Calculation of the TL parameters;
  - B4.1 Resistance;
  - B4.2 Inductance;
  - B4.3 Capacitance.
- B5- Calculation of the TL sequence parameters;
- B6- TL series impedance;
- B7- Admittance TL shunt;
- B8- Characteristic impedance;
- B9- Propagation constant;
- B10- Calculation of the sending-end voltage, current, power, and power-factor;
- B11- The sending-end voltage drop in %;
- B12- The sending-end active power;
- B13- The sending-end reactive power;
- B14- Calculation of line losses including crown;
- B15- Voltage regulation;

B16- TL efficiency;

B17- Calculation of shunt compensation;

B18- Calculation of the series compensation.

The calculations of item B refer to the long-length lines and should be adapted for the medium-length and short-length lines.

All the items described above have already been duly detailed in Chapters 2, 3, and 4; thus, we can exercise all the items through Problem 4.17.10.

4.17 Problems

4.17.1 A 230 kV, 60 Hz, 300 km power transmission line has the following parameters:

$$R = 0.031 \frac{\Omega}{km}$$

$$L = 0.8435\ mH/km$$

$$C = 13.61\ nF/km$$

$$G = 0.0125\ \mu\ S/km$$

The TL meets a load of 230 MW with a power factor of 0.85 inductive. Calculate:

a. The characteristic impedance.
b. The propagation constant.
c. The attenuation constant.
d. The phase constant.
e. The wavelength.
f. The propagation velocity.
g. The sending-end voltage and current.

Answers: a) $Z_C = 249.5398 \angle -2.7141°$, b) $\gamma = 0.06375 + j1.2788\ 1/m$, c) $\alpha = 0.06375\ Np/m$, d) $\beta = 1.2788\ rad/m$, e) $\lambda = 4913.4716\ km$, f) $V_0 = 294808.2979\ km/s$, g) $V_S = 161.8970 \angle 15.7519°\ kV$; $I_S = 464.1260 \angle -9.8171°$.

4.17.2 Consider the data in question 4.17.1. Disregarding the parallel conductance of the TL, make a program in MATLAB language to calculate the sending-end voltage and a current, using the voltage and current equations in the hyperbolic form.

Answer: A function of type * .m must be programmed.

4.17.3 A short-length three-phase transmission line has 16 km at 69 kV. The TL has an impedance of Z = 0.125 + j0.4375 Ω/km. Calculate: (a) the sending-end voltage, (b) the voltage regulation, (c) the receiving-end power, and (d) the efficiency of the line when it supplies:

1. 70 MVA, pf = 0.8 inductive at 69 kV.
2. 120 MW, pf = 1 at 69 kV.

Answers: 1a) $V_S = 43.311 \angle 3.4113° \ kV$, 1b) $VR\ (\%) = 8.7203$, 1c) $S = 58.058 + j49.204\ MVA$, 1d) $\eta = 96.4546\%$, 2a) $V_S = 42.432 \angle 9.5348° \ kV$, 2b) $VR\ (\%) = 6.5124$, 2c) $S = 12.605 + j21.172\ MVA$, 2d) $\eta = 95.2010\%$.

4.17.4 A three-phase line of 230 kV, 200 km, 60 Hz has a phase resistance of 0.031 Ω/km and the phase inductance is 0.8435 mH/km. The shunt capacitance is 13.91 nF/km. The receiving-end load is 250 MVA with pf = 0.8 inductive at 230 kV. Use the medium-length model and calculate:

a. The sending-end voltage.
b. The receiving-end and the sending-end complex power.
c. The voltage regulation.
d. The transmission line efficiency.

Answers: a) $V_S = 112.95 \angle 17.9254° \ kV$, b) $VR\ (\%) = 3.4572$, c) $S_R = 200 + j150\ MVA$; $S_S = 206.44 + j149.01\ MVA$, d) $\eta = 95.9736\%$.

4.17.5 A three-phase transmission line with 350 km at 60 Hz supplies a load of 600 MVA, pf = 0.85 inductive at 500 kV. The line constants are R = 0.025 Ω/km, L = 0.8541 mH/km, and C = 13.95 nF/km. Calculate:

a. The attenuation and phase constants.
b. The constants A, B, C, and D.
c. The sending-end voltage, current, power, and power factor.
d. Line losses.
e. The efficiency of the line.
f. Voltage regulation.
g. The sending-end charging current at the line no-load.
h. The value of the voltage increases without load if the sending-end voltage is kept constant.
i. The equivalent circuit of the TL.

Answers: a) $\alpha = 0.0504\frac{Np}{m}$; $\beta = 1.3023\ rad/m$, b) $A = 0.8981 + j0.0078$; $B = 8.1543 + j108.8626$; $C = -0.0000048 + j0.0018$; $D = A$, c) $V_S = 526.1601 + j109.7739\ kV$; $I_S = 530.3023 + j189.9924\ A$; $S_S = 519.4076 - j72.3185\ MVA$; $FP = 0.9904\ ind$, d) $P_L = 9.4075\ MW$, e) $\eta = 98.1888$, f) $VR(\%) = -30.8931\%$, g) $I_C = 614.2502\ \angle -78.5555°$, h) $\Delta V = 88.5212\ kV$, i) $Z' = 109.1676\ \angle 85.7163°\ \Omega$; $\frac{Y'}{2} = 0.001979\ \angle 89.6598°\ S$.

4.17.6 Talk about reactive compensation.

Answer: See item 4.14.

4.17.7 Describe a TCR/TSC compensator.

Answer: See TCR/TSC compensator.

4.17.8 A 500 kV TL is fed by a constant voltage bus equal to 500 kV and supplies passive loads. The TL has a length of 250 km and the following parameters:

$$R = 0.025\ \Omega/km$$

$$L = 0.322\ mH/km$$

$$C = 13.65\ nF/km$$

Make a program using MATLAB language to determine:

a. The value of the voltage at the end of the TL when it operates in no-load.
b. Assuming that in no-load operation the voltage at the receiver is at most equal to 515 kV, what power of the reactors should be placed in the TL?
c. What are the constants of the compensated TL?

Answer: A function of type * .m must be programmed.

4.17.9 Make a program in MATLAB language to perform the calculations listed in item 4.16 (Introduction to the Electrical Design of a TL) for a TL with the following characteristics:
Three-phase power in the receiving bus: 10 MW.
Line voltage in the busbar: 13.8 kV.
Frequency of the TL: 60 Hz.
Receiving bus power factor: 0.8 inductive.
Number of TL circuits: 1.
TL length: 130 km.
Arrangement of conductors: horizontal.
Distances: a to b = 3 m; b to c = 3 m.
Conductors gauge: 300 MCM – 19 copper wires.
Temperature: 50°C.

Answer: A function of type * .m must be programmed.

4.17.10 The transmission lines in Brazil have a length ranging from a few meters to about 1,000 km. A line is classified as short-length when it has:

a. Length ranging from 80 to 200 km.
b. Length less than 80 km.
c. Length ranging from 200 to 400 km.
d. Length ranging from 400 to 500 km.
e. Any length.

Answer: b.

4.17.11 High-voltage AC transmission lines allow the transmission and distribution of electricity. As a result, they have different transmission capacities, characteristics, and lengths. Considering this, judge the following items.

a. Transmission lines at industrial frequency have constant active losses, independent of loading.
b. If a line with a nominal voltage exceeding 230 kV and having a length of 200 km is operating in no-load, the voltage at the unloaded end shall always be less than that on the side where the voltage is being applied.

Answers: a) Wrong, b) Wrong.

4.17.12 Regarding the sizing and performance criteria of a power transmission line, judge the following items:

a. High-voltage extra-high voltage transmission lines generate negative and zero sequence voltage unbalance.
b. The limiting factor for the transmission capacity of lines with a capacity greater than or equal to 230 kV is the maximum level of audible noise internal to the line of service of the line.
c. A long-length line, at no-load can generate reactive power higher than that required for its own consumption.

Answers: a) Right, b) Wrong, c) Right.

4.17.13 Consider the following propositions regarding transmission lines:

I. The typical characteristic impedance value for a single-circuit transmission line is 1,500 Ω.
II. The inductive reactance per unit length can be obtained by adding the calculated inductive reactance to a defined spacing plus the corresponding spacing factor.
III. The transposition of conductors aims to maintain the inductive reactance approximately equal in the three phases.

It is correct what is stated in:

a. I, only.

b. I and II, only.

c. I and III, only.

d. II and III, only.

e. I, II, and III.

Answer: d.

4.17.14 Consider an AC power transmission line, of which positive sequence parameters (capacitance, inductance, and resistance) are given per phase and per unit length. The characteristic impedance or impedance of this transmission line, considering its parameters of the positive sequence, is calculated knowing:

I. Capacitance.

II. Resistance.

III. Inductance.

IV. Rated line voltage.

Only certain items are true:

a. I and III.

b. I and IV.

c. II and III.

d. II and IV.

Answer: a.

4.17.15 A three-phase transmission line has a positive sequence impedance per unit of length z = j0.04 Ω and a positive sequence admittance per unit length y = j10-4S. The length of the TL is 100 km and the parameters *A*, *B*, *C*, and *D* of the quadrupole are given by:

$$\begin{bmatrix} V_S \\ I_S \end{bmatrix} = \begin{bmatrix} A & B \\ C & D \end{bmatrix} \begin{bmatrix} V_R \\ I_R \end{bmatrix}$$

Where: V_s and I_s are respectively the voltage and current at the sending-end terminal, V_r and I_r are, respectively, the receiving-end voltage and current. The values of the quadrupole array elements for this line are given by:

a. *A*=1,02; *B*=j 4; *C*=j 0,01 *S*; *D*=*A*.

b. *A*=1,02; *B*=j 4; *C*=j 0,01; *D*=*A*.

c. A=1,02; B=j 4; C=j 0,01 S; D=0,98.

d. A=0,98; B=j 4; C=j 0,01 S; D=A.

e. A=0,98; B=j 0,001; C=j 4; D=A.

Answer: d.

4.17.16 Judge the following items about steady-state power transmission at industrial frequency.

a. Short-length transmission line models, used for studies of electric power transmission, are usually represented by an electric circuit containing a resistor and a series inductor.

b. In models of long-length transmission lines, the capacitive effect of the line should be considered.

c. Extra-high voltage lines use aluminum and copper cables as conducting elements.

Answers: a) Right, b) Right, c) Right.

4.17.17 A no-load distribution line is energized from its emitter terminal. After reaching the steady state, the RMS value of the voltage at the open terminal is equal to twice the value of the voltage applied at the other terminal.
Considering the lossless line, the electric length of this line, in radians, is:

a. $\frac{\pi}{12}$

b. $\frac{\pi}{6}$

c. $\frac{\pi}{4}$

d. $\frac{\pi}{3}$

e. $\frac{\pi}{2}$

Answer: d.

4.17.18 Considering that in the transmission and distribution systems of electric power in AC, the most diverse types of load are found, consider the next items, about electrical equipment and electric charges.

a. In order for two three-phase transmission lines to be connected in parallel, they must have the same electrical parameters (resistance, inductance, and capacitance).

b. The transmission capacity of an electrical system can be improved by replacing old components with similar ones of newer technologies. This is called a retrofit. An example of this practice is the replacement of transmission line towers, which increases the capacity of a line by over 50% of its nominal value.

Answers: a) Wrong, b) Wrong.

4.17.19 In a steady-state study on the energization of a transmission line, it shall be determined:

a. The profile of the magnitude voltage on the line.

b. The rise of temperature of the conductors.

c. Damping of electromechanical oscillations.

d. The ohmic loss in the line.

e. Overvoltage achieved by turbines.

Answer: a.

5

Transmission Lines Operation at Transient State

5.1 Introduction

This chapter deals with transient phenomena from the concepts of electromagnetic wave propagation in TLs.

5.2 Transients in Single-Phase Lines

Transient overvoltages in power systems may originate internally or externally to the system. For example, lightning strikes originate from outside and switching operations originate from the power system itself. The insulation level of the TLs and equipment is a function of the operating voltage, as shown in Table 5.1.

5.3 Traveling Waves

When a lightning strike strikes the ground wire or TL conductors, it causes a split-current injection which travels in opposite directions, as shown in Figure 5.1.

The TL will be represented here by the distributed parameter model developed in Chapter 4.

A lossless line is a good representation for high frequency lines, where wL and wC are much larger compared to R and G of the line, respectively. For outbreaks caused by lightning strikes on a TL, the study of a lossless line is a simplification that allows us to understand some phenomena more objectively, although it is an approximate study.

A lossless line is shown in Figure 5.2.

As we are interested in studying the traveling voltage and current waves along the TL, we select the direction in which x increases from the sending-end ($x = 0$) toward the receiving-end ($x = l$).

TABLE 5.1
Causes for Insulation Level

Voltage Magnitude (kV)	TL Insulation Level Determinant
$V \leq 230$	Lightning discharges
$230 < V \leq 700$	Lightning and switching operations
$V > 700$	Switching operations

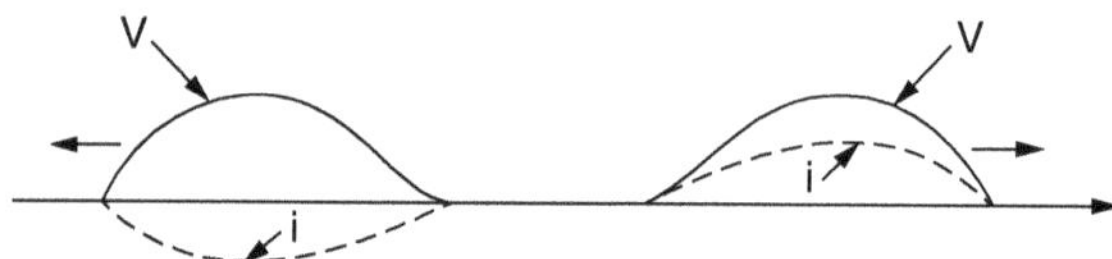

FIGURE 5.1
Traveling waves.

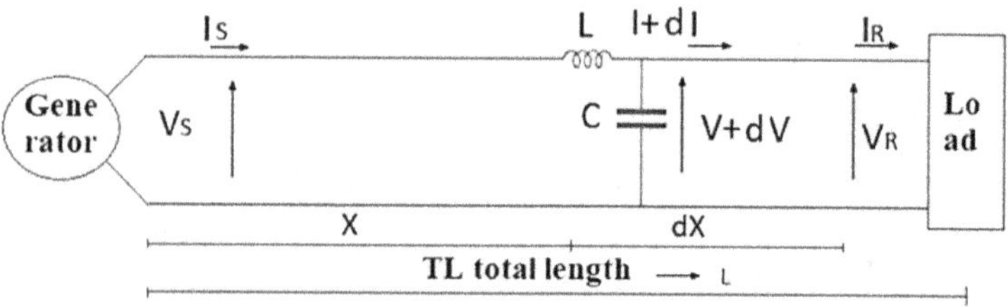

FIGURE 5.2
Lossless TL.

The equations of the telegraphists, (4.14) and (4.15), developed in Chapter 4, without considering losses, are given by:

$$\frac{\partial^2 v}{\partial x^2} = LC\frac{\partial^2 v}{\partial t^2} \tag{5.1}$$

$$\frac{\partial^2 i}{\partial x^2} = LC\frac{\partial^2 i}{\partial t^2} \tag{5.2}$$

Equations (5.1) and (5.2) are called a traveling wave equation for a lossless TL.

Equations (5.1) and (5.2) are solved using the Laplace transform and since under these conditions s is only one parameter, the partial derivatives at x are replaced by total derivatives. Recalling that the Laplace transform of $f''(t)$ is$s^2 f(s) - sf(0) - f'(0)$, where null initial conditions are assumed, we have:

$$\frac{d^2V}{dx^2} = s^2LCV(x,s) \tag{5.3}$$

$$\frac{d^2I}{dx^2} = s^2LCI(x,s) \tag{5.4}$$

The solutions are similar to those obtained in the previous chapter, with the inverted x direction reference, and given by:

$$V(x,s) = A_1(s)e^{-\left(\frac{sx}{v}\right)} + A_2(s)e^{\left(\frac{sx}{v}\right)} = V_i(x,s) + V_r(x,s) \tag{5.5}$$

$$I(x,s) = \frac{A_1(s)}{Z(s)}e^{-\left(\frac{sx}{v}\right)} - \frac{A_2(s)}{Z(s)}e^{\left(\frac{sx}{v}\right)} = I_i(x,s) + I_r(x,s) \tag{5.6}$$

where:

v = the wave propagation speed

$V_i(x,s)$ and $I_i(x,s)$ = waves propagating in the direction of the growth of progressive or incident x waves

$V_r(x,s)$ and $I_r(x,s)$ = waves propagating in the negative direction of x—regressive or reflected waves

From equations (5.5) and (5.6), we get:

$$Z(s) = \frac{V_i(x,s)}{I_i(x,s)} = Z_C = \sqrt{\left(\frac{L}{C}\right)} \tag{5.7}$$

$$Z(s) = \frac{V_r(x,s)}{I_r(x,s)} = -\sqrt{\left(\frac{L}{C}\right)} \tag{5.8}$$

If for the current I_r we chose the positive direction to be the backward wave, we would not have the negative signs in equations (5.6) and (5.8).

5.4 Transient Analysis: Reflections on a Discontinuity

When two traveling waves meet in a discontinuity (short circuit, open circuit, one cable, another TL, winding of a machine, and others) propagating in opposite directions, they add up and, after passing each other, the waves continue with the original shape and magnitude (lossless lines) and can therefore be analyzed separately, following the principle of superposition.

In discontinuity, the wave that goes toward it is the incident wave. A part of the wave is reflected back (reflected wave) and a part of the wave is transmitted (refracted wave).

Figure 5.3 shows a progressive wave of voltage $V_i(x,s)$ accompanied by a progressive wave of current $I_i(x,s)$, focusing on a concentrated impedance

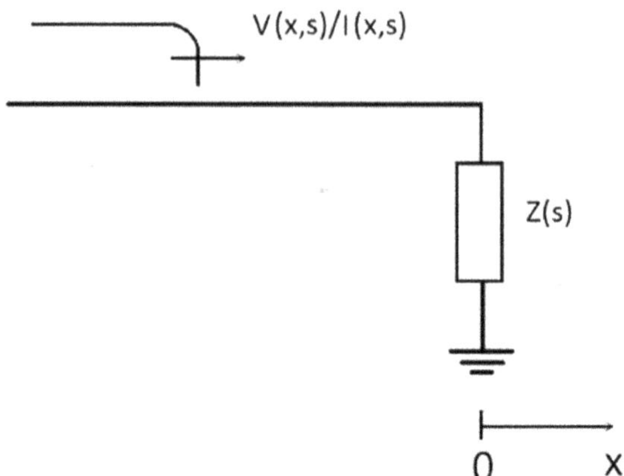

FIGURE 5.3
Z(s) impedance discontinuity point.

Z(s) that represents a discontinuity located at $x = 0$. The ratio of voltage to current at point $x = 0$ is:

$$\frac{V(0,s)}{I(0,s)} = Z(s) \tag{5.9}$$

The incident and reflected waves of voltage and current at the point of discontinuity are given by:

$$V_i(0,s) = A_1(s) \tag{5.10}$$

$$I_i(0,s) = \frac{A_1(s)}{Z_C} \tag{5.11}$$

$$V_r(0,s) = A_2(s) \tag{5.12}$$

$$I_r(0,s) = -\frac{A_2(s)}{Z_C} \tag{5.13}$$

Therefore:

$$V(0,s) = V_i(0,s) + V_r(0,s) = A_1(s) + A_2(s) \tag{5.14}$$

$$I(0,s) = I_i(0,s) + I_r(0,s) = \frac{A_1(s)}{Z_C} - \frac{A_2(s)}{Z_C} \tag{5.15}$$

Dividing equation (5.14) by equation (5.15) and taking the value of $A_2(s)$, we have:

$$A_2(s) = \frac{Z(s) - Z_C}{Z(s) + Z_C} A_1(s) \tag{5.16}$$

$$V_r(0,s) = \frac{Z(s) - Z_C}{Z(s) + Z_C} V_i(0,s) = \rho_{RV} V_i(0,s) \tag{5.17}$$

where:

$$\rho_{RV} = \frac{V_r(0,s)}{V_i(0,s)} = \frac{Z(s) - Z_C}{Z(s) + Z_C} \tag{5.18}$$

is the voltage reflection coefficient.

From equation (5.4), we have:

$$V(0,s) = A_1(s) + A_2(s) = \left(1 + \frac{Z(s) - Z_C}{Z(s) + Z_C}\right) A_1(s) \tag{5.19}$$

$$V(0,s) = \left(\frac{2Z(s)}{Z(s) + Z_C}\right) A_1(s) = \rho_{ReV} V_i(0,s) \tag{5.20}$$

where:

$$\rho_{ReV} = \frac{V(0,s)}{V_i(0,s)} = \left(\frac{2Z(s)}{Z(s) + Z_C}\right) \tag{5.21}$$

is the voltage refraction coefficient.

For the current, the reflection and refraction coefficients are:

$$\rho_{RI} = \frac{I_r(0,s)}{I_i(0,s)} = \frac{-\frac{A_2(s)}{Z_C}}{A_1(s)} = -\frac{(Z(s) - Z_C)}{Z(s) + Z_C} \tag{5.22}$$

$$\rho_{ReI} = \frac{I(0,s)}{I_i(0,s)} = \frac{\left(\frac{A_1(s)}{Z_C} - \frac{A_2(s)}{Z_C}\right)}{\left(\frac{A_1(s)}{Z_C}\right)} = \frac{\left(A_1(s) - \frac{(Z(s) - Z_C)}{Z(s) + Z_C} A_1(s)\right)}{A_1(s)} = \left(\frac{2Z_C}{Z(s) + Z_C}\right) \tag{5.23}$$

Table 5.2 summarizes the reflection and refraction coefficients for a discontinuous point ending in $Z(s)$.

TABLE 5.2

Reflection and Refraction Coefficients: Termination $\mathbf{Z}(s)$

	Reflection Coefficient	Refraction Coefficient
Voltage	$\frac{Z(s)-Z_C}{Z(s)+Z_C}$	$\left(\frac{2Z(s)}{Z(s)+Z_C}\right)$
Current	$-\frac{(Z(s)-Z_C)}{Z(s)+Z_C}$	$\left(\frac{2Z_C}{Z(s)+Z_C}\right)$

5.4.1 Resistive Termination

When the termination is resistive, Table 5.2 may be modified such that $Z(s)=R$ (Table 5.3).

When at the point of discontinuity, we have an open circuit ($R\rightarrow\infty$), the reflection and refraction coefficients are modified as follows:

$$\rho_{RV_{R\rightarrow\infty}}=\frac{1-Z_C/R}{1+Z_C/R}=1 \tag{5.24}$$

$$\rho_{ReV_{R\rightarrow\infty}}=\left(\frac{2}{1+Z_C/R}\right)=2 \tag{5.25}$$

$$\rho_{RI_{R\rightarrow\infty}}=-\frac{\left(1-\frac{Z_C}{R}\right)}{1+\frac{Z_C}{R}}=-1 \tag{5.26}$$

$$\rho_{ReV_{R\rightarrow\infty}}=\left(\frac{2Z_C/R}{1+Z_C/R}\right)=0 \tag{5.27}$$

The modified table is shown below as Table 5.4.

For the case where the termination is a short circuit $R\rightarrow 0$, we have:

$$\rho_{RV_{R\rightarrow 0}}=\frac{1-Z_C/R}{1+Z_C/R}=-1 \tag{5.28}$$

$$\rho_{ReV_{R\rightarrow 0}}=\left(\frac{2}{1+Z_C/R}\right)=0 \tag{5.29}$$

TABLE 5.3

Reflection and Refraction Coefficients: Termination $\mathbf{R}$

	Reflection Coefficient	Refraction Coefficient
Voltage	$\frac{R-Z_C}{R+Z_C}$	$\left(\frac{2R}{R+Z_C}\right)$
Current	$-\frac{(R-Z_C)}{R+Z_C}$	$\left(\frac{2Z_C}{R+Z_C}\right)$

TABLE 5.4

Reflection and Refraction Coefficients: Open-Circuit Termination

	Reflection Coefficient	Refraction Coefficient
Voltage	1	2
Current	–1	0

$$\rho_{RI_{R\to 0}} = -\frac{\left(1-\frac{Z_C}{R}\right)}{1+\frac{Z_C}{R}} = 1 \tag{5.30}$$

$$\rho_{ReV_{R\to 0}} = \left(\frac{2Z_C/R}{1+Z_C/R}\right) = 2 \tag{5.31}$$

The modified table is shown below as Table 5.5.

Here (short circuit), the doubled value after incidence is the refracted current.

Figure 5.4 shows the behavior of the voltage wave after traveling on a TL toward the R, ended open and short-circuit terminations. Figure 5.4 (a) shows that the terminal voltage is the same as the incident voltage because of the terminal resistance being the surge impedance ($R = Z_C$). Figure 5.4 (b) shows that the voltage at the open terminal is twice the incident voltage. Figure 5.4 (c) shows that the reflected voltage is the same and has the opposite signal to the incident wave when the terminal is shorted.

5.4.2 Inductive Termination

When the termination is inductive, as shown in Figure 5.5, the reflected and refracted waves will have deformations, as the reflection and refraction coefficients are now frequency dependent, as shown below.

$$V = L\frac{dI}{dt} \tag{5.32}$$

Doing the Laplace transform of equation (5.32), with initial conditions equal to zero, we have:

$$V(s) = sLI(s) \tag{5.33}$$

TABLE 5.5

Reflection and Refraction Coefficients: Short-Circuit Termination

	Reflection Coefficient	Refraction Coefficient
Voltage	–1	0
Current	1	2

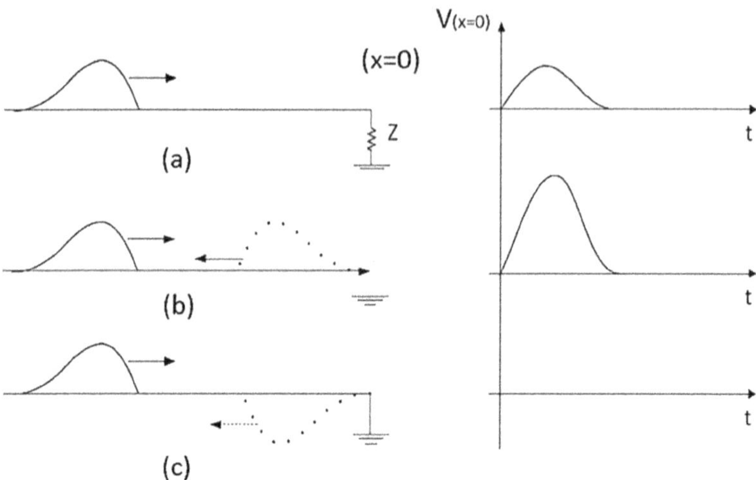

FIGURE 5.4
(a), (b), (c) Voltage wave at terminations.

Therefore, the impedance $Z(s) = sL$. Table 5.6 shows the reflection and refraction coefficients for inductive termination.

Assuming a unit step focusing on the inductive termination at ($t = 0$), the reflected voltage and current waves can be written as:

$$V_r(x,s) = A_2(s)e^{\left(\frac{sx}{v}\right)} = \left(\frac{sL - Z_C}{sL + Z_C}\right)A_1(s)e^{\left(\frac{sx}{v}\right)} = \left(\frac{sL - Z_C}{sL + Z_C}\right)\left(\frac{1}{s}e^{\left(\frac{sx}{v}\right)}\right) \quad (5.34)$$

$$I_r(x,s) = -\left(\frac{A_2(s)}{Z_C}\right)e^{\left(\frac{sx}{v}\right)} = -\left(\frac{sL - Z_C}{sL + Z_C}\right)\frac{1}{Z_C}\left(\frac{1}{s}e^{\left(\frac{sx}{v}\right)}\right) \quad (5.35)$$

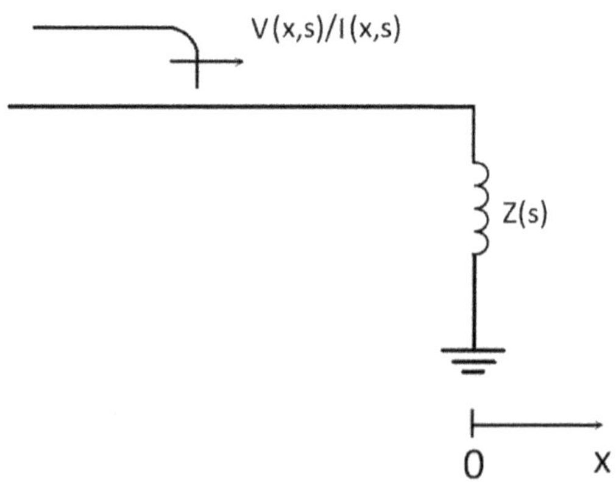

FIGURE 5.5
Inductive termination.

TABLE 5.6

Reflection and Refraction Coefficients: Termination *L*

	Reflection Coefficient	Refraction Coefficient
Voltage	$\frac{sL - Z_C}{sL + Z_C}$	$\left(\frac{2sL}{sL + Z_C}\right)$
Current	$-\frac{(sL - Z_C)}{sL + Z_C}$	$\left(\frac{2Z_C}{sL + Z_C}\right)$

The voltage $V_r(x,s)$ can be placed as follows:

Dividing the term of the second member by *L*, we get:

$$V_r(x,s) = \left(\frac{s - \frac{Z_C}{L}}{s + \frac{Z_C}{L}}\right)\left(\frac{1}{s}e^{\left(\frac{sx}{v}\right)}\right) = \left(\frac{(2s - s) - \frac{Z_C}{L}}{s + \frac{Z_C}{L}}\right)\left(\frac{1}{s}e^{\left(\frac{sx}{v}\right)}\right) = \left(\frac{2}{s + \frac{Z_C}{L}} - \frac{1}{s}\right)e^{\left(\frac{sx}{v}\right)} \quad (5.36)$$

Doing the inverse Laplace transform, we have:

$$V_r(x,t) = 2e^{\left(-\frac{Z_C}{L}\left(t + \frac{x}{v}\right)\right)}u\left(t + \frac{x}{v}\right) - u\left(t - \frac{x}{v}\right) \quad (5.37)$$

The line is located in the range $(-\infty, 0)$; therefore, the voltage reflected at any point on the line $(x = -X, 0,\ \textit{with}\ X > 0)$ is:

$$V_r(-X,t) = 2e^{\left(-\frac{Z_C}{L}\left(t - \frac{X}{v}\right)\right)}u\left(t - \frac{X}{v}\right) - u\left(t + \frac{X}{v}\right) \quad (5.38)$$

The graph in Figure 5.6 shows the voltage reflected in the inductive termination.

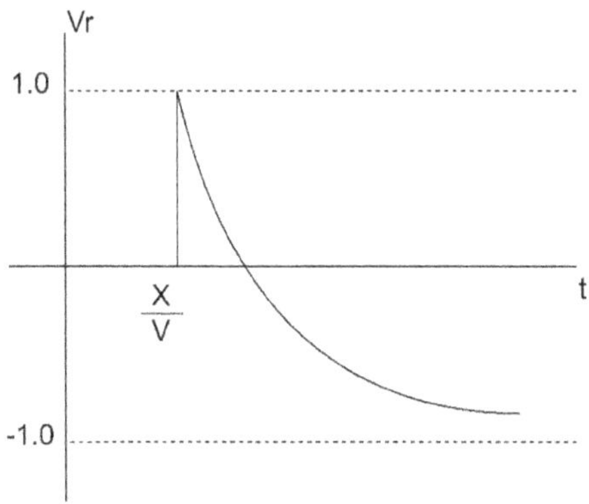

FIGURE 5.6
Reflected voltage wave: inductive termination.

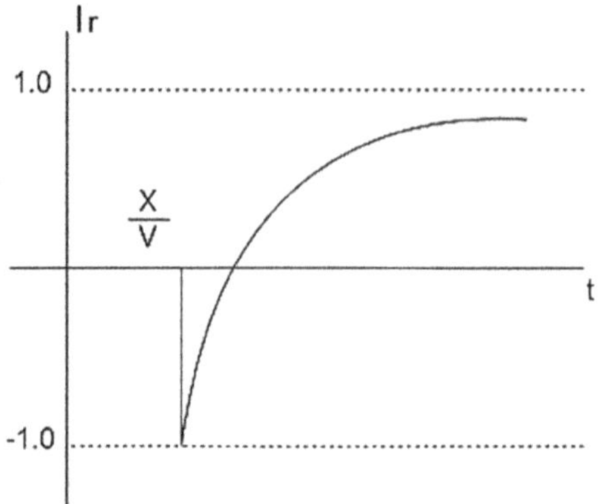

FIGURE 5.7
Reflected current wave: inductive termination.

The reflected current is shown in the graph in Figure 5.7.

The voltage on the inductor is the refracted voltage and can be determined by:

$$V(0,s) = V_i(0,s) + V_r(0,s) = A_1(s) + A_2(s) = \left(1 + \frac{sL - Z_C}{sL + Z_C}\right) A_1(s)$$
$$= \left(\frac{2sL}{sL + Z_C}\right)\frac{1}{s} = \left(\frac{2}{s + \frac{Z_C}{L}}\right) \tag{5.39}$$

Doing the Laplace inverse of equation (5.39) gives:

$$V(t) = 2e^{-\left(\frac{Z_C}{L}t\right)}u(t) \tag{5.40}$$

This expression shows that initially the inductor behaves as an open circuit $t \to 0$ and then as a short circuit for $t \to \infty$, as shown in Figure 5.8.

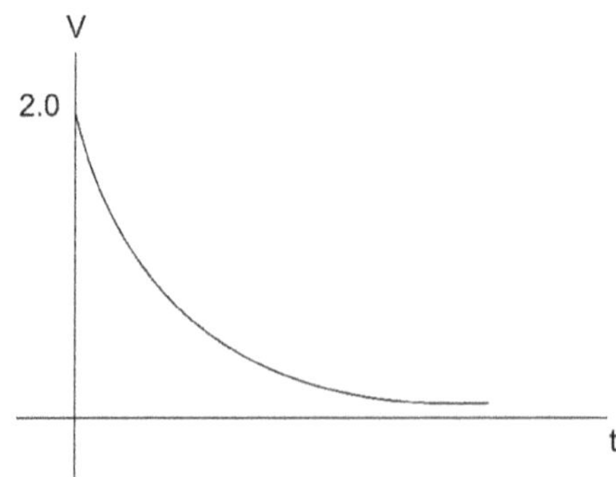

FIGURE 5.8
Inductor voltage.

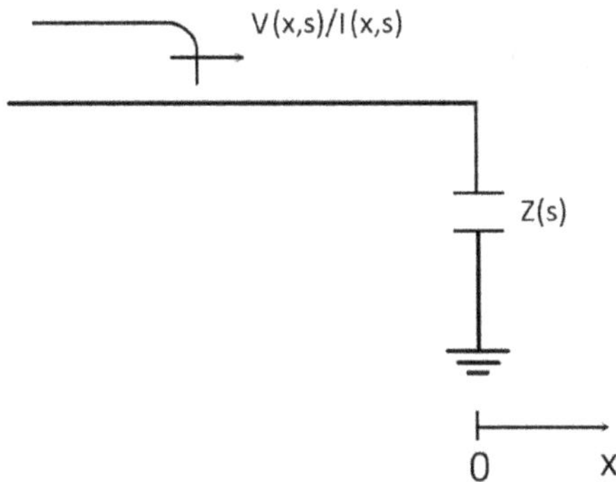

FIGURE 5.9
Capacitive termination.

5.4.3 Capacitive Termination

Using an inductor-like development, one can analyze the behavior of incident waves at a point of discontinuity that has a capacitor, as shown in Figure 5.9.

The equation that relates voltage and current in a capacitor is given by:

$$I = C\frac{dV}{dt} \tag{5.41}$$

Doing the Laplace transform of equation (5.32), with initial conditions equal to zero, we have:

$$I(s) = CsV(s) \tag{5.42}$$

Therefore:

$$Z(s) = \frac{V(s)}{I(s)} = \frac{1}{sC} \tag{5.43}$$

Table 5.7 shows the reflection and refraction coefficients for capacitive termination.

TABLE 5.7
Reflection and Refraction Coefficients: Termination *C*

	Reflection Coefficient	Refraction Coefficient
Voltage	$\frac{\frac{1}{sC} - Z_C}{\frac{1}{sC} + Z_C}$	$\left(\frac{2\frac{1}{sC}}{\frac{1}{sC} + Z_C}\right)$
Current	$-\frac{\left(\frac{1}{sC} - Z_C\right)}{\frac{1}{sC} + Z_C}$	$\left(\frac{2Z_C}{\frac{1}{sC} + Z_C}\right)$

Assuming a unit step focusing on the capacitive termination at ($t = 0$), the reflected voltage and current waves can be written as:

$$V_r(x,s) = A_2(s)e^{\left(\frac{sx}{v}\right)} = \left(\frac{\frac{1}{sC} - Z_C}{\frac{1}{sC} + Z_C}\right)A_1(s)e^{\left(\frac{sx}{v}\right)} = \left(\frac{\frac{1}{sC} - Z_C}{\frac{1}{sC} + Z_C}\right)\left(\frac{1}{s}e^{\left(\frac{sx}{v}\right)}\right) \tag{5.44}$$

$$I_r(x,s) = -\left(\frac{A_2(s)}{Z_C}\right)e^{\left(\frac{sx}{v}\right)} = -\left(\frac{\frac{1}{sC} - Z_C}{\frac{1}{sC} + Z_C}\right)\frac{1}{Z_C}\left(\frac{1}{s}e^{\left(\frac{sx}{v}\right)}\right) \tag{5.45}$$

The voltage $V_r(x,s)$ can be placed as follows:

$$V_r(x,s) = \left(\frac{1}{s} - \frac{2}{2 + \frac{1}{CZ_C}}\right)e^{\left(\frac{sx}{v}\right)} \tag{5.46}$$

Doing the inverse Laplace transform, we get:

$$V_r(x,t) = u\left(t + \frac{x}{v}\right) - 2e^{-\left(\frac{1}{CZ_C}\left(t+\frac{x}{v}\right)\right)}u\left(t + \frac{x}{v}\right) \tag{5.47}$$

The line is located in the range $(-\infty, 0)$; therefore, the voltage reflected at any point on the line ($x = -X, 0$, with $X > 0$) is:

$$V_r(-X,t) = u\left(t - \frac{X}{v}\right) - 2e^{-\left(\frac{1}{CZ_C}\left(t-\frac{X}{v}\right)\right)}u\left(t - \frac{X}{v}\right) \tag{5.48}$$

The graph in Figure 5.10 shows the voltage reflected in the capacitive termination.

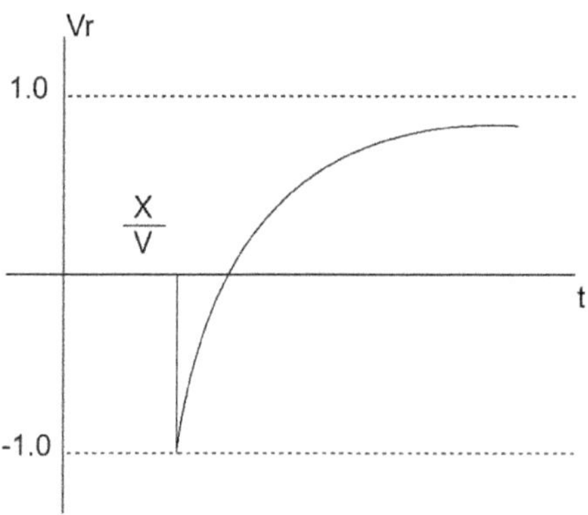

FIGURE 5.10
Reflected voltage wave: capacitive termination.

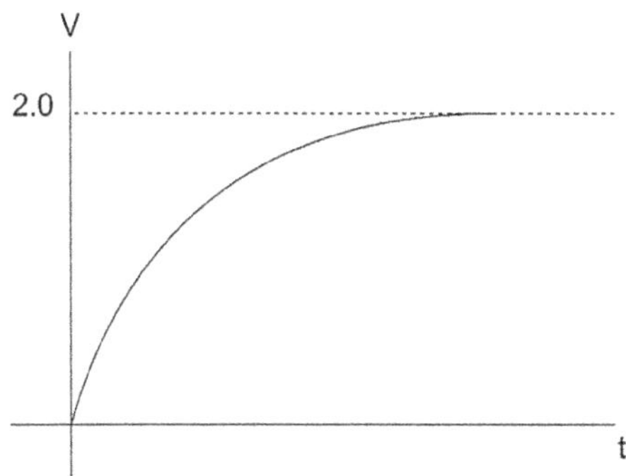

FIGURE 5.11
Capacitor voltage.

The voltage on the capacitor is the refracted voltage and can be determined by:

$$V(0,s)=V_i(0,s)+V_r(0,s)=A_1(s)+A_2(s)=\left(1+\frac{\frac{1}{sC}-Z_C}{\frac{1}{sC}+Z_C}\right)A_1(s)$$
$$=\left(\frac{2\frac{1}{sC}}{\frac{1}{sC}+Z_C}\right)\frac{1}{s}=2\left(\frac{1}{s}-\frac{1}{s+CZ_C}\right) \tag{5.49}$$

Doing the Laplace inverse of equation (5.49), we get:

$$V(t)=2u(t)-2e^{-\left(\frac{t}{CZC}\right)}u(t) \tag{5.50}$$

This expression shows that initially the inductor behaves as a circuit $t \to 0$ and then as an open short for $t \to \infty$, as shown in Figure 5.11.

5.5 Bewley Lattice Diagram

A lattice diagram developed by L. V. Bewley conveniently organizes the reflections that occur during TL transients.

Example 5.1

A 150 V DC source with negligible resistance is connected via a switch to a lossless TL that has a characteristic impedance of 60 Ω. The TL ends with a 120 Ω resistance. Plot the voltage at the resistor terminals up to five times the time required for the voltage wave to travel the full length of the TL. TL transit time is *T* (Figure 5.12).

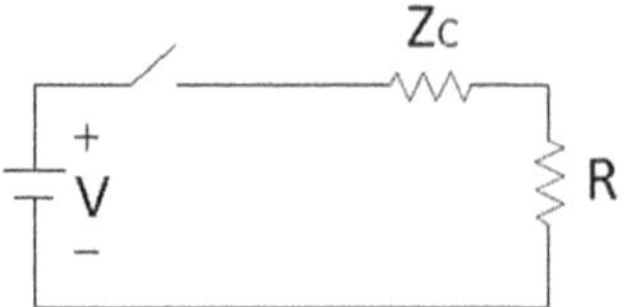

FIGURE 5.12
Circuit for example 5.1.

SOLUTION:

Calculation of reflection coefficients:

Reflection coefficient on resistance:

$$V_r = \frac{R - Z_C}{R + Z_C} V_i = \rho_{RV} V_i$$

$$\rho_{RV} = \frac{120 - 60}{120 + 60} = \frac{1}{3}$$

Coefficient of reflection at source:

$$\rho_{SV} = \frac{0 - 60}{0 + 60} = -1$$

The Bewley diagram (Figures 5.13 to 5.18) shows the voltages reflected in the sending-end and receiving-end terminals.

Calculation of reflected voltage at receiving-end:

$$V_r = \frac{1}{3} V_i = \frac{1}{3} 150 = 50\ V$$

The total voltage at the receiving terminal is shown in Figure 5.14.

$$V_{rT} = 150 + 50 = 200\ V$$

The voltage reflected in the sending-end:

$$V_{ri} = -V_i = -50\ V$$

Figure 5.15 shows the total voltage at the sending-end for time 2*T*.

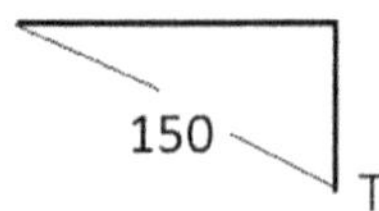

FIGURE 5.13
Voltage in time *T*.

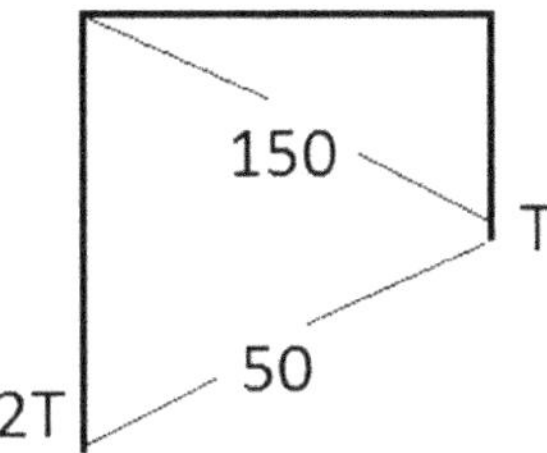

FIGURE 5.14
Total receiving-end voltage.

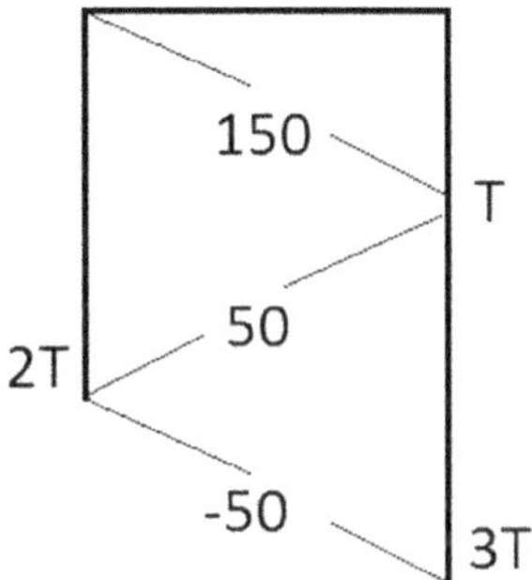

FIGURE 5.15
Voltage reflected on sending-end.

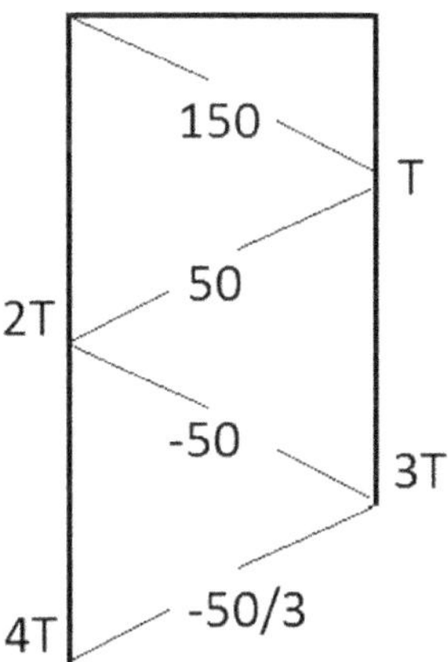

FIGURE 5.16
Receiving-end total voltage for 3*T*.

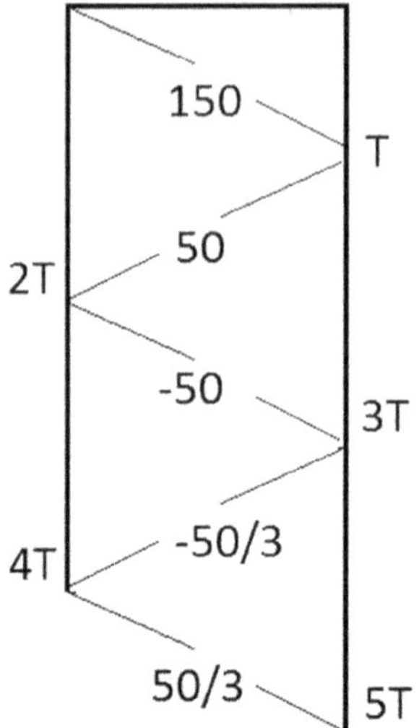

FIGURE 5.17
Total sending-end voltage for 4*T*.

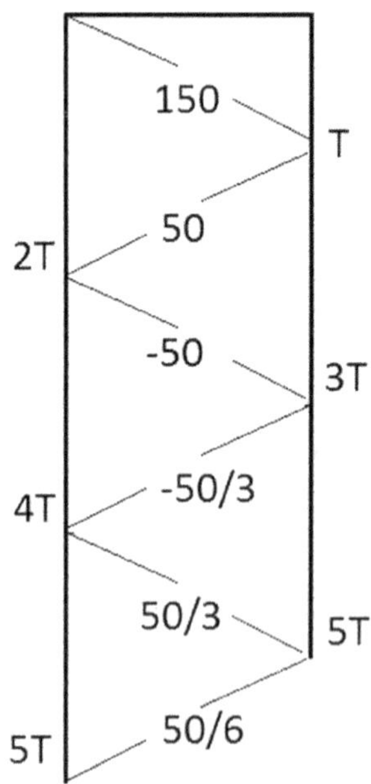

FIGURE 5.18
Receiving-end total voltage for 5*T*.

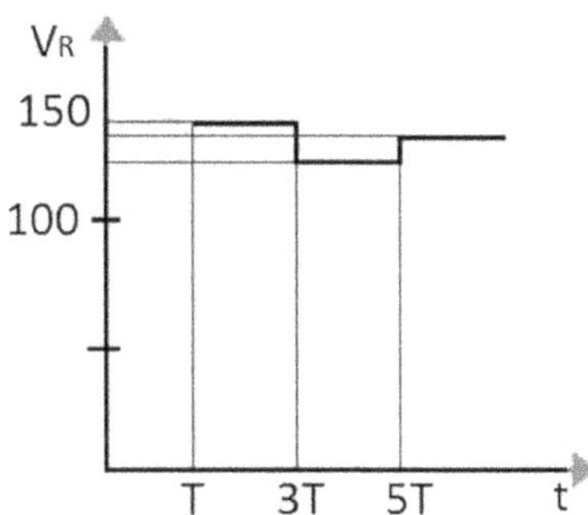

FIGURE 5.19
Receiving-end voltage graph.

The total voltage at the receiving-end is shown in Figure 5.16.

$$V_{r3T} = 200 - 50 - \frac{50}{3} = 133.3333\ V$$

The total voltage at the sending-end for $4T$ is shown in Figure 5.17.

The total voltage at the receiving terminal for $5T$ is shown in Figure 5.18.

$$V_{r5T} = 133.3333 + \frac{50}{6} = 141.6666\ V$$

Obviously, the voltage at the receiving-end tends to:

$$V_R = 150 - (60)x\frac{150}{60+120} = 100\ V$$

The graph of the receiving-end voltage is shown in Figure 5.19.

5.6 Lightning Surges

There are three types of lightning surges that can occur: within clouds, between clouds, and between clouds and ground, the latter being the lightning that matters to the power system, as shown in Figure 5.20.

Several studies on lightning surge modeling in the electrical system use the exponential impulse.

The standardized wave of the impulse isolation test is generally shown in Figure 5.21.

This representation can be seen in the form of current or voltage impulses.

The rise time is 1.2 μs and the tail time is 50 μs until it is reduced to 50% of the peak value.

FIGURE 5.20
Radius between cloud and ground hitting a transmission tower.

For the standardization of studies and tests, the standards establish the existing variables in the curves and values for them.

According to IEC 60060-1, for a voltage wave, the virtual origin O1 is the point where the imaginary line crossing the 30% and 90% points of the peak value reaches the time axis. For a current wave, the imaginary line crosses the points of 10% and 90% of the peak value and, at the intersection with the time axis, has the virtual origin. These points can be seen in Figures 5.22 and 5.23.

The wave front time $T_1 = 1.67 \times T = 1.2\ \mu s \pm 30\%$

The tail time of the wave $T_2 = 50\ \mu s \pm 20\%$

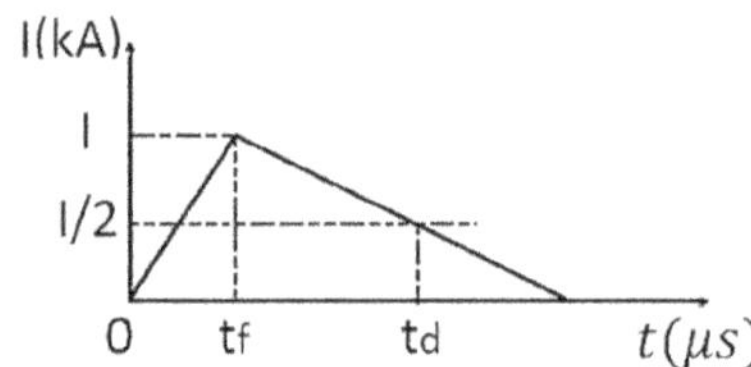

FIGURE 5.21
Standard lightning impulse wave.

where:

- I is the current intensity;
- t_f is the wave front time;
- t_d is the tail time of the wave.

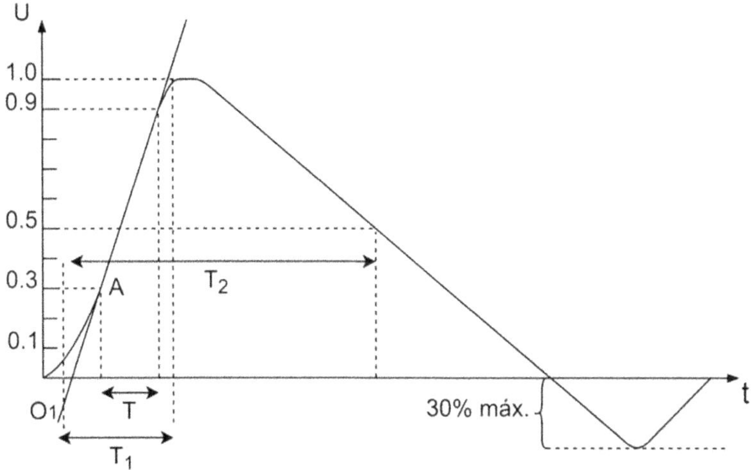

FIGURE 5.22
Standard surge voltage waveform (1.2/50 μs).

The wave front time $T_1 = 1.25 \times T = 8\ \mu s \pm 20\%$
The tail time of the wave $T_2 = 20\ \mu s \pm 20\%$
Also, according to this standard, the *T*1 wavefront time is, for a voltage surge, defined as 1.67 times the time interval between 30% and 90% of the peak value. For a current surge, it is defined as 1.25 times the time interval between 10% and 90% of the peak value.
Tail time *T*2 is the time interval between the virtual origin and the instant when the voltage or current values reach half of the peak value.

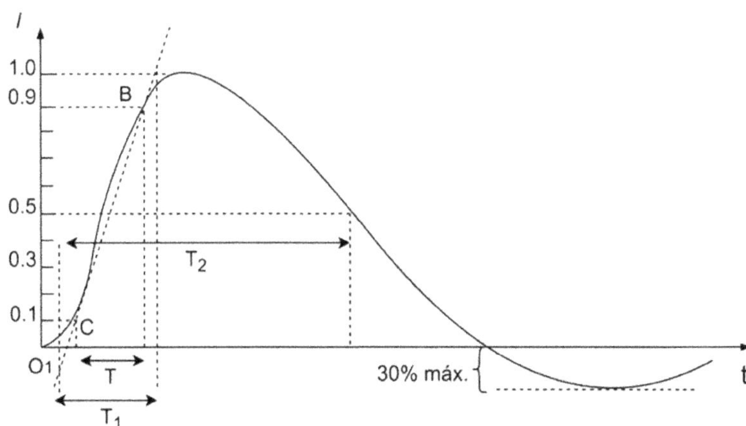

FIGURE 5.23
Standard surge current waveform (1.2/50 μs).

TABLE 5.8

IEC Standard Exponential Impulses 60060-1

Wave Shape	Front Time	Tail Time	Peak Value	Reverse Polarity
1/20	$1\ \mu s \pm 10\%$	$20\ \mu s \pm 10\%$	±10%	20%
4/10	$4\ \mu s \pm 10\%$	$10\ \mu s \pm 10\%$	±10%	20%
8/20	$8\ \mu s \pm 10\%$	$20\ \mu s \pm 10\%$	±10%	20%
30/80	$30\ \mu s \pm 10\%$	$80\ \mu s \pm 10\%$	±10%	20%

The default lead time for a voltage wave is 1.2 μs and the tail time is 50 μs. Standard waveforms for atmospheric impulses follow the exponential form surge wave theory, as shown in Figure 5.22.

The wave front time of a standard surge is 8 μs and the tail time is 20 μs, as shown in Figure 5.23.

As this is a natural phenomenon, it is not possible to establish universal rules for wave front and tail times. However, there is a valid behavior for most of these events. Most of these electrical events have shorter wave front times than tail times by a ratio of two or more times. In fact, the process of standardization of waveforms is appropriate to the technical possibilities presented by the various test laboratories.

Values are then set for exponential current pulses relative to a time, peak values, reverse polarity, energy, and threshold tolerances.

The IEC 60099-4 and IEC 60060-1 standards differ within the limits of each variable according to Tables 5.8 and 5.9.

The differences between the two standards are in the tolerance values.

Example 5.2

Using ATPdraw, simulate a lightning surge over the TL shown in Figure 5.24 using the JMarti model. Then put a surge arrester before the TL and check the voltage after the line.

TABLE 5.9

IEC Standard Exponential Impulses 60099-4

Wave Shape	Front Time	Tail Time	Peak Value	Reverse Polarity
$\frac{1}{20};20\ kA$	$0.9\ \mu s \le T_1 \le 1.1\ \mu s$	$20\ \mu s$	±5%	-
$\frac{4}{10};100\ kA$	$3.5\ \mu s \le T_1 \le 4.5\ \mu s$	$9\ \mu s \le T \le 11\ \mu s$	±10%	±20%
$\frac{8}{20};20\ kA$	$7\ \mu s \le T_1 \le 9\ \mu s$	$20\ \mu s \pm 10\%$	±10%	-
$\frac{30}{80};40\ kA$	$25\ \mu s \le T_1 \le 35\ \mu s$	$70\ \mu s \le T \le 90\ \mu s$	±10%	±20%

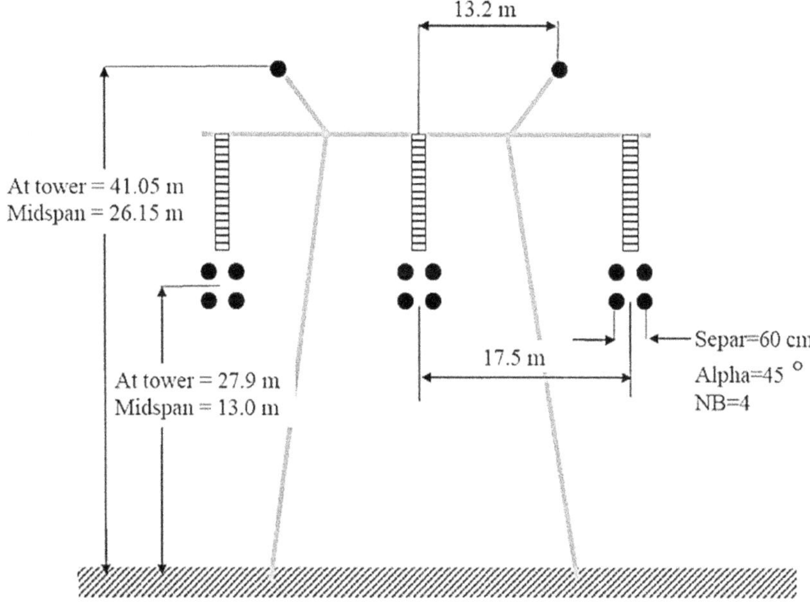

FIGURE 5.24
TL for example 5.2.

TL data are:

Resistance DC = 0.0585 Ω/km;
Outside diameter of the conductors = 3.105 cm;
Inner radius of the tube = 0.55 cm.

The conductors of the ground wire are of reinforced steel and the data are:

Resistance DC = 0.304 Ω/km;
Outside diameter of the ground wire = 1.6 cm;
Inner radius of the tube = 0.3 cm;
The resistivity of the soil equals to 20 Ω-m. The conductor separation in the bundle is 60 cm.

SOLUTION:

The source of lightning surge impulse used was that of Heidler.
The circuit in ATPdraw is shown in Figure 5.25.
The voltage at phase A of the LT terminal is shown in Figure 5.26.
The current in phase A is shown in Figure 5.27.
The circuit with the arrester is as in Figure 5.28.
The voltage in phase A after TL is shown in Figure 5.29.

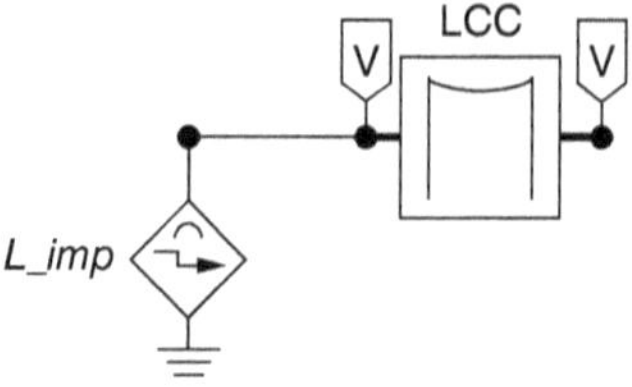

FIGURE 5.25
ATPdraw circuit for example 5.2.

5.7 Switching Surges

It is important that overvoltages caused by energizing and reclosing TLs with operating voltages greater than 345 kV are minimized to reduce the insulation level of TL and open terminal equipment where high overvoltages are recorded. The reduction of these overvoltages is usually accomplished by inserting pre-insertion resistors installed next to a circuit breaker camera. The wiring diagram of such circuit breakers is shown in Figure 5.30, where the TL is initially energized via resistor R by closing the auxiliary contact A of the circuit breaker. After a short time (6 to 10 ms), the main breaker contact B closes, shorting the resistor and energizing the

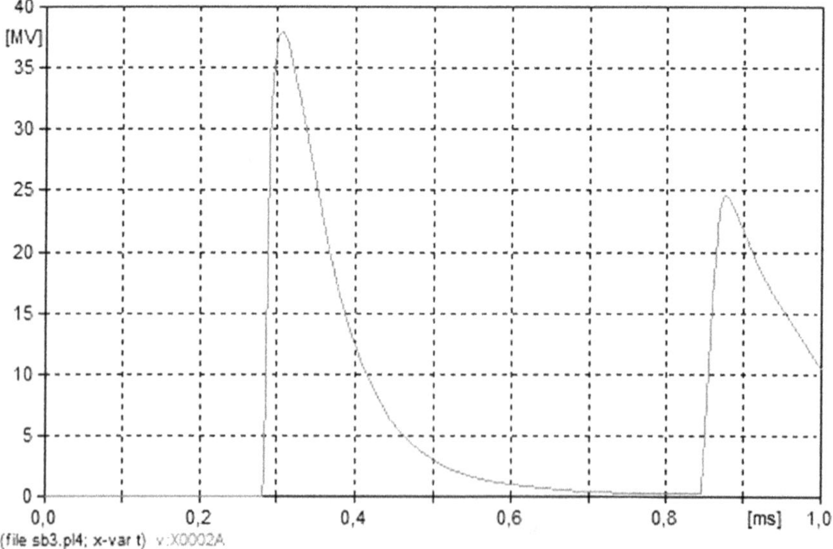

FIGURE 5.26
Phase A overvoltage.

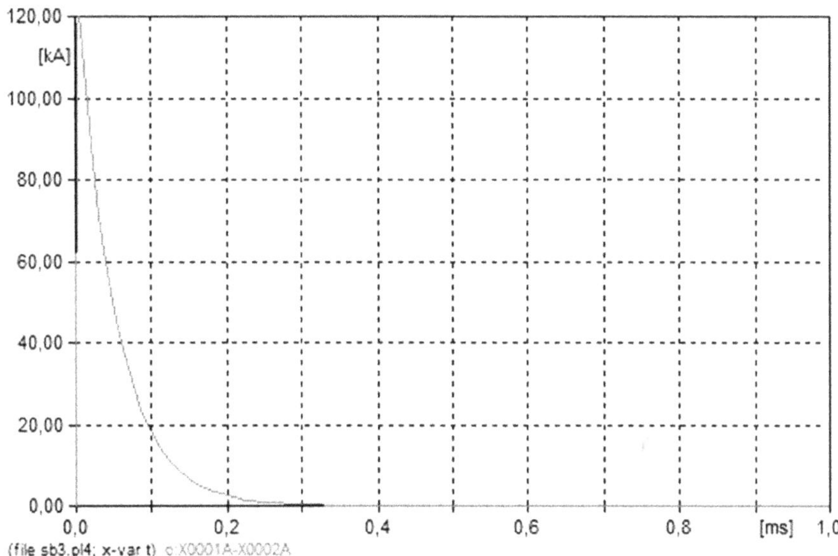

FIGURE 5.27
Phase A overcurrent.

TL with the source voltage. Proper choice of resistor reduces the level of switching overvoltages.

The pre-insertion resistor is ineffective if it is shorted before returning to the source of the first wave reflected by the remote terminal. The pre-insertion resistor is only effective if it is inserted into the circuit for more than twice the TL transit time. In addition, as the resistor is shorted, new overvoltages are introduced and they increase with the resistor value. In general, the value of the pre-insertion resistor ranges from 250 to 450 Ω (characteristic impedance range of TLs).

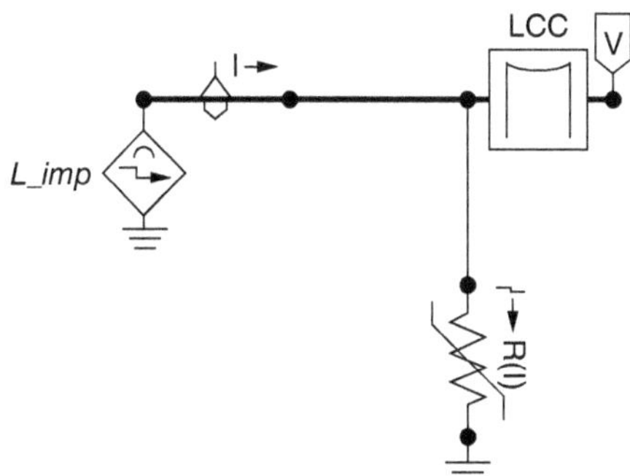

FIGURE 5.28
Ground wire before TL.

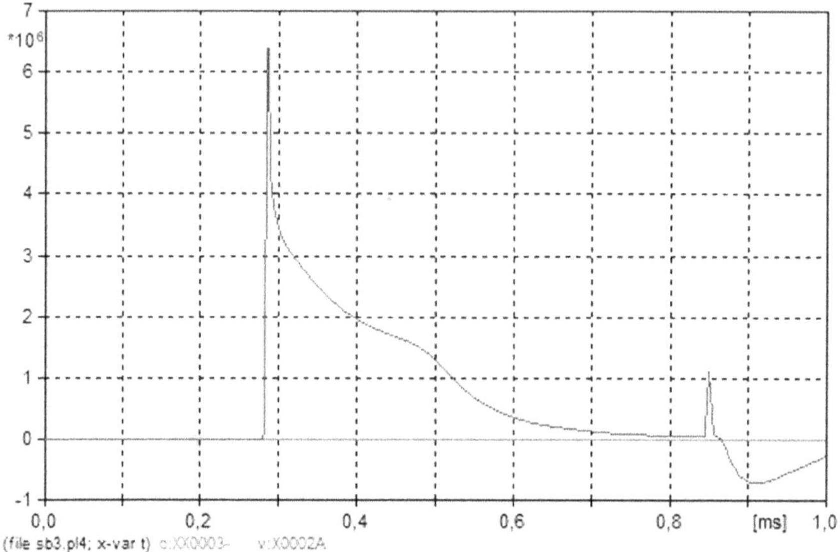

FIGURE 5.29
Phase A voltage after surge arrester placement.

Example 5.3

Simulate switching from one 100 Ω and one 251 Ω resistor to a three-phase TL using ATPdraw. TL has the following characteristics:

Voltage: 500 kV;
Operating frequency: 60 Hz;
Ground resistivity: 1.000 Ω-m;
Surge impedance: 251 Ω;
ACSR conductors: 636 MCM 26/7;
Number of subconductors per phase: 4;
Number of ground wires: 2;
Ground wires conductors: Steel EHS 3/8″ 7 wires;

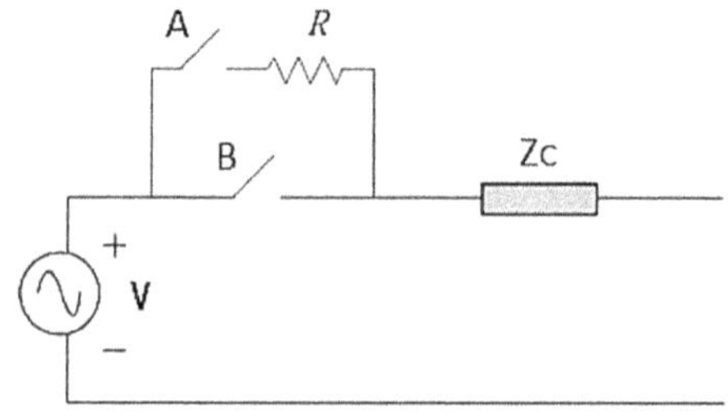

FIGURE 5.30
Pre-insertion resistor switching.

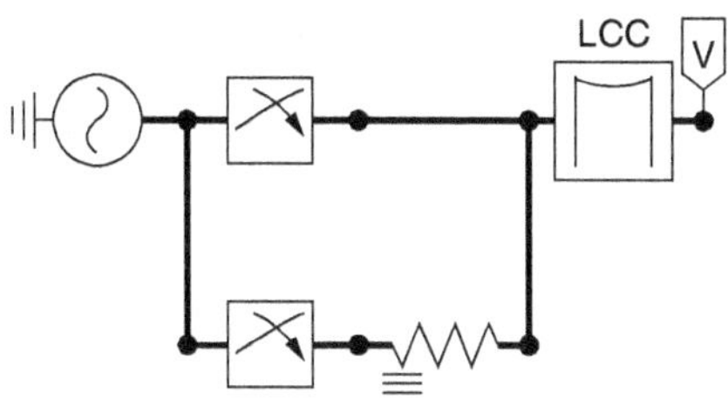

FIGURE 5.31
ATPdraw circuit for example 5.3.

Conductor arrangement: horizontal;
Height of conductors above ground: 26.32 m;
Height of ground conductor cables from ground: 36.91 m;
Phase distance: a-b 11 m; b-c 11 m; and c-a 22 m;
Distance between ground wires: 16.78 m;
Conductor sags: 2 m;
Ground wire sags: 1.5 m.

SOLUTION:

The TL model used was that of Bergeron. ATPdraw Circuit (Figure 5.31):

100-Ω resistor input (Figure 5.32).
251-Ω resistor input (Figure 5.33).
Overvoltage is lower with 251-Ω resistor input.

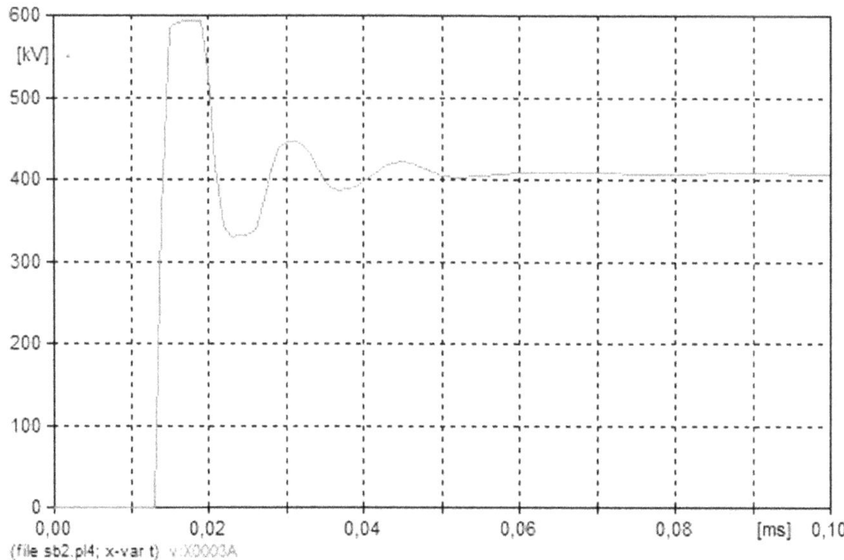

FIGURE 5.32
Phase A line voltage for 100 Ω resistor insertion.

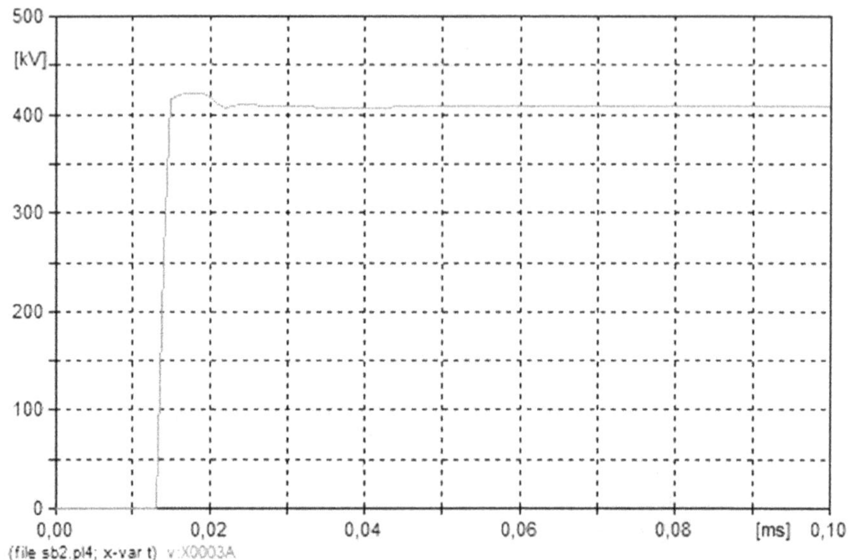

FIGURE 5.33
Phase A line voltage for 251 Ω resistor insertion.

5.8 Problems

5.8.1 Explain what traveling waves are.
Answer: See item 5.3.

5.8.2 Are switching overvoltages larger than overvoltages caused by lightning surges?
Answer: See item 5.2.

5.8.3 A 150 V DC source with negligible resistance is connected via a switch to a lossless TL which has a characteristic impedance of 130 Ω. The TL ends with a 120-Ω resistance. Run a MATLAB language program to calculate the voltage on the receiving-end after five times the time required for the voltage wave to travel the full length of the TL. TL transit time is *T*.
Answer: A function of type * .m must be programmed.

5.8.4 Simulate the switching of a 50 Ω resistor with a three-phase TL using ATPdraw and comment on the results. TL has the following characteristics:

Voltage: 500 kV;

Operating frequency: 60 Hz;

Ground resistivity: 1.000 Ω-m;

Surge impedance: 251 Ω;

ACSR conductors: 636 MCM 26/7;

Number of subconductors per phase: 4;

Number of ground wires: 2;

Ground wires conductors: Steel EHS 3/8″ 7 wires;

Conductor arrangement: horizontal;

Height of conductors above ground: 26.32 m;

Height of ground conductor cables from ground: 36.91 m;

Phase distance: a-b 11 m; b-c 11 m; and c-a 22 m;

Distance between ground wires: 16.78 m;

Conductor sags: 2 m;

Ground wire sags: 1.5 m.

Answer: Use the ATPdraw program.

5.8.5 Switching a no-load transmission line is equivalent to a capacitor bank maneuver, but there are differences in the relationships between voltages and currents in both situations. On this subject, judge the item below.

Factors influencing energizing overvoltages and open line openings include line length, grid grounding conditions, and pre-maneuver voltage.

Answer: Right.

5.8.6 Regarding power substations and transmission lines, judge the following item.

Maneuvering overvoltages resulting from switching in the electrical system are characterized by a wave front lasting from a few microseconds to a few dozen microseconds.

Answer: Wrong.

6

Modal Propagation Theory

6.1 Introduction

In the studies done in previous chapters, the main assumption made was that transmission lines (TLs) could be considered as fully homogeneous circuits operating in a balanced regime. Thus, a single-phase representation of TL was sufficient. This simplification is valid and many of the problems that arise in steady-state studies can be satisfactorily solved with this representation.

Although cyclic conductor transposition allows the TL parameters to be calculated as average values corresponding to the various positions occupied, even in this case, if the line sections between two transpositions are electrically long, the resulting imbalance can be significant.

The definition of an electrically long line depends not only on the physical length of the TL, but also on the working frequencies. At frequencies of 50 Hz or 60 Hz, most TLs can be considered short, and this justifies the use of symmetrical components under normal or abnormal operating conditions.

However, there may be extra-high voltage lines that, even at industrial frequency, may be considered long lines. If the number of transpositions on one of these lines is small, the effect of the imbalance should be analyzed considering the mutual coupling between sequence circuits, which largely negates the analysis using symmetrical components.

The study of a single-phase protection scheme requires a treatment that takes into consideration the coupling between phases, because the impedances seen by the distance relays depend on the distance to the fault and the existing unbalance in the TL, and if this is not obtained erroneous results are inevitable. In addition, when studying the various overvoltages (TL energization, fault reset voltages, etc.), TL should be considered as an electrically long multiconductor system.

In this chapter, TL is treated as a multiconductor system through modal propagation theory.

6.2 Voltage Equation for a Single Conductor System

The value of the effective voltage across a conductor in the presence of perfect earth is calculated using equation (4.41) developed in Chapter 4:

$$V_S = \frac{V_R + I_R Z_C}{2} e^{l\gamma} + \frac{V_R - I_R Z_C}{2} e^{-l\gamma} \tag{6.1}$$

And it is expressed in terms of the incident and reflected voltages seen in Chapter 5 as:

$$V = V_i e^{-x\gamma} + V_r e^{x\gamma} \tag{6.2}$$

Assigning this voltage to the phase 1 conductor and modifying equation (6.2), we have:

$$V_1^{(F)} = V_{1i} e^{-x\gamma} + V_{1r} e^{x\gamma} \tag{6.3}$$

Equation (6.3) can be written as follows:

$$V_1^{(F)} = Q_{11}\left(V_1^{(M+)} e^{-x\gamma} + V_1^{(M-)} e^{x\gamma}\right) \tag{6.4}$$

Or:

$$V_1^{(F)} = Q_{11} V_1^{(M)} \tag{6.5}$$

where:
$V_1^{(F)}$ = the voltage at phase (1)
$V_1^{(M)}$ = the modal voltage (1)
$V_1^{(M+)}$ = the modal voltage (1), incident wave
$V_1^{(M-)}$ = the modal voltage (1), reflected wave
Q_{11} = a constant that transforms a modal voltage into a phase voltage

The transformation of a phase voltage into a modal voltage and the transformation of a modal voltage into a phase voltage are the basis for understanding modal propagation theory.

6.3 The Modal Matrix, Eigenvalues and Eigenvectors

The modal matrix of a TL considered as a homogeneous set of multiconductors is calculated from the concepts of eigenvalues and eigenvectors.

6.3.1 The Eigenvalues of a Matrix

Consider a square matrix $[A]$, of order (nxn), and a column vector $[X]$, of order $(nx1)$. The product of $[A]$ and $[X]$ produces a new vector $[Y]$ of the same order.

$$[A][X]=[Y] \tag{6.6}$$

It is also considered that $[Y]$ has the same direction as $[X]$,

$$[Y]=k[X] \tag{6.7}$$

Where k is a scalar quantity.

Therefore:

$$[A][X]=k[X] \tag{6.8}$$

Manipulating the previous equation comes:

$$[A][X]-k[X]=[0] \tag{6.9}$$

Finally:

$$\left[k[U]-[A][X]\right]=[0] \tag{6.10}$$

where:

$[U]$ is a unitary matrix of order (nxn).

The matrix $\left[k[U]-[A][X]\right]$ is called the characteristic matrix of A.

The matrix equation (6.10) only has a different solution than the trivial one ($[X]=[0]$) if the determinant of $\left[k[U]-[A][X]\right]$ is null, that is:

$$f(k)=|kU-A|=0 \tag{6.11}$$

$$f(k)=k^n+C_1k^{n-1}+C_2k^{n-2}+\ldots\ C_{n-1}k+C_n=0 \tag{6.12}$$

Equation (6.12) is called the characteristic equation. The roots of this equation can be real or complex single or multiple and they are the eigenvalues of the matrix $[A]$.

6.3.2 The Eigenvectors of a Matrix

For each of the roots (eigenvalues) $k=k_i$, $i=1, 2, 3, \ldots, n$ of equation (6.12), it is possible to determine a vector $[X_i]$ solution of the matrix equation (6.10). Thus, we can write:

$$\left[[k_i][U]-[A][X]\right][X_i]=[0]\ i=1, 2, 3, \ldots, n \tag{6.13}$$

Vectors $[X_i]$ that satisfy equation (6.13) are called eigenvectors of matrix $[A]$.

6.3.3 The Modal Matrix

The matrix $[M]$, whose columns comprise the eigenvectors of the original matrix $[A]$, is called the modal matrix of A. That is,

$$[M]=[[X_1]\,[X_2]\,[X_3]\,\ldots\,[X_{n-1}]\,[X_n]] \tag{6.14}$$

The set of equations:

$$\big[[k_i][U]-[A][X]\big][X_i]\;i=1,\,2,\,3,\,\ldots,\,n \tag{6.15}$$

can be written compactly as:

$$[A][M]=[M][D] \tag{6.16}$$

where:
$[M]$ = the modal matrix of $[A]$
$[D]$ = a diagonal matrix containing the eigenvalues of the matrix $[A]$, that is, $D_{ii}=k_i$

From equation (6.16):

$$[D]=[M]^{-1}[A][M] \tag{6.17}$$

Equation (6.17) shows that the pre-multiplication and post-multiplication of a matrix $[A]$ by the inverse of the modal matrix and the modal matrix, respectively, transform matrix $[A]$ into a diagonal matrix, elements of which are the eigenvalues of $[A]$.

6.4 Propagation in a N-Conductor System

Second-order differential equations describing a polyphase transmission line are difficult to solve because of coupling between phases. An important tool for analysis of polyphase systems is the decoupling technique.

Thus, a system that has N phases coupled can be represented by N single-phase systems that are mathematically identical to the original system. To decouple the phases of an LT, phase voltages must be transformed into component voltages. For this, we need to calculate eigenvalues and eigenvectors to get the transformation matrix of phase components for modal components.

For a generic polyphase system, the matrix with the matrix product eigenvectors [Z] [Y] decouples the line phases. For a single product [Z] [Y], there are several sets of eigenvectors that decouple the line. The eigenvalues are unique.

At the end of the study, the inverse process is performed and the quantities of the three-phase system are again obtained.

Consider a line with N conductors.

Frequency domain telegraph equations, in matrix form are written as follows:

$$\frac{d^2[V^F]}{dx^2} = [Z][Y][V^F] \tag{6.18}$$

$$\frac{d^2[I^F]}{dx^2} = [Y][Z][I^F] \tag{6.19}$$

The longitudinal impedance [Z] and transverse admittance [Y] matrices of the line are:

$$[Z] = \begin{bmatrix} Z_{aa} & Z_{ab} & \cdots & Z_{an} \\ Z_{ba} & Z_{bb} & \ddots & Z_{bn} \\ \vdots & & \cdots & \vdots \\ Z_{na} & Z_{nb} & \ldots & Z_{nn} \end{bmatrix} (\Omega) \tag{6.20}$$

$$[Y] = \begin{bmatrix} Y_{aa} & Y_{ab} & \cdots & Y_{an} \\ Y_{ba} & Y_{bb} & \ddots & Y_{bn} \\ \vdots & & \cdots & \vdots \\ Y_{na} & Y_{nb} & \ldots & Y_{nn} \end{bmatrix} (S) \tag{6.21}$$

From equation (4.29), we can write the following equations, in matrix form:

$$\frac{d[V^F]}{dx} = [Z][I^F] \tag{6.22}$$

$$\frac{d[I^F]}{dx} = [Y][V^F] \tag{6.23}$$

Using the concept of component voltage and, by analogy, component current developed in 6.2, we have:

$$[V^F] = [Q][V^M] \tag{6.24}$$

$$[I^F] = [S][I^M] \tag{6.25}$$

Substituting in equations (6.18) and (6.19), we have:

$$\frac{d^2[Q][V^M]}{dx^2}=[Z][Y][Q][V^M] \tag{6.26}$$

$$\frac{d^2[V^M]}{dx^2}=[Q]^{-1}[Z][Y][Q][V^M] \tag{6.27}$$

$$\frac{d^2[S][I^M]}{dx^2}=[Z][Y][S][I^M] \tag{6.28}$$

$$\frac{d^2[I^M]}{dx^2}=[S]^{-1}[Y][Z][S][I^M] \tag{6.29}$$

Comparing equations (6.27) and (6.29) with (6.17), we can conclude that:

The matrix $[Q]$ is the modal matrix of $[Z][Y]$, formed by the eigenvectors of $[Z][Y]$;
The matrix $[S]$ is the modal matrix of $[Y][Z]$, formed by the eigenvectors of $[Y][Z]$.

Substituting (6.24) and (6.25) in the previous equations, we have:

$$\frac{d[Q][V^M]}{dx}=[Z][S][I^M] \tag{6.30}$$

$$\frac{d[S][I^M]}{dx}=[Y][Q][V^M] \tag{6.31}$$

Therefore:

$$\frac{d[V^M]}{dx}=[Q]^{-1}[Z][S][I^M]=[Z]^{modo}[I^M] \tag{6.32}$$

$$\frac{d[I^M]}{dx}=[S]^{-1}[Y][Q][V^M]=[Y]^{modo}[V^M] \tag{6.33}$$

with:

$$[Z]^{modo}=[Q]^{-1}[Z][S]=\begin{bmatrix} Z_{modo1} & 0 & 0 \\ 0 & Z_{modo2} & 0 \\ 0 & 0 & Z_{modo3} \end{bmatrix} \tag{6.34}$$

$$[Y]^{modo}=[S]^{-1}[Y][Q]=\begin{bmatrix} Y_{modo1} & 0 & 0 \\ 0 & Y_{modo2} & 0 \\ 0 & 0 & Y_{modo3} \end{bmatrix} \tag{6.35}$$

6.5 Karrenbauer's Transformation

H. Karrenbauer proposed a more suitable transformation for the solution of electromagnetic transients. This transformation diagonalizes balanced matrices (they are matrices got from transposed lines, where the main diagonal elements are equal to each other and the non-diagonal elements are equal to each other) N × N with the following advantages:

a. The transformed matrices are real and therefore preferable over symmetrical components for the solution of electromagnetic transients;
b. Its structure is very simple (therefore preferable over components α, β, 0);
c. The transformation is valid for N-phase systems;
d. The transformation is valid for any frequency.

The Karrenbauer transformation is defined by:

$$[V^F]=[T][V^M] \tag{6.36}$$

$$[I^F]=[T][I^M] \tag{6.37}$$

where:

$$[T]=\begin{bmatrix} 1 & 1 & 1 & \cdots & 1 \\ 1 & 1-N & 1 & \vdots & 1 \\ 1 & 1 & 1-N & \vdots & 1 \\ \cdots & \cdots & \cdots & \cdots & \cdots \\ 1 & 1 & 1 & \vdots & 1-N \end{bmatrix} \tag{6.38}$$

N is the number of phases.

The inverse transformation:

$$[V^M]=[T]^{-1}[V^F] \tag{6.39}$$

$$[I^M]=[T]^{-1}[I^F] \tag{6.40}$$

has the matrix:

$$[T]^{-1}=\frac{1}{N}\begin{bmatrix} 1 & 1 & 1 & \cdots & 1 \\ 1 & -1 & 0 & \vdots & 0 \\ 1 & 0 & -1 & \vdots & 0 \\ \cdots & \cdots & \cdots & \cdots & \cdots \\ 1 & 0 & 0 & \vdots & -1 \end{bmatrix} \tag{6.41}$$

From equation (6.41), it can be seen that the first mode describes the loop formed with all phases in parallel with the return through earth and lightning conductor cables. This mode is identical to what we commonly call zero sequence mode. The second mode, ..., N-mode, describes the loop formed by the first phase with the second mode, ..., N-mode, respectively. The matrices of equations (6.38) and (6.41) are not normalized.

The following transformation can get the mode impedance matrix:

$$\left[Z^M\right]=[T]^{-1}\left[Z^F\right][T]=\begin{bmatrix} Z^M_{zero} & 0 & 0 & \dots & 0 \\ 0 & Z^M_{positiva} & 0 & \vdots & 0 \\ 0 & 0 & Z^M_{positiva} & \vdots & 0 \\ \dots & \dots & \dots & \dots & \dots \\ 0 & 0 & 0 & \vdots & Z^M_{positiva} \end{bmatrix} \tag{6.42}$$

6.6 Clarke's Transformation

The α, β, 0 components because of E. Clarke can also diagonalize balanced three-phase matrices got from a TL. The transformation is defined as:

$$[V^F]=[T1][V_{0\alpha\beta}]\text{e}[V_{0\alpha\beta}]=[T1]^{-1}[V^F] \tag{6.43}$$

$$[I^F]=[T1][I_{0\alpha\beta}]\text{e}[I_{0\alpha\beta}]=[T1]^{-1}[I^F] \tag{6.44}$$

where:

$$[V_{\alpha\beta 0}]=\begin{bmatrix} V_0 \\ V_\alpha \\ V_\beta \end{bmatrix},\ [I_{\alpha\beta 0}]=\begin{bmatrix} I_0 \\ I_\alpha \\ I_\beta \end{bmatrix} \tag{6.45}$$

with:

$$[T1]=\begin{bmatrix} 2/\sqrt{6} & 0 & 1/\sqrt{3} \\ -1/\sqrt{6} & 1/\sqrt{2} & 1/\sqrt{3} \\ -1/\sqrt{6} & -1/\sqrt{2} & 1/\sqrt{3} \end{bmatrix} \tag{6.46}$$

Example 6.1

Calculate the modal components for the impedances of the line in example 4.1, considering the transposed and not transposed TL.

SOLUTION:

Transposed TL:

For the three-phase line, the Karrenbauer transformation is given by:

$$[T]=\begin{bmatrix}1 & 1 & 1\\ 1 & -2 & 1\\ 1 & 1 & -2\end{bmatrix}$$

$$[T1]^{-1}=\frac{1}{3}\begin{bmatrix}1 & 1 & 1\\ 1 & -1 & 0\\ 1 & 0 & -1\end{bmatrix}$$

Using equation (6.42), we have:

$$\left[Z^M\right]=[T]^{-1}\left[Z^F\right][T]=$$

$$\frac{1}{3}\begin{bmatrix}1 & 1 & 1\\ 1 & -1 & 0\\ 1 & 0 & -1\end{bmatrix}\begin{bmatrix}0.2410+j0.8499 & 0.0926+j0.4106 & 0.0926+j0.4106\\ 0.0926+j0.4106 & 0.2410+j0.8499 & 0.0926+j0.4106\\ 0.0926+j0.4106 & 0.0926+j0.4106 & 0.2410+j0.8499\end{bmatrix}\times$$

$$\begin{bmatrix}1 & 1 & 1\\ 1 & -2 & 1\\ 1 & 1 & -2\end{bmatrix}=\begin{bmatrix}0.4262+j1.6711 & 0 & 0\\ 0 & 0.1484+j0.4393 & 0\\ 0 & 0 & 0.1484+j0.4393\end{bmatrix}$$

Applying Clarke transformation:

$$\left[Z^M\right]=[T1]^{-1}\left[Z^F\right][T1]$$

$$=\begin{bmatrix}0.1484+j0.4393 & 0 & 0\\ 0 & 0.1484+j0.4393 & 0\\ 0 & 0 & 0.4262+j1.6711\end{bmatrix}$$

which gets the same results as before, with the positions of the changed modes.

TL not transposed:

For not transposed TL, we need to calculate the impedance matrix eigenvectors. For this, we use the MATLAB command [*V*, *D*] = eig (*Z*).

[*V*, *D*] = eig (*A*) produces eigenvalue matrices (*D*) and eigenvectors (*V*) of matrix *A*, so that *A* * *V* = *V* * *D*. The matrix *D* is the canonical form of *A*—a matrix diagonal with eigenvalues of A on the main diagonal. Matrix *V* is the modal matrix, and its columns are the eigenvectors.

$$[Z]=\begin{bmatrix}0.2455+j0.8486 & 0.0937+j0.4407 & 0.0937+j0.3895\\ 0.0937+j0.4407 & 0.2388+j0.8505 & 0.0905+j0.4017\\ 0.0937+j0.3895 & 0.0905+j0.4017 & 0.2388+j0.8505\end{bmatrix}\Omega/km$$

$$[V, D] = eig(Z)$$

$$D = \begin{bmatrix} 0.4262 + j1.6711 & 0 & 0 \\ 0 & 0.1484 + j0.4393 & 0 \\ 0 & 0 & 0.1484 + j0.4393 \end{bmatrix}$$

$$V = \begin{bmatrix} 0.5816 - j0.0056 & -0.6624 - j0.0049 & -0.4721 + j0.0016 \\ 0.5867 & 0.7436 & -0.3207 + j0.0010 \\ 0.5635 - j0.0050 & -0.0905 - j0.0024 & 0.8211 \end{bmatrix}$$

6.7 Problems

6.7.1 Calculate the modal transformation matrix for the not transposed TL of problem 3.11.3.

Answer:

$$V = \begin{bmatrix} 0.5695 - j0.0031 & 0.7071 - j0.0000 & -0.4191 + j0.0023 \\ 0.5927 & 0.0000 & 0.8054 + j0.0000 \\ 0.5695 - j0.0031 & -0.7071 - j0.0000 & -0.4191 - j0.0023 \end{bmatrix}$$

$$D = \begin{bmatrix} 0.2000 + \text{j}1.5627 & 0 & 0 \\ 0 & 0.0253 + \text{j}0.3565 & 0 \\ 0 & 0 & 0.0254 + \text{j}0.2864 \end{bmatrix}$$

Bibliography

Cable & Conductor Manufacturers Catalogs, 2018.

Carson, J. R. Wave propagation in overhead wires with ground return. *Bell System Technical Journal*, v. 5, 1926, 539–554.

Central Station Engineers of the Westinghouse Electric Corporation. *Electrical transmission and distribution reference book*. 4th edition. East Pittsburgh, PA, 1984. 832 p.

Dommel, H. W. Overhead line parameters from handbook formulas and computer programs. *IEEE Transactions on PAS*, v. PAS-104, n. 2, February 1985, 366–372.

Elgerd, O. I. *Electric energy systems: an introduction*. 2nd edition. McGraw Hill, NY, 1998. 533 p.

EPRI. *AC transmission line reference book–200 kV and Above*. 3rd edition. Palo Alto, CA, 2005. 1069 p.

Fuchs, R. D. *Power transmission*, v. 1. 3rd edition. Uberlândia-Brazil: EDUFU, 2015. 244 p.

Fuchs, R. D. *Power transmission*. v. 2. 3rd edition. Uberlândia-Brazil: EDUFU, 2015. 550 p.

Galloway, R. H. *et al*. Calculation of electrical parameters for short and long polyphase transmission lines. *Proceedings IEE*, v. 111, December 1964, 2051–2059.

Glover, J. D.; Sarma, M. S.; Overbye, T. J. *Power system analysis and design*. 6th edition. Cengage Learning, 2016. 818 p.

Gonen, T. *Electrical power transmission system engineering: analysis and design*. 3rd edition. New York: CRC Press, 2014. 719 p.

Grainger, J. J.; Stevenson JR., W. D. *Power system analysis*. New York: McGraw Hill Ed., 1994. 787 p.

Gross, C. A. *Power systems analysis*. New York: Wiley, 1982. 593 p.

Haginomori, E.; Arai, T. K. J.; Ikeda, H. *Power system transient analysis: theory and practice using simulation program (ATP-EMTP)*. UK: John Wiley & Sons, Ltd., 2016. 277 p.

Leuven EMTP Center. *ATP—Alternative Transient Program—Rule Book*. Belgium: Herverlee, 1987.

Stevenson JR., W. D. *Elements of power system analysis*. 4th edition. New York: McGraw Hill, 1982. 436 p.

Appendix: Conductors Data

TABLE A1

Characteristics of COPPER Conductors, Hard Drawn, 97.3 Percent

Size of Conductor MCM –AWG	Number of Strands	Approx. Current-Carrying Capacity (A)	Geometric Mean Radius at 60 (Hz) (m)	Resistance DC at 25°C and 60 Hz Ω/*km*	Resistance AC at 25°C and 60 Hz Ω/*km*	Resistance DC at 50°C and 60 Hz Ω/*km*	Resistance AC at 50°C and 60 Hz Ω/*km*	Outside Diameter (mm)
1000	37	300	0.01122	0.03635	0.0394	0.0397	0.0425	29.235
900	37	220	0.01064	0.04039	0.0431	0.0441	0.0467	27.7368
800	37	130	0.01003	0.04543	0.0479	0.0497	0.0520	26.1366
750	37	90	0.009723	0.04847	0.0508	0.0530	0.0551	25.3238
700	37	40	0.009388	0.05195	0.0541	0.0568	0.0588	24.4602
600	37	940	0.008687	0.07059	0.0625	0.0662	0.0680	22.6314
500	37	840	0.007925	0.07270	0.0743	0.0795	0.0809	20.6756
500	19	840	0.007803	0.07270	0.0743	0.0795	0.0809	20.5994
450	19	780	0.007406	0.08070	0.0822	0.0883	0.0896	19.558
400	19	730	0.006980	0.09080	0.0922	0.0994	0.1006	18.4404
350	19	670	0.006523	0.1038	0.1050	0.1136	0.1146	17.2466
350	12	670	0.006858	0.1038	0.1050	0.1136	0.1146	18.034
300	19	610	0.006056	0.1211	0.1221	0.1323	0.1336	15.9766

(Continued)

TABLE A1 (*Continued*)

Characteristics of COPPER Conductors, Hard Drawn, 97.3 Percent

Size of Conductor MCM –AWG	Number of Strands	Approx. Current-Carrying Capacity (A)	Geometric Mean Radius at 60 (Hz) (m)	Resistance DC at 25°C and 60 Hz Ω/*km*	Resistance AC at 25°C and 60 Hz Ω/*km*	Resistance DC at 50°C and 60 Hz Ω/*km*	Resistance AC at 50°C and 60 Hz Ω/*km*	Outside Diameter (mm)
300	12	610	0.006340	0.1211	0.1221	0.1323	0.1336	16.6878
250	19	540	0.005526	0.1454	0.1460	0.1591	0.1597	14.5796
250	12	540	0.005727	0.1454	0.1460	0.1591	0.1597	15.24
211.6 –4/0	19	480	0.005084	0.1715	0.1727	0.1876	0.1883	13.4112
211.6 –4/0	12	490	0.005334	0.1715	0.1727	0.1876	0.1883	14.0208
211.6 –4/0	7	480	0.004813	0.1715	0.1727	0.1876	0.1883	13.2588
167.8 –3/0	12	420	0.004752	0.2169	0.2175	0.2367	0.2374	12.496
167.8 –3/0	7	420	0.004279	0.2169	0.2175	0.2367	0.2374	11.7856
133.1 –2/0	7	360	0.003816	0.2734	0.2734	0.2989	0.2989	10.5156
105.5 –1/0	7	310	0.003392	0.3449	0.3449	0.3766	0.3772	9.3472
83.69 –1	7	270	0.003023	0.4344	0.4344	0.4654		8.3312
83.69 –1	3	270	0.003097	0.4300	0.4300	0.4704		9.1440
66.37 –2	7	230	0.002691	0.5475	0.5481	0.5992		7.4168

(*Continued*)

TABLE A1 (*Continued*)

Characteristics of COPPER Conductors, Hard Drawn, 97.3 Percent

Size of Conductor MCM –AWG	Number of Strands	Approx. Current-Carrying Capacity (A)	Geometric Mean Radius at 60 (Hz) (m)	Resistance DC at 25°C and 60 Hz Ω/km	Resistance AC at 25°C and 60 Hz Ω/km	Resistance DC at 50°C and 60 Hz Ω/km	Resistance AC at 50°C and 60 Hz Ω/km	Outside Diameter (mm)
66.37 –2	3	240	0.002752	0.5425		0.5935		8.128
66.37 –2	1	220	0.002548	0.5369		0.5873		6.5532
52.63 –3	7	200	0.002398	0.6911		0.7557		6.604
52.63 –3	3	200	0.002453	0.6842		0.7482		7.239
52.63 –3	1	190	0.002270	0.6774		0.7408		5.8166
41.74 –4	3	180	0.002185	0.8626		0.9434		66.4516
41.74 –4	1	170	0.002020	0.8539		0.9341		5.1816
33.1 –5	3	150	0.001944	1.0876		1.1895		5.7404
33.1 –5	1	140	0.001798	1.0770		1.1777		4.6202
26.25 –6	3	130	0.001731	1.3735		0.1497		5.1054
26.25 –6	1	120	0.001603	1.3548		0.1485		4.1148
20.82 –7	1	110	0.001426	1.7091		0.1870		3.665
16.51 –8	1	90	0.001271	2.1566		0.2361		3.2639

TABLE A2

Characteristics of Aluminum Cable Steel Reinforced (ACSR)

Code	MCM–AWG	Strands /Layers	Approx. Current-Carrying Capacity* (A)	Geometric Mean Radius at 60 (Hz) (m)	Resistance DC at 25°C and 60 Hz Ω/km	Resistance AC at 25°C and 60 Hz Ω/km	Resistance DC at 50°C and 60 Hz Ω/km	Resistance AC at 50°C and 60 Hz Ω/km	Inside Radius (cm)	Outside Radius (cm)
Joree	2515	76/19	1750	0.01894				0.02797	0.54	2.3878
Thrasher	2312	76/19	1235	0.01815				0.02996	0.5175	2.2895
Kiwi	2167	72/7	1147	0.01739				0.03176	0.441	2.205
Bluebird	2156	84/19	1092.44	0.17934				0.03145	0.610	2.238
Chukar	1781	84/19	901.93	0.01629				0.03717	0.555	2.035
Falcon	1590	54/19	1380	0.01584	0.03648	0.0367	0.04014	0.0425	0.655	1.963
Parrot	1510.5	54/19	1340	0.01545	0.03840	0.0386	0.04226	0.0447	0.6375	1.9125
Plover	1431	54/19	1300	0.01502	0.04052	0.0407	0.04462	0.0472	0.620	1.862
Martin	1321.5	54/19	1250	0.01459	0.04294	0.0431	0.04729	0.0499	0.6025	1.8085
Pheasant	1272	54/19	1200	0.01417	0.04561	0.0458	0.05021	0.0528	0.585	1.755
Grackel	1192.5	54/19	1160	0.01371	0.04866	0.0489	0.05357	0.0563	0.5675	1.6985
Finch	1113	54/19	1110	0.01325	0.05214	0.0524	0.05742	0.0602	0.5475	1.6425
Curlew	1033.5	54/7	1060	0.01280	0.05612	0.0564	0.06177	0.0643	0.5265	1.5795

(Continued)

TABLE A2 (*Continued*)

Characteristics of Aluminum Cable Steel Reinforced (ACSR)

Code	MCM–AWG	Strands /Layers	Approx. Current-Carrying Capacity* (A)	Geometric Mean Radius at 60 (Hz) (m)	Resistance DC at 25°C and 60 Hz Ω/*km*	Resistance AC at 25°C and 60 Hz Ω/*km*	Resistance DC at 50°C and 60 Hz Ω/*km*	Resistance AC at 50°C and 60 Hz Ω/*km*	Inside Radius (cm)	Outside Radius (cm)
Cardinal	954	54/7	1010	0.01228	0.06084	0.0610	0.06699	0.0701	0.507	1.521
Canary	900	54/7	970	0.01191	0.06463	0.0646	0.07116	0.0736	0.492	1.476
Crane	874.5	54/7	950	0.01176	0.06650	0.0671	0.07321	0.0763	0.4845	1.4535
Condor	795	54/7	900	0.01121	0.07271	0.0739	0.08004	0.0856	0.462	1.386
Drake	795	26/7	900	0.01143	0.07271	0.0727	0.08004	0.0800	0.5175	1.4055
Mallard	795	30/19	910	0.01197	0.07271	0.0727	0.08004	0.0800	0.620	1.448
Crow	715.5	54/7	830	0.01063	0.08141	0.0820	0.08962	0.0921	0.438	1.314
Starling	715.5	26/7	840	0.01082	0.08141	0.0814	0.08962	0.0896	0.492	1.334
Redwing	715.5	30/19	840	0.01133	0.08141	0.0814	0.08962	0.0896	0.5875	1.3715
Flamingo	666.6	54/7	800	0.01028	0.08701	0.08763	0.095774	0.099503	0.423	1.269
Rook	636	54/7	770	0.01003	0.09136	0.09198	0.10056	0.10491	0.414	1.242
Gull	666.6	54;7	800	0.01027	0.08701	0.0876	0.09577	0.0995	0.423	1.269
Goose	636	54/7	770	0.01002	0.09136	0.0919	0.10055	0.1049	0.414	1.242
Grosbeak	636	26/7	780	0.01021	0.09136	0.0913	0.10055	0.1005	0.4635	1.2575
Egret	636	30/19	780	0.01069	0.09136	0.0913	0.10055	0.1005	0.555	1.295
Peacock	605	54/7	750	0.009791	0.095712	0.096333	0.105345	0.110317	0.4035	1.2095

(Continued)

TABLE A2 (*Continued*)

Characteristics of Aluminum Cable Steel Reinforced (ACSR)

Code	MCM–AWG	Strands /Layers	Approx. Current-Carrying Capacity* (A)	Geometric Mean Radius at 60 (Hz) (m)	Resistance DC at 25°C and 60 Hz Ω/km	Resistance AC at 25°C and 60 Hz Ω/km	Resistance DC at 50°C and 60 Hz Ω/km	Resistance AC at 50°C and 60 Hz Ω/km	Inside Radius (cm)	Outside Radius (cm)
Duck	605	54/7	750	0.00978	0.09571	0.0963	0.10534	0.1103	0.4035	1.210
Squab	605	26/7	760	0.00996	0.09571	0.0957	0.10565	0.1068	0.4515	1.2255
Dove	556.5	26/7	730	0.00954	0.10411	0.1044	0.11491	0.1155	0.4335	1.1775
Eagle	556.5	30/7	730	0.00999	0.10441	0.1044	0.11491	0.1155	0.519	1.211
Heron	500	30/7	690	0.00947	0.11622	0.1162	0.12802	0.12802	0.492	1.1475
Hawk	477	26/7	670	0.00885	0.12181	0.12181	0.13424	0.13424	0.402	1.090
Hen	477	30/7	670	0.00926	0.12181	0.1218	0.13424	0.13424	0.480	1.120
Ibis	397.5	26/7	590	0.00807	0.14605	Equal CC	0.16096	Equal CC	0.366	0.994
Lark	397.5	30/7	600	0.00847	0.14605	Equal CC	0.16096	Equal CC	0.438	1.022
Linnet	336.5	26/7	530	0.00743	0.17277	Equal CC	0.19017	Equal CC	0.3375	0.9155
Oriole	336.4	30/7	530	0.00777	0.17277	Equal CC	0.19017	Equal CC	0.4035	0.9415
Ostrich	300	26/7	490	0.00701	0.19328	Equal CC	0.21255	Equal CC	0.318	0.8635

(Continued)

TABLE A2 (*Continued*)

Characteristics of Aluminum Cable Steel Reinforced (ACSR)

Code	MCM–AWG	Strands /Layers	Approx. Current-Carrying Capacity* (A)	Geometric Mean Radius at 60 (Hz) (m)	Resistance DC at 25°C and 60 Hz Ω/*km*	Resistance AC at 25°C and 60 Hz Ω/*km*	Resistance DC at 50°C and 60 Hz Ω/*km*	Resistance AC at 50°C and 60 Hz Ω/*km*	Inside Radius (cm)	Outside Radius (cm)
Piper	300	30/7	500	0.00734	0.19328	Equal CC	0.21255	Equal CC	0.381	0.889
Partridge	266.8	26/7	460	0.00661	0.21752	Equal CC	0.23917	Equal CC	0.3	0.814
Owl	266.8	6/7	460	0.00208	0.21814	0.2187	0.23989	0.3430	0.2685	0.8045
Peguin		4/0	340	0.00248	0.27408	0.2765	0.30142	0.3679	0.2385	0.6375
Pigeon		3/0	300	0.00182	0.34555	0.3480	0.38035	0.4493	0.2125	0.6375
Quail		2/0	270	0.00155	0.43629	0.4387	0.48041	0.5562	0.189	0.567
Raven		1/0	230	0.00135	0.55002	0.5518	0.60534	0.6960	0.1685	0.5055
Robin		1	200	0.00127	0.69608	0.6960	0.76444	0.8576	0.15	0.45
Sparrow		2	180	0.00127	0.87631	0.8763	0.96332	1.0503	0.1335	0.4005
-		2	180	0.00153	0.87631	0.8763	0.96332	1.0254		
Swallow		3	160	0.00131	1.10627	1.1062	1.21192	1.2865	0.119	0.357
Swan		4	140	0.00133	1.39216	1.3921	1.53510	1.5972	0.106	0.318
-		4	140	0.00137	1.39216	1.3921	1.53510	1.5848		
Thush		5	120	0.00126	1.75263	1.7526	1.92675	1.9763	0.0945	0.2835
Turkey		6	100	0.00120	2.21245	2.2125	2.43628	2.4735	0.34	0.252

* For conductors at 25°C, wind 2.2526 km/h, frequency 60 Hz. Data were transcribed from tables of cable manufacturers.

TABLE A3

Galvanized Steel Cables for Ground Wires

	Outside Diameter		Resistance at 60 Hz (Ω/km)		
Type (7 Strands)	**Inches**	**cm**	**I=0 A**	**I=30 A**	**I=60 A**
Common	1/4	0.6350	5.903	7.084	7.022
	9/32	0.7144	4.412	5.717	5.592
	5/16	0.7937	3.355	4.660	4.847
	3/8	0.9525	2.672	4.039	4.101
	1/2	1.2700	1.420	2.672	3.107
HS	1/4	0.6350	4.971	7.456	6.276
	9/32	0.7144	3.728	6.214	5.406
	5/16	0.7937	3.045	4.971	4.305
	3/8	0.9525	2.299	4.350	3.915
	1/2	1.2700	1.305	3.045	3.107
EHS	1/4	0.6350	4.350	7.954	6.773
	9/32	0.7144	3.355	6.773	5.406
	5/16	0.7937	2.485	5.592	4.225
	3/8	0.9525	2.175	4.909	3.728
	1/2	1.2700	1.243	3.542	2.920

TABLE A4

Fiber Optic Guard Cables—OPGW Cables (Aluminum Foil-Clad Steel Core)

	Diameter (cm)			
Designation	**Layer**	**Cable**	**Max Resistance at 20°C Ω/km**	**Maximum Fault Current kA**
92-AL3/28 48 fo	0.90	1.5	0.3230	116
92-AL3/35 24 fo	0.90	1.5	0.3230	127
91-AL2/38 40 fo	0.93	1.55	0.3190	127
91-AL2/45 16 fo	0.93	1.55	0.3080	194
92-AL2/57 48 fo	1.06	1.62	0.2930	165
100-AL3/50 24 fo	0.98	1.63	0.2800	170
125-AL3/48 24 fo	1.05	1.75	0.2310	321
204-AL5/34 36 fo	0.882	2.06	0.1450	504

Index

D

E

F

For Product Safety Concerns and Information please contact our EU representative GPSR@taylorandfrancis.com
Taylor & Francis Verlag GmbH, Kaufingerstraße 24, 80331 München, Germany

www.ingramcontent.com/pod-product-compliance
Lightning Source LLC
Chambersburg PA
CBHW060335220326
41598CB00023B/2715

* 9 7 8 1 0 3 2 3 3 6 2 3 7 *